I0488593

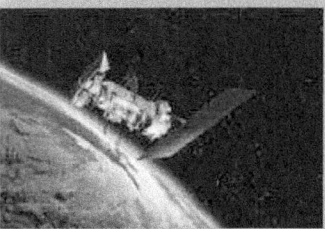

Decision-Support Experiments and Evaluations using Seasonal-to-Interannual Forecasts and Observational Data: *A Focus on Water Resources*

Synthesis and Assessment Product 5.3
Report by the U.S. Climate Change Science Program
and the Subcommittee on Global Change Research

EDITED BY:
Executive Editor: Nancy Beller-Simms,
Contributing Editors: Helen Ingram, David Feldman, Nathan Mantua,
Katharine L. Jacobs, and Anne Waple

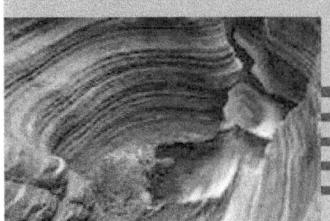

November, 2008

Members of Congress:

On behalf of the National Science and Technology Council, the U.S. Climate Change Science Program (CCSP) is pleased to transmit to the President and the Congress this Synthesis and Assessment Product (SAP) *Decision-Support Experiments and Evaluations using Seasonal-to-Interannual Forecasts and Observational Data: A Focus on Water Resources*. This is part of a series of 21 SAPs produced by the CCSP aimed at providing current assessments of climate change science to inform public debate, policy, and operational decisions. These reports are also intended to help the CCSP develop future program research priorities.

The CCSP's guiding vision is to provide the Nation and the global community with the science-based knowledge needed to manage the risks and capture the opportunities associated with climate and related environmental changes. The SAPs are important steps toward achieving that vision and help to translate the CCSP's extensive observational and research database into informational tools that directly address key questions being asked of the research community.

This SAP evaluates decision support experiments that have used seasonal-to-interannual forecasts and observational data. It was developed with broad scientific input and in accordance with the Guidelines for Producing CCSP SAPs, the Information Quality Act (Section 515 of the Treasury and General Government Appropriations Act for Fiscal Year 2001 (Public Law 106-554)), and the guidelines issued by the Department of Commerce and the National Oceanic and Atmospheric Administration pursuant to Section 515.

We commend the report's authors for both the thorough nature of their work and their adherence to an inclusive review process.

Sincerely,

Carlos M. Gutierrez
Secretary of Commerce
Chair, Committee on Climate Change
Science and Technology Integration

Samuel W. Bodman
Secretary of Energy
Vice Chair, Committee on Climate
Change Science and Technology
Integration

John H. Marburger III
Director, Office of Science and
Technology Policy
Executive Director, Committee
on Climate Change Science and
Technology Integration

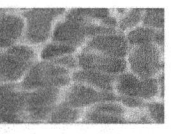

TABLE OF CONTENTS

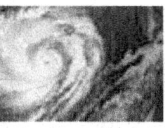

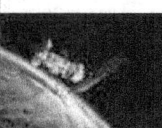

Contents:

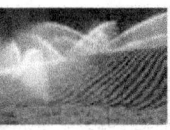
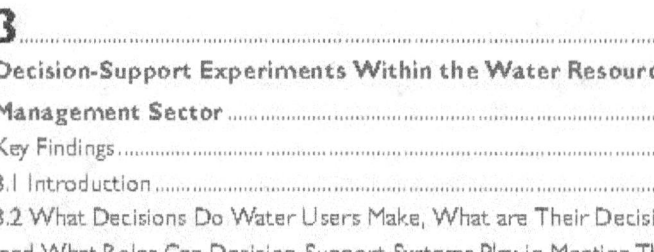
TABLE OF CONTENTS

TABLE OF CONTENTS

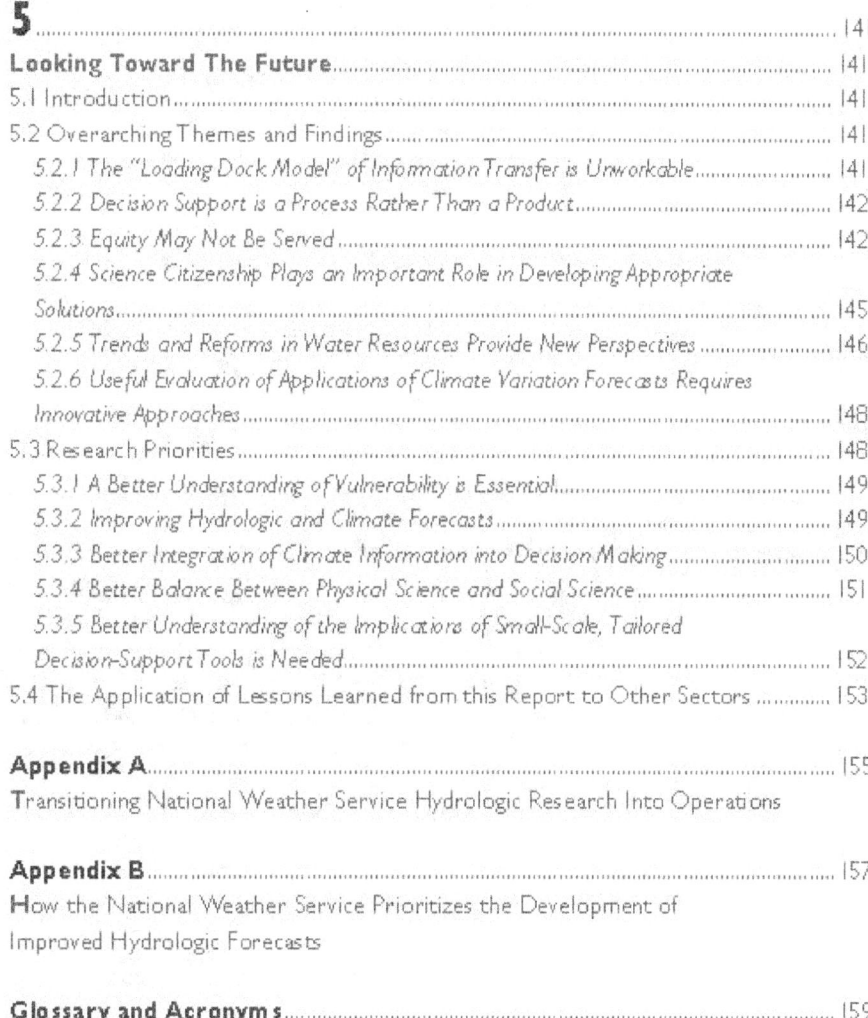

AUTHOR TEAM FOR THIS REPORT

Preface
Convening Lead Author: Nancy Beller-Simms, NOAA
Lead Authors: Helen Ingram, Univ. of Arizona; David Feldman, Univ. of California, Irvine; Nathan Mantua, Climate Impacts Group, Univ. of Washington; Katharine L. Jacobs, Arizona Water Institute

Executive Summary
Convening Lead Author: Helen Ingram, Univ. of Arizona
Lead Authors: David Feldman, Univ. of California, Irvine; Nathan Mantua, Climate Impacts Group, Univ. of Washington; Katharine L. Jacobs, Arizona Water Institute
Contributing Authors: Nancy Beller-Simms, NOAA; Anne M. Waple, STG, Inc.

Chapter 1
Convening Lead Author: Helen Ingram, Univ. of Arizona
Lead Authors: David Feldman, Univ. of California, Irvine; Nathan Mantua, Climate Impacts Group, Univ. of Washington; Katharine L. Jacobs, Arizona Water Institute; Denise Fort, Univ. of New Mexico
Contributing Authors: Nancy Beller-Simms, NOAA; Anne M. Waple, STG, Inc.

Chapter 2
Convening Lead Author: Nathan Mantua, Climate Impacts Group, Univ. of Washington
Lead Authors: Michael D. Dettinger, U.S. Geological Survey, Scripps Institution of Oceanography; Thomas C. Pagano, National Water and Climate Center, NRCS/USDA; Andrew W. Wood, 3TIER™, Inc / Dept. of Civil and Environmental Engineering, Univ. of Washington; Kelly Redmond, Western Regional Climate Center, Desert Research Institute
Contributing Author: Pedro Restrepo, NOAA

Chapter 3
Convening Lead Authors: David L. Feldman, Univ. of California, Irvine; Katharine L. Jacobs, Arizona Water Institute
Lead Authors: Gregg Garfin, Univ. of Arizona; Aris Georgakakos, Georgia Institute of Technology; Barbara Morehouse, Univ. of Arizona; Robin Webb, NOAA; Brent Yarnal, Penn. State Univ.
Contributing Authors: John Kochendorfer, Riverside Technology, Inc.; Cynthia Rosenzweig, NASA; Michael Sale, Oak Ridge National Laboratory; Brad Udall, NOAA; Connie Woodhouse, Univ. of Arizona

Chapter 4
Convening Lead Authors: David L. Feldman, Univ. of California, Irvine; Katharine L. Jacobs, Arizona Water Institute
Lead Authors: Gregg Garfin, Univ. of Arizona; Aris Georgakakos, Georgia Institute of Technology; Barbara Morehouse, Univ. of Arizona; Pedro Restrepo, NOAA; Robin Webb, NOAA; Brent Yarnal, Penn. State Univ.
Contributing Authors: Dan Basketfield, Silverado Gold Mines Inc.; Holly C. Hartmann, Univ. of Arizona; John Kochendorfer, Riverside Technology, Inc.; Cynthia Rosenzweig, NASA; Michael Sale, Oak Ridge National Laboratory; Brad Udall, Univ. of Colorado; Connie Woodhouse, Univ. of Arizona

Chapter 5
Convening Lead Author: Helen Ingram, Univ. of Arizona;
Lead Authors: David L. Feldman, Univ. of California, Irvine; Katharine L. Jacobs, Arizona Water Institute; Nathan Mantua, Climate Impacts Group, Univ. of Washington; Maria Carmen Lemos, Univ. of Michigan; Barbara Morehouse, Univ. of Arizona
Contributing Authors: Nancy Beller-Simms, NOAA; Anne M. Waple, STG, Inc.

Appendix A See Chapter 2 Author List
Appendix B See Chapter 2 Author List

ACKNOWLEDGEMENTS

The authors would like to thank the members of the CCSP Synthesis and Assessment 5.3 Working Committee, who developed the original Prospectus for this Product. Members included from the National Oceanic and Atmospheric Administration: Nancy Beller-Simms, Claudia Nierenberg, Michael Brewer, and Pedro Restrepo; from National Aeronautics and Space Administration: Shahid Habib; from the US Environmental Protection Agency: Janet Gamble; from US Geological Survey: Ron Berenknopf; and from the National Science Foundation: L. Douglas James.

The authors would also like to thank the stakeholders and experts who attended and testified at our open meeting in January 2007: Eric Kuhn, General Manager - Colorado River Water Conservation District; Brett Rosenberg, United States Council of Mayors; Michael J. Sale, Oak Ridge National Laboratory; Joe Grindstaff, Director, The CALFED Bay-Delta Program; Jayantha Obeysekera, Director - Office of Modeling, South Florida Water Management District; Jan Dutton, AWS Convergence Technologies, Inc.; Ed O' Lenic, Chief, Climate Operations Branch, NOAA Climate Prediction Center; and Arun Kumar Chief, Development Branch NOAA Climate Prediction Center. Their participation provided the writing team with new insights into the successes and failures of the use of climate information in decision making.

The authors are also indebted to Joseph Wilder, Director of the University of Arizona Southwest Center who took care of us during our author team meeting in January 2008 by providing meeting rooms, transportation, and warm hospitality.

CCSP Synthesis and Assessment Product 5.3 (SAP 5.3) was developed with the benefit of a scientifically rigorous, first draft peer review conducted by a committee appointed by the National Research Council (NRC). Prior to their delivery to the SAP 5.3 Author Team, the NRC review comments, in turn, were reviewed in draft form by a second group of highly qualified experts to ensure that the review met NRC standards. The resultant NRC Review Report was instrumental in shaping the revised version of SAP 5.3, and in improving its completeness, sharpening its focus, communicating its conclusions and recommendations, and improving its general readability. We wish to thank the members of the NRC Review Committee: Soroosh Sorooshian (Chair), Departments of Civil and Environmental Engineering and of Earth System Science, University of California, Irvine; Kirstin Dow, Department of Geography, University of South Carolina, Columbia; John A. Dracup, Department of Civil and Environmental Engineering, University of California, Berkeley; Lisa Goddard, International Research Institute for Climate and Society, Columbia University; Michael Hanemann, Department of Agricultural and Resource Economics, University of California, Berkeley; Denise Lach, Department of Sociology, Oregon State University, Corvallis; Doug Plasencia, Michael Baker, Jr., Inc., Phoenix, Arizona; Paul C. Stern, Study Director; Jennifer F. Brewer, Staff Officer; and Linda Depugh, Administrative Assistant.

We also thank the individuals who reviewed the NRC Report in its draft form: Richard G. Lawford, Global Energy and Water Cycle Experiment (GEWEX), Silver Spring, MD; Kathleen A. Miller, Institute for the Study of Society and Environment, National Center for Atmospheric Research; and Alex Rothman, Department of Psychology, University of Minnesota, and Roger E. Kasperson, George Perkins Marsh Institute, Clark University, who oversaw the report review.

The review process for SAP 5.3 also included a public and interagency review of the Second Draft, and we would like to thank John D. Wiener, Program on Environment and Society, Institute of Behavioral Science, University of Colorado; Jerry Elwood, Department of Energy; and Samuel P. Williamson who participated in this cycle. The Author Team carefully considered all comments submitted, and a substantial number resulted in further improvements and clarity of SAP 5.3.

Finally, it should be noted that the respective review bodies were not asked to endorse the final version of SAP 5.3, as this was the responsibility of the National Science and Technology Council.

ABSTRACT

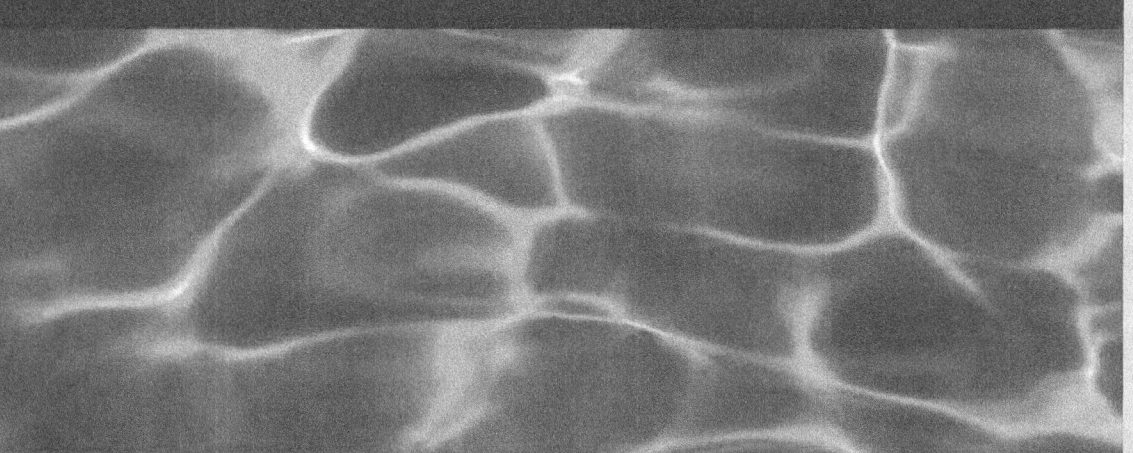

Faced with mounting pressures from a changing climate, an increasing population, a transitory populace, and varying access to available natural resources, decision makers, scientists, and resource managers have an immediate need to understand, obtain, and better integrate climate forecasts and observational data in near- and long-term planning. Reducing our societal vulnerability to variabilities and changes in climate depends upon our ability to bridge the gap between climate science and the implementation of scientific understanding in our management of critical resources, arguably the most important of which is water. Our ability to adapt and respond to climate variability and change depends, in large part, on our understanding of the climate and how to incorporate this understanding into our resource management decisions. This Product focuses on the connection between the scientific ability to predict climate on seasonal scales and the opportunity to incorporate such understanding into water resource management decisions. It directly addresses decision support experiments and evaluations that have used seasonal-to-interannual forecasts and observational data, and is expected to inform (1) decision makers about the relative success of experiences of others who have experimented with these forecasts and data in resource management; (2) climatologists, hydrologists, and social scientists on how to advance the delivery of decision-support resources that use the most recent forecast products, methodologies, and tools; and (3) science and resource managers as they plan for future investments in research related to forecasts and their role in decision support. It is important to note, however, that while the focus of this Product is on the water resources management sector, the findings within this Synthesis and Assessment Product may be directly transferred to other sectors.

RECOMMENDED CITATIONS

For the Report as a whole:

CCSP, 2008: *Decision-Support Experiments and Evaluations using Seasonal-to-Interannual Forecasts and Observational Data: A Focus on Water Resources.* A Report by the U.S. Climate Change Science Program and the Subcommittee on Global Change Research [Nancy Beller-Simms, Helen Ingram, David Feldman, Nathan Mantua, Katharine L. Jacobs, and Anne M. Waple (eds.)]. NOAA's National Climatic Data Center, Asheville, NC, 192 pp.

For the Preface:

Beller-Simms, N., H. Ingram, D. Feldman, N. Mantua, and K.L. Jacobs, 2008: Preface. In: *Decision-Support Experiments and Evaluations using Seasonal-to-Interannual Forecasts and Observational Data: A Focus on Water Resources.* A Report by the U.S. Climate Change Science Program and the Subcommittee on Global Change Research [Nancy Beller-Simms, Helen Ingram, David Feldman, Nathan Mantua, Katharine L. Jacobs, and Anne M. Waple (eds.)]. NOAA's National Climatic Data Center, Asheville, NC, pp. XI–XII.

For the Executive Summary:

Ingram, H., D. Feldman, N. Mantua, K.L. Jacobs, A. M. Waple, and N. Beller-Simms, 2008: Executive Summary. In: *Decision-Support Experiments and Evaluations using Seasonal-to-Interannual Forecasts and Observational Data: A Focus on Water Resources.* A Report by the U.S. Climate Change Science Program and the Subcommittee on Global Change Research [Nancy Beller-Simms, Helen Ingram, David Feldman, Nathan Mantua, Katharine L. Jacobs, and Anne M. Waple (eds.)]. NOAA's National Climatic Data Center, Asheville, NC, pp. 1-6.

For Chapter 1:

Ingram, H., D. Feldman, N. Mantua, K.L. Jacobs, D. Fort, N. Beller-Simms, and A. M. Waple, 2008: The Changing Context. In: *Decision-Support Experiments and Evaluations using Seasonal-to-Interannual Forecasts and Observational Data: A Focus on Water Resources.* A Report by the U.S. Climate Change Science Program and the Subcommittee on Global Change Research [Nancy Beller-Simms, Helen Ingram, David Feldman, Nathan Mantua, Katharine L. Jacobs, and Anne M. Waple (eds.)]. NOAA's National Climatic Data Center, Asheville, NC, pp. 7-28.

For Chapter 2:

Mantua, N., M.D. Dettinger, T.C. Pagano, A.W. Wood, K. Redmond, and P. Restrepo, 2008: A Description and Evaluation of Hydrologic and Climate Forecast and Data Products that Support Decision-Making for Water Resource Managers. In: *Decision-Support Experiments and Evaluations using Seasonal-to-Interannual Forecasts and Observational Data: A Focus on Water Resources.* A Report by the U.S. Climate Change Science Program and the Subcommittee on Global Change Research [Nancy Beller-Simms, Helen Ingram, David Feldman, Nathan Mantua, Katharine L. Jacobs, and Anne M. Waple (eds.)]. NOAA's National Climatic Data Center, Asheville, NC, pp. 29-64.

For Chapter 3:

Feldman, D.L., K.L. Jacobs, G. Garfin, A. Georgakakos, B. Morehouse, R. Webb, B. Yarnal, J. Kochendorfer, C. Rosenzweig, M. Sale, B. Udall, and C. Woodhouse, 2008: Decision-Support Experiments Within the Water Resource Management Sector. In: *Decision-Support Experiments and Evaluations using Seasonal-to-Interannual Forecasts and Observational Data: A Focus on Water Resources.* A Report by the U.S. Climate Change Science Program and the Subcommittee on Global Change Research [Nancy Beller-Simms, Helen Ingram, David Feldman, Nathan Mantua, Katharine L. Jacobs, and Anne M. Waple (eds.)]. NOAA's National Climatic Data Center, Asheville, NC, pp. 65-100.

For Chapter 4:

Feldman, D.L., K.L. Jacobs, G. Garfin, A. Georgakakos, B. Morehouse, P. Restrepo, R. Webb, B. Yarnal, D. Basketfield, H.C. Hartmann, J. Kochendorfer, C. Rosenzweig, M. Sale, B. Udall, and C. Woodhouse, 2008: Making Decision-Support Information Useful, Useable, and Responsive to Decision-Maker Needs. In: *Decision-Support Experiments and Evaluations using Seasonal-to-Interannual Forecasts and Observational Data: A Focus on Water Resources.* A Report by the U.S. Climate Change Science Program and the Subcommittee on Global Change Research [Nancy Beller-Simms, Helen Ingram, David Feldman, Nathan Mantua, Katharine L. Jacobs, and Anne M. Waple (eds.)]. NOAA's National Climatic Data Center, Asheville, NC, pp. 101-140.

For Chapter 5

Ingram, H., D.L. Feldman, K.L. Jacobs, N. Mantua, M.C. Lemos, B. Morehouse, A. M. Waple, and N. Beller-Simms, 2008: Looking Toward the Future. In: *Decision-Support Experiments and Evaluations using Seasonal-to-Interannual Forecasts and Observational Data: A Focus on Water Resources.* A Report by the U.S. Climate Change Science Program and the Subcommittee on Global Change Research [Nancy Beller-Simms, Helen Ingram, David Feldman, Nathan Mantua, Katharine L. Jacobs, and Anne M. Waple (eds.)]. NOAA's National Climatic Data Center, Asheville, NC, pp. 141-154.

For Appendix A

Mantua, N., M.D. Dettinger, T.C. Pagano, A.W. Wood, K. Redmond, and P. Restrepo, 2008: Transitioning the National Weather Service Hydrologic Research into Operations. In: *Decision-Support Experiments and Evaluations using Seasonal-to-Interannual Forecasts and Observational Data: A Focus on Water Resources.* A Report by the U.S. Climate Change Science Program and the Subcommittee on Global Change Research [Nancy Beller-Simms, Helen Ingram, David Feldman, Nathan Mantua, Katharine L. Jacobs, and Anne M. Waple (eds.)]. NOAA's National Climatic Data Center, Asheville, NC, pp. 155-156.

For Appendix B

Mantua, N., M.D. Dettinger, T.C. Pagano, A.W. Wood, K. Redmond, and P. Restrepo, 2008: How the National Weather Service Prioritizes the Development of Improved Hydrologic Forecasts. In: *Decision-Support Experiments and Evaluations using Seasonal-to-Interannual Forecasts and Observational Data: A Focus on Water Resources.* A Report by the U.S. Climate Change Science Program and the Subcommittee on Global Change Research [Nancy Beller-Simms, Helen Ingram, David Feldman, Nathan Mantua, Katharine L. Jacobs, and Anne M. Waple (eds.)]. NOAA's National Climatic Data Center, Asheville, NC, pp. 157-158.

PREFACE

Report Motivation and Guidance for Using this Synthesis/Assessment Report

Convening Lead Author: Nancy Beller-Simms, NOAA

Lead Authors: Helen Ingram, Univ. of Arizona; David Feldman, Univ. of California, Irvine; Nathan Mantua, Climate Impacts Group, Univ. of Washington; Katharine L. Jacobs, Arizona Water Institute

Editor: Anne M. Waple, STG, Inc.

P.1 MOTIVATION AND GUIDANCE FOR USING THIS SYNTHESIS AND ASSESSMENT PRODUCT

The core mission of the U.S. Climate Change Science Program (CCSP) is to "Facilitate the creation and application of knowledge of the Earth's global environment through research, observations, decision support, and communication". To accomplish this goal, the CCSP has commissioned 21 Synthesis and Assessment Products to summarize current knowledge and evaluate the extent and development of this knowledge for future scientific explorations and policy planning.

These Products fall within five goals, namely:
1. Improve knowledge of the Earth's past and present climate and environment, including its natural variability, and improve understanding of the causes of observed variability and change;
2. Improve quantification of the forces bringing about changes in the Earth's climate and related systems;
3. Reduce uncertainty in projections of how the Earth's climate and environmental systems may change in the future;
4. Understand the sensitivity and adaptability of different natural and managed ecosystems and human systems to climate and related global changes; and
5. Explore the uses and identify the limits of evolving knowledge to manage risks and opportunities related to climate variability and change.

CCSP Synthesis and Assessment Product 5.3 is one of three products to be developed for the final goal.

This Product directly addresses decision-support experiments and evaluations that have used seasonal-to-interannual forecasts and observational data, and is expected to inform (1) decision makers about the experiences of others

who have experimented with these forecasts and data in resource management; (2) climatologists, hydrologists, and social scientists on how to advance the delivery of decision-support resources that use the most recent forecast products, methodologies, and tools; and (3) science and resource managers as they plan for future investments in research related to forecasts and their role in decision support.

P.2 BACKGROUND

Gaining a better understanding of how to provide better decision support to decision and policy makers is of prime importance to the CCSP, and it has put considerable effort and resources towards achieving this goal. For example, within its Strategic Plan, the CCSP identifies decision support as one of its four core approaches to achieving its mission[1]. The plan endorses the transfer of knowledge gained from science in a format that is usable and understandable, and indicates levels of uncertainty and confidence. CCSP expects that the resulting tools will promote the development of new models, tools, and methods that will improve current economic and policy analyses as well as advance environmental management and decision making.

CCSP has also encouraged the authors of the 21 Synthesis and Assessment Products to support informed decision making on climate variability and change. Most of the Synthesis and Assessment Products' Prospectuses have outlined efforts to involve decision makers, including a broad group of stakeholders, policy makers, resource managers, media, and the general public, as either writers or as special workshop/meeting participants. Inclusion of decision makers in the Synthesis and Assessment Products also helps to fulfill the requirements of the Global Change Research Act (GCRA) of 1990 (P.L. 101-606, Section 106), which directs the program

---

[1] The four core approaches of CCSP include science, observations, decision support, and communications.

to "produce information readily usable by policymakers attempting to formulate effective strategies for preventing, mitigating, and adapting to the effects of global change" and to undertake periodic science "assessments".

In November 2005, the CCSP held a workshop to address the potential of those working in the climate sciences to inform decision and policy makers. The workshop included discussions about decision-maker needs for scientific information on climate variability and change. It also addressed future steps, including the completion of this and other Synthesis and Assessment Products, for research and assessment activities that are necessary for sound resource management, adaptive planning, and policy formulation. The audience included representatives from academia; governments at the state, local, and national levels; non-governmental organizations (NGOs); decision makers, including resource managers and policy developers; members of Congress; and the private sector.

P.3 FOCUS OF THIS SYNTHESIS AND ASSESSMENT PRODUCT

In response to the 2003 Strategic Plan for the Climate Change Science Program Office, which recommended the creation of a series of Synthesis and Assessment Product reports, the National Oceanic and Atmospheric Administration (NOAA) took responsibility for this Product. An interagency group comprised of representatives from NOAA, National Aeronautics and Space Administration, U.S. Environmental Protection Agency, U.S. Geological Survey and National Science Foundation wrote the Prospectus[2] for this Product and recommended that this Synthesis and Assessment Product should concentrate on the water resource management sector. This committee felt that focusing on a single sector would allow for a detailed synthesis of lessons learned in decision-support experiments within that sector. These lessons, in turn, would be relevant, transferable, and essential to other climate-sensitive resource management sectors. Water resource management was selected, as it was the most relevant of the sectors proposed and would be of interest to all agencies participating in this process. The group wrote a Prospectus and posed a series of questions that they felt the CCSP 5.3 Product authors should address in this Report. Table 1.2 lists these questions and provides the location within the Synthesis and Assessment Product where the authors addressed them.

P.4 THE SYNTHESIS AND ASSESSMENT WRITING TEAM

This study required an interdisciplinary team that was able to integrate scientific understandings about forecast and data products with a working knowledge of the needs of water resource managers in decision making. As a result, the team included researchers, decision makers, and federal government employees with varied backgrounds in the social sciences, physical sciences, and law. The authors were identified based on a variety of considerations, including their past interests and involvements with decision-support experiments and their knowledge of the field as demonstrated by practice and/or involvement in research and/or publications in refereed journals. In addition, the authors held a public meeting, in January 2007, in which they invited key stakeholders to discuss their decision support experiments with the committee. Working with authors and stakeholders with such varied backgrounds presented some unique challenges including preconceived notions of other disciplines, as well as the realization that individual words have different meanings in the diverse disciplines. For example, those with a physical science background understood a more quantifiable definition for the words 'confidence' and 'uncertainty' than the more qualitative (*i.e.*, behavioral) view of the social scientists.

The author team for this Product was constituted as a Federal Advisory Committee in accordance with the Federal Advisory Committee Act of 1972 as amended, 5 U.S.C. App.2. The full list of the author team, in addition to a list of lead authors provided at the beginning of each Chapter, is provided on page 3 of this Report. The editorial staff reviewed the scientific and technical input and managed the assembly, formatting, and preparation of the Product.

---

[2]   The Prospectus is posted on the Climate Change Science Program website at: http://www.climatescience.gov.

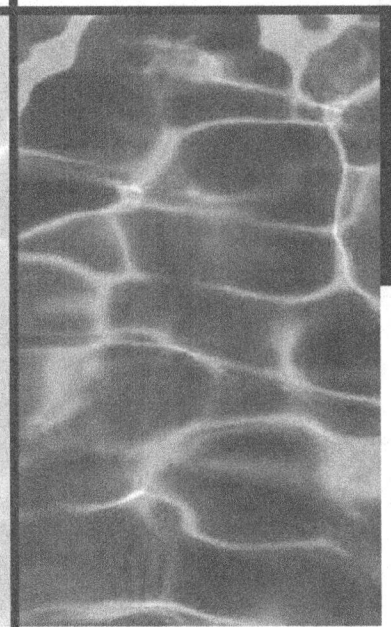

Convening Lead Author: Helen Ingram, Univ. of Arizona

Lead Authors: David Feldman, Univ. of California, Irvine; Nathan Mantua, Climate Impacts Group, Univ. of Washington; Katharine L. Jacobs, Arizona Water Institute

Contributing Authors: Anne M. Waple, STG, Inc.; Nancy Beller-Simms, NOAA

EXECUTIVE SUMMARY

ES.1 WHAT IS DECISION SUPPORT AND WHY IS IT NECESSARY?

Earth's climate is naturally varying and also changing in response to human activity. Our ability to adapt and respond to climate variability and change depends, in large part, on our understanding of the climate and how to incorporate this understanding into our resource management decisions. Water resources, in particular, are directly dependent on the abundance of rain and snow, and how we store and use the amount of water available. With an increasing population, a changing climate, and the expansion of human activity into semi-arid regions of the United States, water management has unique and evolving challenges. This Product focuses on the connection between the scientific ability to predict climate on seasonal scales and the opportunity to incorporate such understanding into water resource management decisions. Reducing our societal vulnerability to changes in climate depends upon our ability to bridge the gap between climate science and the implementation of scientific understanding in our management of critical resources, arguably the most important of which is water. It is important to note, however, that while the focus of this Product is on the water resources management sector, the findings within this Synthesis and Assessment Product may be directly transferred to other sectors.

The ability to predict many aspects of climate and hydrologic variability on seasonal-to-interannual time scales is a significant success in Earth systems science. Connecting the improved understanding of this variability to water resources management is a complex and evolving challenge. While much progress has been made, conveying climate and hydrologic forecasts in a form useful to real world decision making introduces complications that call upon the

skills of not only climate scientists, hydrologists, and water resources experts, but also social scientists with the capacity to understand and work within the dynamic boundaries of organizational and social change.

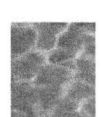

Up until recent years, the provision of climate and hydrologic forecast products has been a producer-driven rather than a user-driven process. The momentum in product development has been largely skill-based rather than a response to demand from water managers. It is now widely accepted that there is considerable potential for increasing the use and utility of climate information for decision support in water resources management even without improving the skill level of climate and hydrologic forecasts. The outcomes of "experiments" intended to deliver climate-related decision support through "knowledge-to-action networks" in water resource related problems are encouraging.

Linkages between climate and hydrologic scientists are getting stronger as they now more frequently collaborate to create forecast products. A number of complex factors influence the rate at which seasonal water supply forecasts and climate-driven hydrologic forecasts are improving in terms of skill level. Mismatches between needs and information resources continue to occur at multiple levels and scales. Currently, there is substantial tension between providing tools at the space and time scales useful for water resources decisions that are also scientifically accurate, reliable, and timely.

The concept of decision support has evolved over time. Early in the development of climate information tools, decision support meant the translation and delivery of climate science information into forms believed to be useful to decision makers. With experience, it became clear that climate scientists often did not know what kind of information would be useful to decision makers. Further, decision makers who had never really considered the possibility of using climate information were not yet in a position to articulate what they needed. It became obvious that user groups had to be involved at the point at which climate information began to be developed. Making climate science useful to decision makers involves a process in which climate scientists, hydrologists, and the potential users of their products engage in an interactive dialogue during which trust and confidence is built at the same time that climate information is exchanged.

The institutional framework in which decision-support experiments are developed has important effects. Currently there is a disconnect between agency-led operational forecasts and experimental hydrologic forecasts being carried out in universities. However, as shown by the experiments highlighted in this Product, it is possible to develop decision-support tools, processes and institutions that are relevant to different geographical scales and are sufficiently flexible to serve a diverse body of users. Such tools and processes can reveal commonalities of interests and shared vulnerabilities that are otherwise obscure. Well-designed tools, institutions, and processes can clarify necessary trade-offs of short- and long-term gains and losses to potentially competing values associated with water allocation and management.

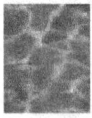

Evidence suggests that many of the most successful applications of climate information to water resource problems occur when committed leaders are poised and ready to take advantage of unexpected opportunities. In evaluating the ways in which science-based climate information is finding its way to users, it is important to recognize that straightforward, goal-driven processes do not characterize the real world. We usually think of planning and innovation as a linear process, but experience shows us that, in practice, it is a nonlinear, chaotic process with emergent properties. This is particularly true when working with climate impacts and resource management. It is clear that we must address problems in new ways and understand how to encourage diffusion of innovations.

The building of knowledge networks is a valuable way to provide decision support and pursue strategies to put knowledge to use. Knowledge networks require widespread, sustained human efforts that persist through time. Collaboration and adaptive management efforts among resource managers and forecast producers with different missions show that mutual learning informed by climate information can occur between scientists with different disciplinary backgrounds and between scientists and managers. The benefits of such linkages and relation-

ships are much greater than the costs incurred to create and maintain them, however, the opportunities to build these associations are often neglected or discouraged. Collaborations across organizational, professional, disciplinary, and other boundaries are often not given high priority; incentives and reward structures need to change to take advantage of these opportunities. In addition, the problem of data overload for people at critical junctions of information networks, and for people in decision-making capacity such as those of resource managers and climate scientists, is a serious impediment to innovation.

Decision-support experiments employing climate related information have had varying levels of success in integrating their findings with the needs of water and other resource managers.

ES.2 CLIMATE AND HYDROLOGIC FORECASTS: THE BASIS FOR MAKING INFORMED DECISIONS

There are a wide variety of climate and hydrologic data and forecast products currently available for use by decision makers in the water resources sector. However, the use of official seasonal-to-interannual (SI) climate and hydrologic forecasts generated by federal agencies remains limited in this sector. Forecast skill, while recognized as just one of the barriers to the use of SI climate forecast information, remains a primary concern among forecast producers and users. Simply put, there is no incentive to use SI climate forecasts when they are believed to provide little additional skill to existing hydrologic and water resource forecast approaches (described in Chapter 2). Not surprisingly, there is much interest in improving the skill of hydrologic and water resources forecasts. Such improvements can be realized by pursuing several research pathways, including:

- Improved monitoring and assimilation of real-time hydrologic observations in land surface hydrologic models that leads to improved estimates for initial hydrologic states in forecast models;
- Increased accuracy in SI climate forecasts; and

- Improved bias corrections in existing forecasts.

Another aspect of forecasts that serves to limit their use and utility is the challenge in interpreting forecast information. For example, from a forecast producer's perspective, confidence levels are explicitly and quantitatively conveyed by the range of possibilities described in probabilistic forecasts. From a forecast user's perspective, probabilistic forecasts are not always well understood or correctly interpreted. Although structured user testing is known to be an effective product development tool, it is rarely done. Evaluation should be an integral part of improving forecasting efforts, but that evaluation should be extended to factors that encompass use and utility of forecast information for stakeholders. In particular, very little research is done on effective SI forecast communication. Instead, users are commonly engaged only near the end of the product development process.

Other barriers to the use of SI climate forecasts in water resources management have been identified and those that relate to institutional issues and aspects of current forecast products are discussed in Chapters 3 and 4 of this Product.

Pathways for expanding the use and improving the utility of data and forecast products to support decision making in the water resources sector are currently being pursued at a variety of spatial and jurisdictional scales in the United States. These efforts include:

- An increased focus on developing forecast evaluation tools that provide users with opportunities to better understand forecast products in terms of their expected skill and applicability;
- Additional efforts to explicitly and quantitatively link SI climate forecast information with SI hydrologic and water supply forecasting efforts;
- An increased focus on developing new internet-based tools for accessing and customizing data and forecast products to support hydrologic forecasting and water resources decision making (e.g., the Advanced Hydrologic Prediction Service [AHPS] described in Chapters 2 and 3); and

- Further improvements in the skill of hydrologic and water supply forecasts.

Many of these pathways are currently being pursued by the federal agencies charged with producing the official climate and hydrologic forecast and data products for the United States, but there is substantial room for increasing these activities.

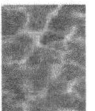

Recent improvements in the use and utility of data and forecast products related to water resources decision making have come with an increased emphasis on these issues in research funding agencies through programs like the National Oceanic and Atmospheric Administration's Regional Integrated Sciences and Assessments (RISA), Sectoral Applications Research Program (SARP), Transition of Research Applications to Climate Services (TRACS) and Climate Prediction Program for the Americas (CPPA) and the World Climate Research Programme's Global Energy and Water Cycle Experiment (GEWEX) programs. Sustaining and accelerating future improvements in the use and utility of official data and forecast products in the water resources sector rests in part on investments in programs focused on improving the skill in forecasts, increasing the access to data and forecast products, identifying processes that influence the creation of knowledge-to-action networks for making climate information useful for decision making, and fostering sustained interactions between forecast producers and consumers.

ES.3 DECISION-SUPPORT EXPERIMENTS IN THE WATER RESOURCE SECTOR

Decision-support experiments that test the utility of SI information for use by water resource decision makers have resulted in a growing set of successful applications. However, there is significant opportunity for expansion of applications of climate-related data and decision-support tools, and for developing more regional and local tools that support management decisions within watersheds. Among the factors as to how and/or whether tools are used depends on:

- The range and complexity of water resources decisions. This is compounded by the numerous organizations responsible for making these decisions and the shared responsibility for implementing them.

- Policies and organizational rules that impact the rate at which innovation occurs. Some larger institutions have historically been reluctant to change practices, in part because of value differences, risk aversion, fragmentation, and sharing of authority. This conservatism impacts how decisions are made as well as whether to use newer, scientifically generated information, including SI forecasts and observational data." However, its not necessarily true that policies and rule inhibit all innovation, or that policies and rules are always inflexible. In fact many policies are specifically developed to advance innovation and the quality of information can promote use even under unfavorable circumstances.

- Different spatial and temporal frames for decisions. Spatial scales for decision making range from local, state, and national levels to international. Temporal scales range from hours to multiple decades impacting policy, operational planning, operational management, and near real-time operational decisions. Resource managers often make multi-dimensional decisions spanning various spatial and temporal frames.

- Communication of risks differs among scientific, political, and mass media elites, each systematically selecting aspects of these issues that are most salient to their conception of risk, and thus, socially constructing and communicating its aspects most salient to a particular perspective.

Decision-support systems are not often well integrated into planning and management activities, making it difficult to realize the full benefits of these tools. Because use of many climate products requires special training or access to data that are not readily available, decision-support products may not equitably reach all audiences. Moreover, over-specialization and narrow disciplinary perspectives make it difficult for information providers, decision makers, and the public to communicate with one another. Three lessons stem from this:

- Decision makers need to understand the types of predictions that can be made, and

the tradeoffs between longer-term predictions of information at the local or regional scale on one hand, and potential decreases in accuracy on the other.

- Decision makers and scientists need to work together in formulating research questions relevant to the spatial and temporal scale of problems the former manage.

- Scientists should aim to generate findings that are accessible and viewed as useful, accurate, and trustworthy by stakeholders.

ES.4 MAKING DECISION-SUPPORT INFORMATION USEFUL, USEABLE, AND RESPONSIVE TO DECISION-MAKER NEEDS

Decision-support experiments that apply SI climate variability information to basin and regional water resource problems serve as testbeds that address diverse issues faced by decision makers and scientists. They illustrate how to articulate user needs, overcome communication barriers, and operationalize forecast tools. They also demonstrate how user participation can be incorporated in tool development.

Five major lessons emerge from these experiments and supporting analytical studies:

- The effective integration of SI climate information in decisions requires long-term collaborative research and application of decision support through identifying problems of mutual interest. This collaboration will require a critical mass of scientists and decision makers to succeed, and there is currently an insufficient number of "integrators" of climate information for specific applications.

- Investments in long-term research-based relationships between scientists and decision makers must be encouraged. In general, progress on developing effective decision-support systems is dependent on additional public and private interest and efforts to facilitate better networking among decision makers and scientists at all levels as well as public engagement in the fabric of decision making.

- Effective decision-support tools must wed national production of data and technologies to ensure efficient, cross-sector useful-

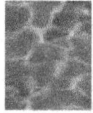

ness with customized products for local users. This requires that tool developers engage a wide range of participants, including those who generate tools and those who translate them, to ensure that specially-tailored products are widely accessible and are immediately adopted by users insuring relevancy and utility.

- The process of tool development must be inclusive, interdisciplinary, and provide ample dialogue among researchers and users. To achieve this inclusive process, professional reward systems that recognize people who develop, use, and translate such systems for use by others are needed within management and related agencies, universities, and organizations. Critical to this effort, further progress in boundary spanning—the effort to translate tools to a variety of audiences—requires considerable organizational skills.

- Information generated by decision-support tools must be implementable in the short term for users to foresee progress and support further tool development. Thus, efforts must be made to effectively integrate public concerns and elicit public information through dedicated outreach programs.

ES.5 LOOKING TOWARD THE FUTURE; RESEARCH PRIORITIES

A few central themes emerge from this Product, and are summarized in this Section. Key research priorities are also highlighted.

ES.5.1 Key Themes
1) The "Loading Dock Model" of Information Transfer is Unworkable.
Skill is a necessary ingredient in perceived forecast value, yet more forecast skill by itself does not imply more forecast value. Lack of forecast skill and/or accuracy may be one of the impediments to forecast use, but there are many other barriers as well. Such improvements must be accompanied by better communication and stronger linkages between forecasters and potential users. In this Product, we have stressed that forecasts flow through knowledge networks and across disciplinary and occupational boundaries. Thus, forecasts need to be useful and relevant in the full range from observations to applications, or "end-to-end useful".

2) Decision Support is a Process Rather Than a Product.
As knowledge systems have come to be better understood, providing decision support has come to be understood not as information products but as a communications process that links scientists with users.

3) Equity May Not Be Served.
Information is power in global society and, unless it is widely shared, the gaps between the rich and the poor, and the advantaged and disadvantaged may widen. Efforts to meet, communicate effectively with, and incorporate the perspectives of the poor and disadvantaged require the ability: to transmit and disseminate information in a clear, non-technical and vernacular language; to embrace the actual concerns of farmers, peasants, villagers, *etc.* (*e.g.*, drought, floods, their effects on crops, livelihoods), and to undertake public outreach that elicits the type of information they need – not just the kind of information scientists are likely to generate.

4) Science Citizenship Plays an Important Role in Developing Appropriate Solutions.
A new paradigm in science is emerging, one that emphasizes science-society collaboration and production of knowledge tailored more closely to society's decision-making needs. Concerns about climate impacts on water resource management are among the most pressing problems that require close collaboration between scientists and decision makers.

5) Trends and Reforms in Water Resources Provide New Perspectives.
Some researchers suggest that, since the 1980s, a "new paradigm" or frame for federal water planning has occurred, although no clear change in law has brought this change about. This new paradigm appears to reflect the ascendancy of an environmental protection ethic among the general public. The new paradigm emphasizes greater stakeholder participation in decision making; explicit commitment to environmentally-sound, socially-just outcomes; greater reliance upon drainage basins as planning units; program management via spatial and managerial flexibility, collaboration, participation, and sound, peer-reviewed science; and,

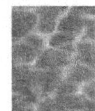

embracing of ecological, economic, and equity considerations.

6) Useful Evaluation of Applications of Climate Variation Forecasts Requires Innovative Approaches.

There can be little argument that SI forecast applications must be evaluated just as most other programs that involve substantial public expenditures are assessed. This Product illustrates many of the difficulties of using standard evaluation techniques.

ES.5.2 Research Priorities

As a result of the findings in this Product, we suggest that a number of research priorities should constitute the focus of attention for the foreseeable future. These priorities (not in order) are:

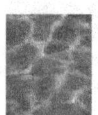

- Improving climate and hydrologic forecasts;
- Improving the communication of uncertainties;
- Enhancing monitoring to better link climate and hydrologic forecasts;
- Expanding our understanding of the decision context within which decision support tools are used,
- Enhancing assessments of decision-maker perceptions of climate risk and vulnerability;
- Understanding the role of public pressures and networks in generating demands for climate information,
- Bettering integration of SI climate science into decision making;
- Improving the generalizability/transferability of case studies on decision-support experiments, and
- Sustaining long-term scientist-decision-maker interactions and collaborations and development of science citizenship and production of knowledge tailored more closely to society's decision-making needs within a variety of natural resource management areas.

CHAPTER 1

The Changing Context

Convening Lead Author: Helen Ingram, Univ. of Arizona

Lead Authors: David Feldman, Univ. of California, Irvine; Nathan Mantua, Climate Impacts Group, Univ. of Washington; Katherine L. Jacobs, Arizona Water Institute; Denise Fort, Univ. of New Mexico

Contributing Authors: Nancy Beller-Simms, NOAA; Anne M. Waple, STG Inc.

1.1 INTRODUCTION

Increasingly frequent headlines such as "UN Calls Water Top Priority" (*The Washington Post, January 25, 2008*), "Drought-Stricken South Facing Tough Choices" (*The New York Times, Oct 15, 2007*), and "The Future is Drying Up" (*The New York Times, October 21, 2007*), coupled with the realities of less-available water, have alerted decision makers, from governors and mayors to individual farmers, that climate information is crucial for future planning. Over the past quarter-century, there have been significant advances in the ability to monitor and predict important aspects of seasonal-to-interannual (SI) variations in climate, especially those associated with variations of the El-Niño Southern Oscillation (ENSO) cycle. Predictions of climate variability on SI time scales are now routine and operational, and consideration of these forecasts in making decisions has become more commonplace. Some water resources decision makers have already begun to use seasonal, interseasonal, and even longer time scale climate forecasts and observational data to assess future options, while others are just beginning to realize the potential of these resources. This Product is designed to show how climate and hydrologic forecast and observational data are being used or neglected by water resources decision makers and to suggest future pathways for increased use of this data.

The Climate Change Science Program (CCSP) included a chapter in its 2003 Strategic Plan that described the critical role of decision support in climate science; previous assessment analyses and case studies have highlighted the importance of assuring that climate information and data would be used by decision makers and not be produced without knowledge of its application. Since that time, there has been increased interest and research in decision-support science focused on organizations using SI forecasts and observational data in future planning. Since the release of the 2003 Strategic Plan, one of the main purposes of CCSP continues to be to "provide information for decision-making through the development of decision-support resources" (CCSP, 2008[1]). As a result, CCSP has charged this author group to produce a Synthesis and Assessment Product (SAP) that directly addresses decision-support experiments and evaluations in the water resources sector. This is that Product.

The authors of this Product concentrated their efforts on discussing SI forecasts and data products. In some cases, however, longer-range forecasts are discussed because they have become a part of the context for decision-making processes. We provided a range of

---

[1] According to this same document, "Decision-support resources, systems, and activities are climate-related products or processes that directly inform or advise stakeholders to help them make decisions".

The impact of climate on water resource management has far-reaching implications for everyone, from the farmer who may need to change the timing of crop planting/ harvesting or the crop type itself, to citizens who may have to relocate because their potable water supply has disappeared.

domestic case study examples, referred to as "experiments and/or evaluations", and have also provided some international examples, where appropriate.

1.2 INCREASING STRESS AND COMPLEXITY IN WATER RESOURCES

Under global warming conditions and an accelerating demand for abundant water supplies, water management may become an increasingly politically charged issue throughout the world in the coming century. Emerging challenges in water quantity, quality, pricing, and water management in relation to seasonal climate fluctuations may increase as the demand for water continues to rise. Though the total volume of water on the planet may be sufficient for societal needs, the largest portion of this water is geographically remote, misallocated, wasted, or degraded by pollution (Whiteley *et al.*, 2008). At the same time, there are shifts in water usage, the societal value of natural water systems, and the laws that govern management of this resource. Accordingly, the impact of climate on water resource management has far-reaching implications for everyone, from the farmer who may need to change the timing of crop planting/harvesting or the crop type itself, to citizens who may have to relocate because their potable water supply has disappeared.

In the United States, water resource decisions are made at multiple levels of government and, increasingly, by the private sector. Water is controlled, guided, governed, or measured by a gamut of federal agencies that oversee various aspects from quality (e.g., U.S. Environmental Protection Agency [EPA]) to quantity (e.g., U.S. Geological Survey [USGS], Bureau of Reclamation [Reclamation], and U.S. Army Corps of Engi-

neers [USACE]). This is complicated by state, regional, and jurisdictional boundaries and responsibilities. Defining a "decision maker" is equally difficult given the complexity of water's use and the types of information that can be used to make decisions. Our challenge in writing this Product is to reflect the various models under which water is managed and the diverse character of decisions that comprise water management. To illustrate, the term "water management" encompasses decisions made by: a municipal water entity regarding when to impose outdoor water restrictions; a federal agency regarding how to operate a storage facility; the United States Congress regarding funding of recovery efforts for an endangered species; and by state governments regarding water purchases necessary to ensure compliance with negotiated compacts.

These types of decisions may be based on multiple factors, such as cost, climate (past trends and future projections), community preferences, political advantage, and strategic concerns for future water decisions. Further, water is associated with many different values including economic security, opportunity, environmental quality, lifestyle, and a sense of place (Blatter and Ingram, 2001). Information about climate variability can be expected to affect some of these decisions and modify some of these values. For other decisions, it may be of remote interest or viewed as entirely irrelevant. For instance, the association of access to water with respect to economic security is relatively fixed while the association of water to lifestyle choices such as a preference for water-based sports may vary with additional information about variability in climate.

The rapidly-closing gap between usable supplies and rising demand is being narrowed by a myriad of factors, including, but not limited to:

- Increasing demand for water with population growth in terms of potable drinking water, agricultural/food requirements, and energy needs.
- Greater political power of recreational and environmental interests that insist on minimum instream flows in rivers.
- Groundwater reserves where development enabled the expansion of agriculture in the

western United States and is the basis for the development of several urban regions. As groundwater reserves are depleted, pressure increases on other water sources.

- Water quality problems that persist in many places, despite decades of regulations and planning.

At the same time, there are some compensating innovations taking place in some areas (see Section 5.2.5).

The best-documented pressure is population growth, which is occurring in the United States as a whole, and especially in the South and Southwest regions where water resources are also among the scarcest. Water rights are afforded to the earliest users in many states, and new users without senior rights often must search for additional supplies. Las Vegas, Nevada is a case study of the measures required to provide water in the desert, but Phoenix, Albuquerque, Denver, Los Angeles and a host of other western cities provide comparable examples. In the southeastern United States, rapid population growth in cities (e.g., Atlanta), combined with poor management and growing environmental concerns that require water to sustain fish and wildlife habitats, have led to serious shortages.

Recreational and environmental interests also have a direct stake in how waters are managed. For example, fishing and boating have increased in importance in recent decades as recreational uses have expanded and the economic basis of our economy has shifted from manufacturing to service.

Groundwater mining is a wild card in national water policy. Water resource allocation is generally a matter of state, not federal, control, and states have different policies with respect to groundwater. Some have no regulation; others permit mining (also referred to as groundwater overdrafting). Because groundwater is not visible and its movement is not well understood, its use is less likely to be regulated than surface water use. The effects of groundwater mining become evident not only in dewatering streams, but also impact regions that must search for alternative sources of water when sources diminish or disappear.

Historically, the solution for a supply-side response to increasing demand has focused on building new reservoirs, new pipelines to import water from distant basins, and new groundwater extraction systems. In the recent past, the United States engaged in an extended period of big dam and aqueduct construction (Worster, 1985) in which most of the appropriate construction sites were utilized. Other options have also been explored such as water reuse. As rivers have become fully appropriated, or over appropriated, there is no longer "surplus" water available for development. Environmental and recreational issues are impacted by further development of rivers, making additional water projects more difficult. Increasing demands for water are not likely to lead to the development of major additional water sources, although additional storage as well as other conservation tools (possibly including but not limited to water reuse, best management practices, and wetland banking) are being considered by water managers; however, it is too early in their evolution and adoption to determine what their impact will be on water supply.

In response to the growing imbalance between demand and supply, water utilities and jurisdictions have been investing in new sources of water and improved system efficiency for decades Reuse of municipal wastewater has become a significant component of the water supply picture in the Southwestern US (California, Arizona, New Mexico, and Texas) and Florida, and is quickly expanding in other regions. It is viewed as a particularly important resource in areas where the population is growing, since production of wastewater generally expands in proportion to the number of households involved as other sources are diminished. Other jurisdictions have tried options such as con-

servation, capturing rainwater for on-site use, improving capture and retention of floodflows, conjunctive management of groundwater and surface water, *etc.*

Many utilities have found that in the absence of a public perception of imminent threat to the adequacy of the water supply, that it is difficult to provide incentives to cause changes in human behavior leading to substantial water conservation because despite its actual value to society, water is relatively inexpensive. Politicians have found that the public does not welcome sharp increases in the price of water, even if the rationale for price increases is well described (Martin, 1984).

Water usage may also be examined by the relative flexibility of each demand. Municipal and industrial demands can be moderated through conservation or temporary restrictions, but these demands are less elastic than agricultural use. Agricultural uses, which comprise the largest users by volume, can be restricted in times of drought without major economic dislocations if properly implemented; however, the increasing connection between water and energy may limit this flexibility. Greater reliance on biofuels both increases competition for scarce water supplies and diverts irrigated agriculture from the production of food to the production of oilseeds such as soybeans, corn, rapeseed, sunflower seed, and sugarcane, among other crops used for biofuel. This changes the pattern of agricultural water use in the United States (Whiteley *et al.*, 2008).

The rationalization of U.S. policies concerning water has been a goal for many decades. Emergent issues of increased climate variability and change may be the agents of transformation for United States water policies as many

Natural disasters, including Hurricane Katrina and recent sustained droughts in the United States, have raised awareness of society's vulnerability to flood, drought, and degradation of water quality.

regions of the country are forced to examine the long term sustainability of water related management decisions (NRC, 1999b; Jacobs and Holway, 2004).

1.2.1 The Evolving Context: The Importance of Issue Frames

In order to fully understand the context in which a decision is made, those in the decision support sciences often look at the "issue frame" or the factors influencing the decision makers, including society's general frame of mind at the time. A common denominator for conceptualizing a frame is the notion that a problem can be understood or conceptualized in different ways (Dewulf *et al.*, 2005). For the purpose of this Product, an issue frame can be considered a tool that allows us to understand the importance of a problem (Weick, 1995). Thus, salience is an important part of framing. Historically low public engagement in water resource decisions was associated with the widespread perception that the adequate delivery of good quality water is within the realm of experts. Further, the necessary understanding and contribution to decisions takes time, commitment, and knowledge that few possess or seek to acquire as water appears to be plentiful and is available when needed. It was understood that considerable variations in water supply and quality can occur, but it was accepted that water resource managers know how to handle variation.

A series of events and disclosures of scientific findings have profoundly changed the framing of water issues and the interaction between such framing and climate variability and change. As illustrated in Figure 1.1, natural disasters, including Hurricane Katrina and recent sustained droughts in the United States, have raised awareness of society's vulnerability to flood, drought, and degradation of water quality. Such extreme events occur as mounting evidence indicates that water quantity and quality, fundamental components of ecological sustainability in many geographical areas, are threatened (*e.g.*, deVilliers, 2003). The February 2007 Intergovernmental Panel on Climate Change, Working Group 1, Fourth Assessment Report (IPCC, 2007a) reinforced the high probability of significant future climate change and more extreme climate variation, which is expected to affect many sectors, including water resources.

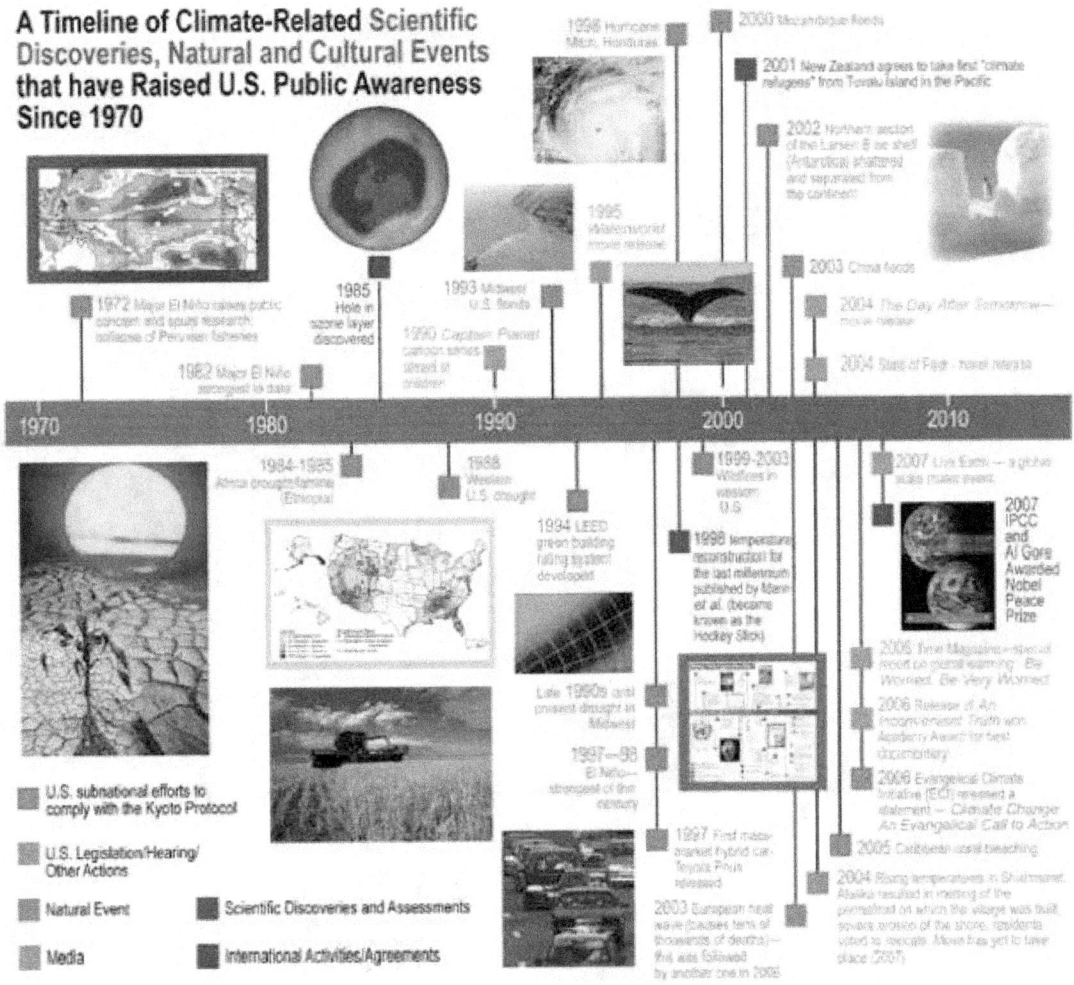

Figure 1.1 Timeline from 1970 to present of key natural and cultural events contributing to a widespread change in context for increasing awareness of climate issues.

The Report received considerable press coverage and generated increased awareness among the public and policy makers. Instead of being a low visibility issue, the issue frame for water resources has become that of attention-grabbing risk and uncertainty about such matters as rising sea levels, altered water storage in snow packs, and less favorable habitats for endangered fish species sensitive to warmer water temperatures. Thus, the effects of global warming have been an emerging issue-frame for water resources management.

Along with greater visibility of water and climate issues has come greater political and public involvement. At the same time, with an increase in discovery and awareness of climate impacts, there has been a deluge of policy actions in the form of new reports and passage of climate-related agreements and legislation

(see Figure 1.2). Higher visibility of climate and water variability has put pressure on water managers to be proactive in response to expected negative effects of climate variability and change (Hartmann, *et al.*, 2002; Carbone and Dow, 2005). Specifically, in the case of water managers in the United States, perception of risk has been found to be a critical variable for the adoption of innovative management in the sector (O'Connor *et al.*, 2005).

Frames encompass expectations about what can happen and what should be done if certain predicted events do occur (Minsky, 1980). The emergent issue frame for water resource management is that new knowledge (about climate change and variability) is being created that warrants management changes. Information and knowledge about climate variability experienced in the recent historical past is no longer

The emergent issue frame for water resource management is that new knowledge (about climate change and variability) is being created that warrants management changes.

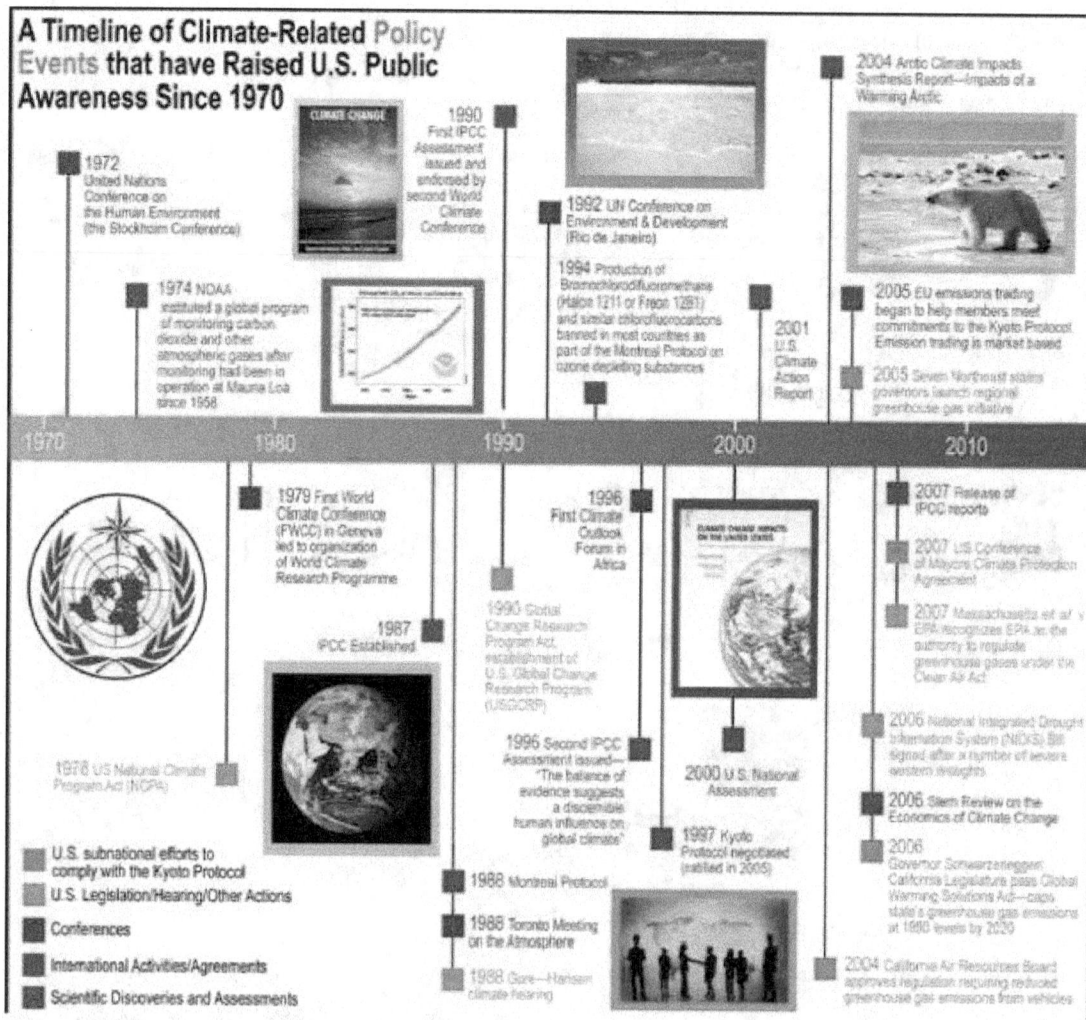

Figure 1.2 Timeline from 1970 to present of key policy events contributing to a widespread change in context for increasing awareness of climate issues.

Only in the last decade or so have climate scientists become able to predict aspects of future climate variations one to a few seasons in advance with better forecast skill than can be achieved by simply using historical averages for those seasons. This is a fundamentally new scientific advance.

as valuable as once it was, and new knowledge must be pursued (Milly *et al.*, 2008). Organizations and individuals face a context today where perceived failure to respond to climate variation and change is more risky than maintaining the status quo.

1.2.2 Climate Forecasting Innovations and Opportunities in Water Resources

Only in the last decade or so have climate scientists become able to predict aspects of future climate variations one to a few seasons in advance with better forecast skill than can be achieved by simply using historical averages for those seasons. This is a fundamentally new scientific advance (NRC, 2008).

It is important to emphasize that SI climate forecasting skill is still quite limited, and varies considerably depending on lead time,

geographic scale, target region, time of year, status of the ENSO cycle, and many other issues that are addressed in Chapter 2. Despite that, the potential usefulness of this new scientific capability is enormous, particularly in the water resources sector. This potential is being harvested through a variety of experiments and evaluations, some of which appear in this Product. For instance, reservoir management changes in the Columbia River Basin in response to SI climate forecast information have the potential to generate an average of $150 million per year more hydropower with little or no loss to other management objectives (Hamlet *et al.*, 2002). Table 1.1 illuminates the potential of SI climate forecasts to influence a wide range of water-related decisions, potentially providing great economic, security, environmental quality, and other gains.

BOX 1.1 Seasonal-to-Interannual Climate Forecasts

Weather forecasts seek to predict the exact state of the atmosphere for a specific time and place at lead-times ranging from nowcasts (e.g., severe weather warnings) out to a maximum of two weeks. Observations that can be used to accurately characterize the initial state of the atmosphere are crucial to the accuracy of these short-term weather forecasts. In contrast, seasonal-to-interannual climate forecasts seek to predict the statistics of the atmosphere for a region over a specified window of time, typically from one month to a few seasons in advance.

Observations of the slowly varying boundary conditions on the atmosphere, including upper ocean temperatures, snow cover, and soil moisture are critical to the accuracy of climate forecasts. Climate forecasts can also address the expected probabilities for extreme events (floods, freezes, blizzards, hurricanes, etc.), and the expected range of climate variability. Much of the skill in seasonal-to-interannual climate forecasts for the United States derives from an ability to monitor and accurately predict the future evolution of El Niño-Southern Oscillation (ENSO), however, the actual skill demonstrated is not yet high. As a general principle, all climate forecasts are probabilistic. They are probabilistic both in the future state of ENSO and in the consequences of ENSO for remotely influenced regions like the United States. For example, a typical ENSO-related climate forecast for the Pacific Northwest region of the United States might be presented as follows:

> *Based on expectations for continued El Niño conditions in the tropical Pacific, we expect increased likelihoods for above average winter and spring temperatures with below average precipitation, with small but non-zero odds for the opposite conditions (i.e., below average likelihoods for below average winter and spring temperatures and above average precipitation) in the Pacific Northwest.*

At lead times of a few decades to centuries, climate change scenarios are based on scenarios for changes in the emissions and concentrations of atmospheric greenhouse gases and aerosols that are important for the Earth's energy budget. Climate change scenarios do not require real-time observations needed to accurately initialize the atmosphere or slowly-evolving boundary conditions (upper ocean temperatures, snow cover, etc.). However, a recent study by Keenleyside et al. (2008) demonstrates that there is potential for improving the forecast skill in decadal climate predictions made within longer-term climate change scenarios by initializing global climate models with ocean observations.

Aside from the potential applications suggested in Table 1.1, there are other overarching opportunities for use of SI climate and hydrologic forecasts recently introduced to the water resources sector. Adaptive Management and Integrated Water Resources Management are examples of reforms that are still in relative infancy (discussed in further detail in Chapters 3 and 4) but could gain considerable momentum through fostering continuous feedback from forecasts to changes in practice and improved performance. Adaptive management embraces the need for continuous monitoring and feedback. Information provided by forecasts can prompt real time adaptations by public and private agencies and water users (NRC 2004). Integrated Water Resources Management provides a more holistic view of water supply or demand and is based around the concepts of flexibility and adaptability, using measures that can be easily reversed or are robust under changing circumstances (IPCC, 2007b). Such potential flexibility and adaptability extends not only to water agencies, but also to the general public. Advances in climate forecast skills and their applications provide an opportunity to give the public a deeper understanding about the relationship of climate variability to increased risk, vulnerability, and uncertainty related to water that now tends to be perceived in terms of a replication of the past. In addition, tuning water management more closely to real time climate prediction allows for reducing the lead time for response to climate variation.

> Adaptive management embraces the need for continuous monitoring and feedback. Information provided by forecasts can prompt real time adaptations by public and private agencies and water users.

Table 1.1 Examples of Water Resource Decisions Related to seasonal-to-interannual Climate Forecasts.

Decision/topic	Agency/organization Responsible	Activities Affected	Climate Forecast Information Relevance
Dam and reservoir management and reservoir allocation	• U.S. Army Corps of Engineers • U.S. DOI*, Bureau of Reclamation • Tennessee Valley Authority • FERC* and its licensed projects • Federal power marketing agencies • State, local, and regional water management entities and utilities, irrigation districts	Distribution of inflows and outflows for: • Agriculture • Public supply • Industry • Power • Flood control • Navigation • Instream flow maintenance • Protecting reserved waters for resources/other needs	• Total reservoir inflow • Long-range precipitation • Long-range temperature • Flow data • Snow melt data • Flood forecasts • Shifts in "phase" in decadal cycles
Irrigation/water allocation for agriculture/ aquaculture	• Federal, state, and regional facility operators • Irrigation districts • Agricultural cooperatives • Farmers	How much water and when and where to allocate it.	• Long/short-range precipitation • Long-range temperature
Ecosystem protection/ ecosystem services	• Federal and state resource agencies*, e.g., • U.S. DOI, Fish and Wildlife Service • U.S. DOA, Forest Service, U.S. DOI, Park Service, U.S. DOI, BLM, U.S. DOC, NMFS, etc. • State, regional and watershed-based protected areas NGOs, e.g., Nature Conservancy, local and regional land trusts	• Instream flow management • Riverine/riparian management • Wildlife management	• Climate cycles • Long-term climate predictions
Public water supply/ wastewater management	• Municipalities • Special water districts • Private water utilities • Water supply/wastewater utilities/ utility districts	• Needs for new reservoirs, dams, wastewater treatment facilities, pumping stations, groundwater management areas, distribution systems; • Needs for long term water supply and demand management plans; • Drought planning.	Changes in temperature/ precipitation effect water demand; reduction in base-flows, increased demands, and greater evaporation rates (Gleick et al., 2000; Clarkson and Smerdon, 1989). Predictive information at multiple scales and multiple time frames.
Coastal zones	• Regional coastal zone management agencies • Corps of Engineers • NMFS, other federal agencies • Local/regional flood control agencies • Public supply utilities	• Impacts to tidal deltas, low lying coastal plains • Changes to fish production/ coastal food systems, salt water intrusion • Erosion; deterioration of marshes • Flood control, water supply and sewage treatment implications	Predicted sea level rise & land subsidence; fluctuation in surface water temperature; tropical storm predictions; change to precipitation patterns; wind & water; storm surges and flood flow circulation patterns (Davidson, 1997).
Navigation	• Harbor managers • River system and reservoir managers, barge operators	• River and harbor channel depth; flow	• Stream flow, seasonality, and flooding potential
Power production	• Federal water and power agencies; FERC; private utilities with licensed hydropower projects; private utilities using power from generation facilities	• Water for hydropower • Water for steam generation in fossil fuel and nuclear plants • Water for cooling	• Temperature (and relation ships to demand for power) • Precipitation • Stream flow and runoff
Flooding/ floodplain management	• Floodplain managers; flood zone agencies; insurance companies; risk managers, land use planners	• Infrastructure needs planning • Emergency management	Short and long-term runoff predictions, especially long term trends in intensity of precipitation, storm surges, etc.

*Abbreviations used in table: BLM: Bureau of Land Management; DOA: Department of Agriculture; DOC: Department of Commerce; DOI: Department of the Interior; FERC: Federal Energy Regulatory Commission; NGO: Non-Governmental Organization; NMFS: National Marine Fisheries Service.

1.2.3 Organizational Dynamics and Innovation

The flow of information among agencies and actors in the complex organizational fields of climate forecasting and water resources is not always effective. Even as skill levels of climate and hydrologic forecasts have improved, resistance to their use in water resources management both exists and persists (O'Connor *et al.*, 1999; Rayner *et al.*, 2005; Yarnal *et al.*, 2006). Such resistance to innovation is to be expected, according to organizational and management literature that addresses the management of information across boundaries of various kinds that include organizations, disciplines, fields, and practices (Carlile, 2004; Feldman *et al.*, 2006). The same specialization that makes organizations effective in meeting internal organizational goals can make them resistant to innovation (Weber, 1947). Creating a product or service requires experience, terminologies, tools, and incentives that are embedded in a specific organization. Because knowledge requires time, resource, and opportunity cost investments, it constitutes a kind of "stake", and therefore significant costs are associated with acquiring new knowledge across boundaries (Carlile, 2002). Further, if the kind of knowledge that needs to be coordinated across boundaries is so different that a bridge of a common language must be created to allow translation, then the barriers are more difficult to overcome. Finally, demands made by sharing information across boundaries may be so novel that an organization must make a fundamental readjustment that challenges everything it knows.

Figure 1.3, adapted from Carlile (2004), depicts the challenges that must be addressed in order to share knowledge across boundaries, and conveys the challenge of innovation through information sharing across different organizations, levels of government, and public and private sectors. The lowest level of the inverted triangle shows information transfer is relatively simple between climate forecasters from different organizations. Forecasters generally share common knowledge, and know each others' language and levels of expertise regardless of organizational ties. Because a common lexicon exists, knowledge transfer is relatively simple. The usual barriers to smooth information flow apply, including information overload, avail-

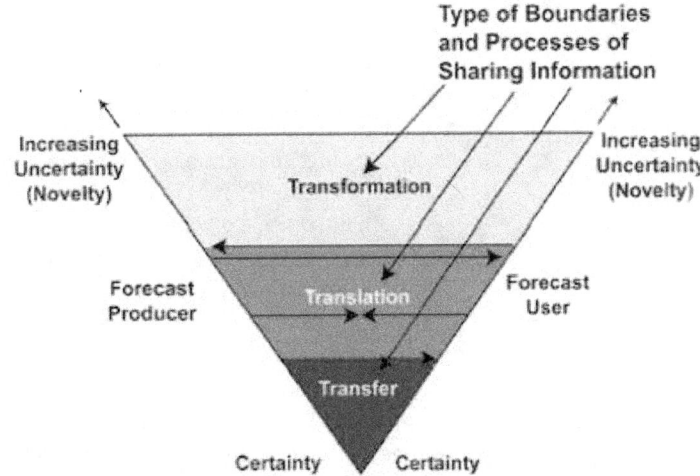

Figure 1.3 Illustration of information sharing processes. At the tip of the triangle forecast producers and forecast users are sharing a common syntax and framework, and therefore knowledge is simply transferred. As the products and uses become increasingly different and novel, a process of learning has to occur for information to be translated (middle of inverted triangle). Finally, information will need to be transformed in order for knowledge to be accessible to very different parties (top of the inverted triangle). Adapted from Carlile, 2004.

ability of storage and retrieval technologies and other information processing challenges. Unfortunately, because agencies tend to use their own terminology and information, because they know and trust the sources, before using terminology and information from outside, the adoption of SI climate forecast information in the water resource sector rarely fits this simple transfer profile.

At the second, or translation, level of information management, language issues become problematic and development of shared information is more difficult. This level of information sharing typifies the relationships between climate forecasters and water resource forecasters who have long predicted water futures using data such as snowpack, soil moisture, and basin and watershed models. Efforts to communicate at this level involve a large expenditure of effort that must be justified within the organization and may encounter resistance unless offset by some considerable worthwhile pay-off. Successful efforts for communication could include the creation of a lexicon with common definitions, the development of shared methodologies, the formulation of cross-organizational teams, the engagement in strategies such as collocation of offices, and the employment of individuals who can act as translators or brokers.

The third, or transformation, level of managing information requires considerable change in the ways in which organizations presently process and use information. Currently, climate forecasters tend to follow what has been termed the "Loading Dock Model", or simply issuing forecasts with little notion of whether they will be used by other organizations (Cash and Buizer, 2005). Knowledge at this third level (ultimately at all levels) must be created collaboratively, that is, coproduced with outside organizations, interests and entities, rather than delivered; and must be clear, credible and legitimate to all engaged actors. Information is likely to be more salient if it comes from known and trusted sources (NRC, 1989, 2008). Credibility is not just credibility of scientists, but also to users; information is more credible if it recognizes and addresses multiple perspectives. Legitimacy relates to even-handedness and the absence of narrow organizational or political agendas (Cash *et al.*, 2003; NRC, 2007, 2008). Almost all of the important applications of SI climate forecasts involve information management at the third level.

1.2.4 Decision Support, Knowledge Networks, Boundary Organizations, and Boundary Objects

A recent National Academy of Sciences Report (2008) observed that decision support is widely used but definitions of what constitutes that support vary. Following the lead of this Product, decision support is defined here as creating conditions that foster the appropriate use of information. This definition presumes that the climate scientists who generate SI climate forecasts often do not know what type of useful information they could provide to water resources managers, and that water managers do not necessarily know how they could apply SI climate forecasts and related information (NRC, 2008). The primary objective of decision-support activities is to foster transformative information exchange that will both change the kind of information that is produced and the way it is used (NRC 1989, 1996, 1999a, 2005, 2006, 2008).

Decision support involves engaging effective two-way communication between the producers and users of climate information (Jacobs *et al.*, 2005; Lemos and Morehouse, 2005; NRC, 1999a, 2006) rather than just the development of tools and products that may also be useful though less functional. This conception of decision support brings into focus human relationships and networks in information utilization. The test of transformed information is that it is trusted and considered reliable, and is fostered by familiarity and repeated interaction between information collaborators and the working and reworking of relationships. A knowledge network is built through such human interactions across organizational boundaries, creating and conveying information that is useful for all participants, ranging from scientists to multiple decision makers.

A variety of mechanisms can be employed to foster the creation of knowledge networks and the coproduction of knowledge that transcends what is already available. Among such mechanisms are boundary organizations that play an intermediary role between different organizations, specializations, disciplines, practices, and functions; including science and policy (Cash, 2001; Guston, 2001). These organizations can play a variety of roles in decision support, such as convening together, collaboration among users and producers, mediation for the various parties and the production of boundary objects. A boundary object is a prototype, model or other artifact through which collaboration can occur across different kinds of boundaries. Collaborative participants may come to appreciate the contribution of other kinds of knowledge, perspectives, expertise or practices and how they may augment or modify their own knowledge through engagement (Star and Griesemer, 1989). For example, a fish ladder is a kind of boundary object since it is an add-on to a dam structure. It must be integrated into the structural design, so hydrologists and engineers must collaborate on design decisions. At the same time, it serves fish species, so the insight of biologists about fish behavior is necessary for the ladder to work as it is intended.

1.3 OUTLINE OF THE REPORT AND WHERE PROSPECTUS QUESTIONS ARE ADDRESSED

This Chapter addresses the types of SI forecast-related decisions that are made in the water resources community and the role that such

Decision support is defined here as creating conditions that foster the appropriate use of information. The primary objective of decision-support activities is to foster transformative information exchange that will both change the kind of information that is produced and the way it is used.

Interannual Forecasts and Observational Data: *A Focus on Water Resources*

forecasts could play. It describes the general contextual opportunities and limitations to innovations that could limit the use of SI forecast information.

Chapter 2 answers the question: What are SI forecast products and how do they evolve from a scientific prototype to an operational product? It also addresses the issue of forecast skill, the impediments to progress in improving skill, and the steps necessary to ensure a product is needed and will be used in decision support. It describes the level of confidence about SI forecast products in the science and decision-making communities.

Chapter 3 focuses on the obstacles, impediments, and challenges in fostering close collaboration between scientists and decision makers in terms of theory and observation. Researchers have documented why and how resource decision makers use information, Chapter 3 addresses the following kinds of questions: How are hazards and risks related to climate variability perceived and managed? What are the challenges related to determining and serving the needs of decision makers, emphasizing

the importance of reliability and trust, and suggesting how decision support could leverage scientific and technological advances?

Chapter 4 provides examples of a range of decision support experiments in the context of SI forecast information. It describes the limitations on the kinds of information available and the need to employ logical inference. It also discusses how decision support tools can be improved.

Chapter 5 provides a summary of this Product, especially identifying overarching themes. It suggests the kinds of research and action needed to improve progress in this area. Finally, it addresses how the knowledge gained in water resources might be useful to other sectors.

The prospectus for this study contained a series of questions that were to direct this study, vetted by the Climate Change Science Program office and by public review. Table 1.2 summarizes the questions and specifies which chapter section they are addressed. Table 1.3 is a summary of the case studies provided in this Product.

Table 1.2 Questions To Be Addressed in Synthesis and Assessment Product 5.3.

Prospectus Question	Report Location where Question is Addressed
What seasonal-to-interannual (e.g., probabilistic) forecast information do decision makers need to manage water resources?	2.1
What are the seasonal-to-interannual forecast/data products currently available and how does a product evolve from a scientific prototype to an operational product?	2.2
What is the level of confidence of the product within the science community and within the decision-making community, who establishes these confidence levels and how are they determined?	2.2
How do forecasters convey information on climate variability and how is the relative skill and level of confidence of the results communicated to resource managers?	2.3
What is the role of probabilistic forecast information in the context of decision support in the water resources sector?	2.3
How is data quality controlled?	2.3
What steps are taken to ensure that this product is needed and will be used in decision support?	2.5
What types of decisions are made related to water resources?	3.2
What is the role that seasonal-to-interannual forecasts play and could play?	3.2
How does climate variability influence water resource management?	3.2
What are the obstacles and challenges decision makers face in translating climate forecasts and hydrology information into integrated resource management?	3.2
What are the barriers that exist in convincing decision makers to consider using risk-based hydrology information (including climate forecasts)?	3.2
What challenges do tool developers have in finding out the needs of decision makers?	3.3
How much involvement do practitioners have in product development?	4.1
What are the measurable indicators of progress in terms of access to information and its effective uses?	4.3
Identify critical components, mechanisms, and pathways that have led to successful utilization of climate information by water managers.	4.4
Discuss options for (a) improving the use of existing forecasts/data products and (b) identify other user needs and challenges in order to prioritize research for improving forecasts and products.	4.4 and 5
Discuss how these findings can be transferred to other sectors.	5

Table 1.3 Summary of Case Studies (*i.e.*, Experiments and Evaluations) presented in this Product.

Study or Experiment	Chapter	Type of Decision Support Information Needed, Used or Delivered	Most Successful Feature(s) or Lesson(s) Learned from Case Study
CPC Seasonal Drought Outlook (DO)	2, Box 2.3	DO is a monthly subjective consensus forecast between several agencies and academic experts, of drought evolution for three months following the forecast date.	Primary drought-related agency forecast produced in US; widely used by drought management and response community from local to regional scales. Research is ongoing for product improvements.
Testbeds	2, Box 2.4	Testbeds are a mix of research and operations, and serve as a conduit between operational, academic and research communities. NOAA currently operates several testbeds (*e.g.*, Hazardous Weather, Climate and Hurricanes).	Testbeds focus on introducing new ideas and data to the existing system and analyzing the results through experimentation and demonstration. Satisfaction with testbeds has been high for operational and research participants alike.
Advanced Hydrologic Prediction Service (AHPS)	2, Box 2.5;3, Section 3.3.1.2	AHPS provides data more quickly and at smaller scale (*i.e.*, local watershed) than previous hydrographic models; directly links to local decision makers.	More accurate, detailed, and visually oriented outputs provide longer-range forecasts than current methods. Also includes a survey process and outreach, training, and educational activities.
NWS Local 3-Month Outlook for Temp & Precip (L3MO)	2, Box 2.6	Designed to clarify and downscale the national-scale CPC Climate Outlook temperature forecast product.	Outlook is new; it became operational in January 2007. The corresponding local product for precipitation is still in development as of this writing.
Southwest drought-climate variability & water management	3, Section 3.2.3.2	Regional studies of: associations between ENSO teleconnections, multi-decadal variations in Pacific Ocean-atmosphere system, and regional climate show potential predictability of seasonal climate and hydrology.	New Mexico and Arizona have been working to integrate new decision support tools and data into their drought plans; Colorado River Basin water managers have commissioned tree ring reconstructions of streamflow to revise estimates of record droughts, and to improve streamflow forecast performance.
Red River of the North —Flooding and Water Management	3, Section 3.2.4	Model outputs to better use seasonal precipitation, snowmelt, *etc.*, are being used in operations decisions; however, the 1997 floods resulted in $4 billion in losses. The River crested 5 feet over the flood height predicted by the North Central River Forecast Center; public blamed National Weather Service for a faulty forecast.	There is a need for (1) improved forecasts (*e.g.*, using recent data in flood rating curves, real-time forecasting); (2) better forecast communication (*e.g.*, warnings when rating curve may be exceeded and including user feedback in improved forecast communication); and (3) more studies (*e.g.*, reviewing data for future events).

Study or Experiment	Chapter	Type of Decision Support Information Needed, Used or Delivered	Most Successful Feature(s) or Lesson(s) Learned from Case Study
Credibility and the Use of Climate Forecasts: Yakima River Basin/El Niño	3, Section 3.2.4	In 1997, USBR issued a faulty forecast for summer runoff to be below an established threshold. Result was increased animosity between water rights holders, loss of confidence in USBR, lawsuits against USBR.	There is a need for greater transparency in forecast methods (including issuing forecast confidence limits), better communication between agencies and the public, and consideration of consequences of actions taken by users in the event of a bad forecast.
Credibility and the Use of Climate Forecasts: Colorado Basin Case Studies	3, Section 3.2.4	In 1997, the USBR issued a forecast, based on snowpack, for summer runoff to be below the legally established threshold, resulting in jeopardized water possibilities for junior water rights holders.	Need to improve transparency in forecast methods (e.g., issuing forecast confidence limits, better communication between agencies and the public, and consideration of users' actions in the event of a bad forecast), would have improved the forecast value and the actions taken by the USBR.
Southeast Drought: Another Perspective on Water Problems in the Southeastern United States	3, Section 3.3.1	A lack of tropical storms/ hurricanes and societal influences such as operating procedures, laws and institutions led to the 2007-2008 Southeast Drought, resulting in impacts to agriculture, fisheries, and municipal water supplies.	Impacts exacerbated by (1) little action to resolve river basin conflicts between GA, AL, and FL; (2) incompatibility of river usage (e.g., protecting in-stream flow while permitting varied off-stream use), (3) conflicts between up- and down-stream demands (i.e., water supply/wastewater discharge, recreational use), and (4) negotiating process (e.g., compact takes effect only when parties agree to allocation formula).
Policy learning and seasonal climate forecasting application in NE Brazil—integrating information into decisions	3, Section 3.3.1.1	In 1992, in response to a long drought, the State of Ceara created several levels of water management including an interdisciplinary group within the state water management agency to develop and implement reforms.	Inclusion of social and physical scientists and stakeholders resulted in new knowledge (i.e., ideas and technologies) that critically affected water reform, including helping poorer communities better adapt to, and build capacity for managing climate variability impacts on water resources; also helped democratize decision making.

Study or Experiment	Chapter	Type of Decision Support Information Needed, Used or Delivered	Most Successful Feature(s) or Lesson(s) Learned from Case Study
Interpreting Climate Forecasts—uncertainties and temporal variability: Use of ENSO based information	3, Section 3.3.2	The Arizona Salt River Project (SRP) made a series of decisions based on the 1997/1998 El Niño (EN) forecast plus analysis of how ENs tended to affect their rivers and reservoirs.	SRP managers reduced groundwater pumping in 1997 in anticipation of a wet winter; storms provided ample water for reservoirs. Success was partly due to availability of climate and hydrology research and federal offices in close proximity to managers. Lack of temporal and geographical variability information in climate processes remains a barrier to adoption/use of specific products; decisions based only on forecasts are risky.
How the South Florida Water Management District (SFWMD) Uses Climate Information	4, Experiment 1	SFWMD established a regulation schedule for Lake Okeechobee that uses climate outlooks as guidance for regulatory release decisions. A decision tree with a climate outlook is a major advance over traditional hydrologic rule curves used to operate large reservoirs. This experiment is the only one identified that uses decadal climate data in a decision-support context.	To improve basin management, modeling capabilities must: improve ability to differentiate trends in basin flows associated with climate variation; gauge skill gained in using climate information to predict basin hydro-climatology; account for management uncertainties caused by climate; and evaluate how climate projections may affect facility planning and operations. Also, adaptive management is effective in incorporating SI variation into modeling and operations decision-making processes.

Study or Experiment	Chapter	Type of Decision Support Information Needed, Used or Delivered	Most Successful Feature(s) or Lesson(s) Learned from Case Study
Long-Term Municipal Water Management Planning— New York City (NYC)	4, Experiment 2	NYC is adapting strategic and capital planning to include the potential effects of climate change (i.e., sea-level rise, higher temperatures, increases in extreme events, and changing precipitation patterns) on the City's water systems. NYC Department of Environmental Protection, in partnership with local universities and private sector consultants, is evaluating climate change projections, impacts, indicators, and adaptation and mitigation strategies to support agency decision making.	This case illustrates (1) plans for regional capital improvements can include measures that reduce vulnerability to sea level rise; (2) the meteorological and hydrology communities need to define and communicate current and increasing risks, with explicit discussion of the inherent uncertainties; (3) more research is needed (e.g., to further reduce uncertainties associated with sea-level rise, provide more reliable predictions of changes in frequency/intensity of tropical and extra-tropical storms, etc.); (4) regional climate model simulations and statistical techniques used to predict long-term climate change impacts could be down-scaled to help manage projected SI climate variability; and (5) decision makers need to build support for adaptive action despite uncertainties. The extent and effectiveness of this action will depend on building awareness of these issues among decision makers, fostering processes of interagency interaction and collaboration, and developing common standards.

Study or Experiment	Chapter	Type of Decision Support Information Needed, Used or Delivered	Most Successful Feature(s) or Lesson(s) Learned from Case Study
Integrated Forecast and Reservoir Management (INFORM)—Northern California	4, Experiment 3	INFORM aims to demonstrate the value of climate, weather, and hydrology forecasts in reservoir operations. Specific objectives are to: (1) implement a prototype integrated forecast-management system for the Northern California river and reservoir system in close collaboration with operational forecasting and management agencies, and (2) demonstrate the utility of meteorological/climate and hydrologic forecasts through near-real-time tests of the integrated system with actual data and management input.	INFORM demonstrated key aspects of integrated forecast-decision systems, i.e., (1) seasonal climate and hydrologic forecasts benefit reservoir management, provided that they are used in connection with adaptive dynamic decision methods that can explicitly account for and manage forecast uncertainty; (2) ignoring forecast uncertainty in reservoir regulation and water management decisions leads to costly failures; and (3) static decision rules cannot take full advantage of and handle forecast uncertainty information. The extent that forecasts help depends on their reliability, range, and lead time, in relation to the management systems' ability to regulate flow, water allocation, etc.
How Seattle Public Utility (SPU) District Uses Climate Information to Manage Reservoirs	4, Experiment 4	Over the past several years SPU has taken steps to improve incorporation of climate, weather, and hydrologic information into the real-time and SI management of its mountain water supply system. They are receptive to new management approaches due to public pressure and the risk of legal challenges related to the protection of fish populations	The SPU case shows: (1) access to skillful SI forecasts enhances credibility of using climate information in the region; (2) monitoring of snowpack moisture storage and mountain precipitation is essential for effective decision making and for detecting long-term trends that can affect water supply reliability; and (3) SPU has significant capacity to conduct in-house investigations/assessments. This provides confidence in the use of information.

Study or Experiment	Chapter	Type of Decision Support Information Needed, Used or Delivered	Most Successful Feature(s) or Lesson(s) Learned from Case Study
Using Paleo-climate Information to Examine Climate Change Impacts	4, Experiment 5	Because of repeated drought, western water managers, through partnerships with researchers in the inter-mountain West have chosen to use paleoclimate records of streamflow and hydroclimatic variability to provide an extended record for assessing the potential impact of a more complete range of natural variability as well as providing a baseline for detecting regional impacts of global climate change.	Partnerships have led to a range of applications evolving from a better understanding of historical drough conditions to assessing drought impacts on water systems using tree ring reconstructed flows. Workshops have expanded applications of the tree ring based streamflow reconstructions for drought planning and water management. Also, an online resource provides water managers access to gage and reconstruction data and a tutorial on reconstruction methods for gages in Colorado and California.
Climate, Hydrology, and Water Resource Issues in Fire-Prone United States Forests	4, Experiment 6	The 2000 experiment, consisting of annual workshops to evaluate the utility of climate information for fire management, was initiated to inform fire managers about climate forecasting tools and to enlighten climate forecasters about the needs of the fire management community.	Fire-climate workshops are now accepted practice by agencies with an annual assessment of conditions and production of pre-season fire-climate forecasts. Scientists and decision makers continue to explore new questions, as well as involve new participants, disciplines and specialties, to make progress in key areas (e.g., lightning climatologies).
The CALFED – Bay Delta Program: Implications of Climate Variability	4 Experiment 7	Delta requirements to export water supplies to southern California are complicated by: managing habitat and water supplies in the region, maintaining endangered fish species, making major long-term decisions about rebuilding flood control levees and rerouting water supply networks through the region.	A new approach has led to consideration of climate change and sea level rise in infrastructure planning; the time horizon for planning has been extended to 200 years. Because of incremental changes in understanding changing climate, this case shows the importance of using adaptive management strategies.

Study or Experiment	Chapter	Type of Decision Support Information Needed, Used or Delivered	Most Successful Feature(s) or Lesson(s) Learned from Case Study
Regional Integrated Science and Assessment Teams (RISAs)—An Opportunity for Boundary Spanning, and a Challenge	Section 4.3.2	The eight RISA teams that are sponsored by NOAA represent a new collaborative paradigm in which decision makers are actively involved in developing research agendas. RISAs explicitly seek to work at the boundary of science and decision making.	RISA teams facilitate engagement with stakeholders and design climate-related decision-support tools for water managers through using: (1) a robust "stakeholder-driven research" approach focusing on both the supply (i.e., information development) and demand side (i.e., the user and her/his needs); (2) an "information broker" approach, both producing new scientific information themselves and providing a conduit for new and old information and facilitating the development of information networks; (3) a "participant/advocacy" or "problem-based" approach, involving a focus on a particular problem or issue and engaging directly in solving it; and (4) a "basic research" approach where researchers recognize gaps in the key knowledge needed in the production of context sensitive, policy-relevant information.
Leadership in the California Department of Water Resources (CDWR)	4, Case Study A	Drought in the Colorado River Basin and negotiations over shortage and surplus guidelines prompted water resources managers to use climate data in plans and reservoir forecast models. Following a 2005 workshop on paleohydrologic data use in resource management, RISA and CDWR scientists developed ties to improve the usefulness of hydroclimatic science in water management.	CDWR asked the NAS to convene a panel to clarify scientific understanding of Colorado River Basin climatology and hydrology, past variations, projections for the future, and impacts on water resources. NAS issued the report in 2007; a new Memorandum of Agreement now exists to improve cooperation with RISAs and research laboratories.

Study or Experiment	Chapter	Type of Decision Support Information Needed, Used or Delivered	Most Successful Feature(s) or Lesson(s) Learned from Case Study
Cooperative extension services, watershed stewardship: the Southeast Consortium	4, Case Studies B and F	The Southeast Climate Consortium RISA (SECC), a confederation of researchers at six universities in Alabama, Georgia, and Florida, has used a top-down approach to develop stakeholder capacity to use climate information in region's $33 billion agricultural sector. Early on, SECC researchers recognized the potential of using ENSO impact on local climate data to provide guidance to farmers, ranchers, and forestry sector stakeholders on yields and changes to risk (e.g, frost occurrence).	SECC determined that (1) benefits from producers use of seasonal forecasts depends on factors that include the flexibility and willingness to adapt farming operations in response to forecasts, and the effectiveness of forecast communication; (2) success in championing integration of new information requires sustained interactions (e.g., with agricultural producers in collaboration with extension agents; and (3) direct engagement with stakeholders provides feedback to improve the design of the tool and to enhance climate forecast communication.
Approaches to building user knowledge and enhancing capacity building—Arizona Water Institute	4, Case Study C	The Arizona Water Institute, initiated in 2006, focuses resources of the State of Arizona's university system on the issue of water sustainability. The Institute was designed as a "boundary organization" to build pathways for innovation between the universities and state agencies, communities, Native American tribal representatives, and the private sector.	The Institute focuses on: capacity building, training students through engagement in real-world water policy issues, providing better access to hydrologic data for decision makers and assisting in visualizing implications of decisions they make, providing workshops and training programs for tribal entities, jointly defining research agendas between stakeholders and researchers, and building employment pathways to train students for jobs requiring special training (e.g, water and wastewater treatment plant operators).

Study or Experiment	Chapter	Type of Decision Support Information Needed, Used or Delivered	Most Successful Feature(s) or Lesson(s) Learned from Case Study
Murray–Darling Basin— sustainable development and adaptive management	4, Case Study D	1985 Murray–Darling Basin Agreement (MDBA), formed by New South Wales, Victoria, South Australia and Commonwealth, provides for integrated management of water and related land resources of world's largest catchment system. MDBA encourages use of climate information for planning and management; seeks to integrate quality and quantity concerns within a single management framework; has a broad mandate to embrace social, economic, environmental and cultural issues in decisions, and authority to implement water & development policies.	According to Newson (1997), while the policy of integrated management has "received wide endorsement", progress towards effective implementation has fallen short—especially in the area of floodplain management. This has been attributed to a "reactive and supportive" attitude as opposed to a proactive one. Despite such criticism, it is hard to find another initiative of this scale and sophistication that has attempted adaptive management based on community involvement.
Adaptive management in Glen Canyon, Arizona and Utah	4, Case Study E	Glen Canyon Dam was constructed in 1963 to provide hydropower, irrigation, flood control, and public water supply—and to ensure adequate storage for upper basin states of Colorado River Compact. When dam's gates closed, the river above and below Glen Canyon was altered. In 1996, USBR created an experimental flood to restore the river ecosystem.	Continued drought in the Southwest is placing increased stress on land and water resources of region, including agriculture. Efforts to restore the river to conditions more nearly approximating the era before the dam was built will require changes in the dam's operating regime to force a greater balance between instream flow, sediment management, power generation and offstream water supply. This will require forecast use to ensure that these various needs can be optimized.
Potomac River Basin	4, Case Study G	The Interstate Commission on the Potomac River Basin (ICPRB) periodically studies the impact of climate change on the supply reliability to the Washington metropolitan area (WMA).	A 2005 study stated that the 2030 demand in the WMA could be 74% to 138% greater than that of 1990. According to the report, with aggressive conservation and operation policies, existing resources should be sufficient through 2030; recommended incorporating potential climate impacts in future planning.

Study or Experiment	Chapter	Type of Decision Support Information Needed, Used or Delivered	Most Successful Feature(s) or Lesson(s) Learned from Case Study
Fire prediction workshops as a model for climate science–water management process to improve water resources decisions	4, Case Study H	Given strong mutual interests in improving the range of tools available to fire management, with goal of reducing fire related damage and loss of life, fire managers and climate scientists have developed long-term process to: improve fire potential prediction; better estimate costs; most efficiently deploy fire fighting resources.	Emphasis on process, as well as product, may be a model for climate science in support of water resources management decision making. Another key facet in maintaining this collaboration and direct application of climate science to operational decision making has been the development of strong professional relationships between the academic and operational partners.
Incentives to Innovate—Climate Variability and Water Management along San Pedro River	4, Case Study I	The highly politicized issue of water management in upper San Pedro River Basin has led to establishment of Upper San Pedro Partnership, whose primary goal is balancing water demands with supply without compromising region's economic viability, much of which is tied to Fort Huachuca Army base.	Studies show growing vulnerability to climate impacts. Climatologists, hydrologists, social scientists, and engineers work with partnership to strengthen capacity/interest in using climate forecast products. A decision-support model being developed by University of Arizona with partnership members will hopefully integrate climate into local decisions.

CHAPTER 2

A Description and Evaluation of Hydrologic and Climate Forecast and Data Products that Support Decision-Making for Water Resource Managers

Convening Lead Author: Nathan Mantua, Climate Impacts Group, Univ. of Washington

Lead Authors: Michael Dettinger, U.S.G.S., Scripps Institution of Oceanography; Thomas C. Pagano, N.W.C.C., NRCS/USDA; Andrew Wood, 3TIER™, Inc./Dept. of Civil and Environmental Engineering, Univ. of Washington; Kelly Redmond, W.R.C.C., Desert Research Institute

Contributing Author: Pedro Restrepo, NOAA

KEY FINDINGS

There are a wide variety of climate and hydrologic data and forecast products currently available for use by decision makers in the water resources sector, ranging from seasonal outlooks for precipitation and surface air temperature to drought intensity, lake levels, river runoff and water supplies in small to very large river basins. However, the use of official seasonal-to-interannual (SI) climate and hydrologic forecasts generated by National Oceanic and Atmospheric Administration (NOAA) and other agencies remains limited in the water resources sector. Forecast skill, while recognized as just one of the barriers to the use of SI climate forecast information, remains a primary concern among forecast producers and users. Simply put, there is no incentive to use SI climate forecasts when they are believed to provide little additional skill to existing hydrologic and water resource forecast approaches. Not surprisingly, there is much interest in improving the skill of hydrologic and water resources forecasts. Such improvements can be realized by pursuing several research pathways, including:

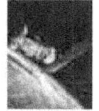

- Improved monitoring and assimilation of real-time hydrologic observations in land surface hydrologic models that leads to improved estimates for initial hydrologic states in forecast models;
- Increased accuracy in SI climate forecasts; and,
- Improved bias corrections in existing forecast.

Because runoff and forecast conditions are projected to gradually and continually trend towards increasingly warmer temperatures as a consequence of human-caused climate change, the expected skill in regression-based hydrologic forecasts will always be limited by having only a brief reservoir of experience with each new degree of warming. Consequently, we must expect that regression-based forecast equations will tend to be increasingly and perennially out of date in a world with strong warming trends. This problem with the statistics of forecast skill in a changing world suggests that development and deployment of more physically-based, less statistically-based, forecast models should be a priority in the foreseeable future.

Another aspect of forecasts that serves to limit their use and utility is the challenge in interpreting forecast information. For example, from a forecast producer's perspective, confidence levels are explicitly and quantitatively conveyed by the range of possibilities described in probabilistic forecasts. From a forecast user's perspective, probabilistic forecasts are not always well understood or correctly interpreted. Although structured user testing is known to be an effective product development tool, it is rarely done. Evaluation should be an integral part of improving forecasting efforts, but that evaluation should be extended to factors that encompass use and utility of forecast information for stakeholders. In particular, very little research is done on effective seasonal forecast communication. Instead, users are commonly engaged only near the end of the product development process.

Other barriers to the use of SI climate forecasts in water resources management have been identified and those that relate to institutional issues and aspects of current forecast products are discussed in Chapters 3 and 4 of this Product.

Pathways for expanding the use and improving the utility of data and forecast products to support decision making in the water resources sector are currently being pursued at a variety of spatial and jurisdictional scales in the United States. These efforts include:

- An increased focus on developing forecast evaluation tools that provide users with opportunities to better understand forecast products in terms of their expected skill and applicability;
- Additional efforts to explicitly and quantitatively link SI climate forecast information with SI hydrologic and water supply forecasting efforts;
- An increased focus on developing new internet-based tools for accessing and customizing data and forecast products to support hydrologic forecasting and water resources decision making; and,
- Further improvements in the skill of hydrologic and water supply forecasts.

Many of these pathways are currently being pursued by the federal agencies charged with producing the official climate and hydrologic forecast and data products for the United States, but there is substantial room for increasing these activities.

An additional important finding is that recent improvements in the use and utility of data and forecast products related to water resources decision-making have come with an increased emphasis on these issues in research funding agencies through programs like the Global Energy and Water Cycle Experiment (GEWEX, a program initiated by the World Climate Research Programme) and NOAA's Regional Integrated Sciences and Assessment (RISA), Sectoral Applications Research Program (SARP), Transition of Research Applications to Climate Services (TRACS) and Climate Prediction Program for the Americas (CPPA) programs. Sustaining and accelerating future improvements in the use and utility of official data and forecast products in the water resources sector rests, in part, on sustaining and expanding federal support for programs focused on improving the skill in forecasts, increasing the access to data and forecast products, and supporting sustained interactions between forecast producers and consumers. One strategy is to support demonstration projects that result in the development of new tools and applications that can then be transferred to broader communities of forecast producers, including those in the private sector, and broader communities of forecast consumers.

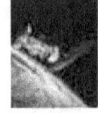

2.1 INTRODUCTION

In the past, water resource managers relied heavily on observed hydrologic conditions such as snowpack and soil moisture to make seasonal-to-interannual (SI) water supply forecasts to support management decisions. Within the last decade, researchers have begun to link SI climate forecasts with hydrologic models (e.g., Kim *et al.*, 2000; Kyriakidis *et al.*, 2001) or statistical distributions of hydrologic parameters (e.g., Dettinger *et al.*, 1999; Sankarasubramanian and Lall, 2003) to improve hydrologic and water resources forecasts. Efforts to incorporate SI climate forecasts into water resources forecasts have been prompted, in part, by our growing understanding of the effects of global-scale climate phenomena, like El Niño-Southern Oscillation (ENSO), on U.S. climate, and the expectation that SI forecasts of hydrologically-significant climate variables like precipitation and temperature provide a basis for predictability that is not currently being exploited. To the extent that climate variables like temperature and precipitation can be forecasted seasons in advance, hydrologic and water-supply forecasts can also be made skillfully well before the end, or even beginning, of the water year[1].

More generally speaking, the use of climate data and SI forecast information in support of water resources decision making has been aided by efforts to develop programs focused on fostering sustained interactions between data and forecast producers and consumers in ways that support co-discovery of applications (e.g. see Miles *et al.*, 2006).

This Chapter focuses on a description and evaluation of hydrologic and climate forecast and data products that support decision making for water resource managers. Because the focus of this CCSP Product is on using SI forecasts and data for decision support in the water resources sector, we frame this Chapter around key forecast and data products that contribute towards improved hydrologic and water sup-

ply forecasts. As a result, this Product does not contain a comprehensive review and assessment of the entire national SI climate and hydrologic forecasting effort. In addition, the reader should note that, even today, hydrologic and water supply forecasting efforts in many places are still not inherently linked with the SI climate forecasting enterprise.

Surveys identify a variety of barriers to the use of climate forecasts (Pulwarty and Redmond, 1997; Callahan *et al.*, 1999; Hartmann *et al.*, 2002), but insufficient accuracy is always mentioned as a barrier. It is also well established that an accurate forecast is a necessary, but in and of itself, insufficient condition to make it useful or usable for decision making in management applications (Table 2.1). Chapters 3 and 4 provide extensive reviews, case studies, and analyses that provide insights into pathways for lowering or overcoming barriers to the use of SI climate

Table 2.1 Barriers to the use of climate forecasts and information for resource managers in the Columbia River Basin (Reproduced from Pulwarty and Redmond, 1997).

a. Forecasts not "accurate" enough.
b. Fluctuation of successive forecasts ("waffling").
c. The nature of what a forecast is, and what is being forecast (e.g., types of El Niño and La Niña impacts, non-ENSO events, what are "normal" conditions?).
d. Non-weather/climate factors are deemed to be more important (e.g., uncertainty in other arenas, such as freshwater and ocean ecology [for salmon productivity]).
e. Low importance is given to climate forecast information because its role is unclear or impacts are not perceived as important enough to commit resources.
f. Other constraints deny a flexible response to the information (e.g., meeting flood control or Endangered Species Act requirements).
g. Procedures for acquiring knowledge and making and implementing decisions which incorporate climate information, have not been clearly defined.
h. Events forecast may be too far in the future for a discrete action to be engaged.
i. Availability and use of locally-specific information may be more relevant to a particular decision.
j. "Value" may not have been demonstrated by a credible reliable organization or competitor.
k. Desired information not provided (e.g., number of warm days, regional detail).
l. There may be competing forecasts or other conflicting information.
m. Lack of "tracking" information; does the forecast appear to be verifying?
n. History of previous forecasts not available. Validation statistics of previous forecasts not available.

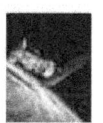

---

[1] The *water year*, or hydrologic year, is October 1st through September 30th. This reflects the natural cycle in many hydrologic parameters such as the seasonal cycle of evaporative demand, and of the snow accumulation, melt, and runoff periods in many parts of the United States.

forecasts in water resources decision making. It is almost impossible to discuss the perceived value of forecasts without also discussing issues related to forecast skill. Many different criteria have been used to evaluate forecast skill (see Wilks, 1995 for a comprehensive review). Some measures focus on aspects of deterministic skill (e.g., correlations between predicted and observed seasonally averaged precipitation anomalies), while many others are based on categorical forecasts (e.g., Heidke skill scores for categorical forecasts of "wet", "dry", or "normal" conditions). The most important measures of skill vary with different perspectives. For example, Hartmann et al. (2002) argue that forecast performance criteria based on "hitting" or "missing" associated observations offer users conceptually easy entry into discussions of forecast quality. In contrast, some research scientists and water supply forecasters may be more interested in correlations between the ensemble average of predictions and observed

measures of water supply like seasonal runoff volume.

Forecast skill remains a primary concern among many forecast producers and users. Skill in hydrologic forecast systems derives from various sources, including the quality of the simulation models used in forecasting, the ability to estimate the initial hydrologic state of the system, and the ability to skillfully predict the statistics of future weather over the course of the forecast period. Despite the significant resources expended to improve SI climate forecasts over the past 15 years, few water-resource related agencies have been making quantitative use of climate forecast information in their water supply forecasting efforts (Pulwarty and Redmond 1997; Callahan et al., 1999).

In Section 2.2 of this Chapter, we review hydrologic data and forecasts products. Section 2.3 provides a parallel discussion of the climate

BOX 2.1: Agency Support

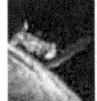

Federal support for research supporting improved hydrologic forecasts and applications through the use of climate forecasts and data has received increasing emphasis since the mid-1990s. The World Climate Research Program's Global Energy and Water Cycle Experiment (GEWEX) was among the first attempts to integrate hydrology/land surface and atmosphere models in the context of trying to improve hydrologic and climate predictability.

There have been two motivations behind this research: understanding scientific issues of land surface interactions with the climate system, and the development or enhancement of forecast applications, e.g., for water, energy and hazard management. Early on, these efforts were dominated by the atmospheric (and related geophysical) sciences.

In the past, only a few U.S. programs have been very relevant to hydrologic prediction: the NOAA Climate Prediction Program for the Americas (CPPA), NOAA predecessors GEWEX Continental-scale International Project (GCIP), GEWEX Americas Prediction Project (GAPP) and the NASA Terrestrial Hydrology Program. The hydrologic prediction and water management focus of NOAA and NASA has slowly expanded over time. Presently, the NOAA Climate Dynamics and Experimental Prediction (CDEP), Transition of Research Applications to Climate Services (TRACS) and Sectoral Applications Research Program (SARP) programs, and the Water Management program within NASA, have put a strong emphasis on the development of both techniques and community linkages for migrating scientific advances in climate and hydrologic prediction into applications by agencies and end use sectors. The longer-standing NOAA Regional Integrated Sciences and Assessments (RISA) program has also contributed to improved use and understanding of climate data and forecast products in water resources forecasting and decision making. Likewise, the recently initiated postdoctoral fellowship program under the Predictability, Predictions, and Applications Interface (PPAI) panel of U.S. CLIVAR aims to grow the pool of scientists qualified to transfer advances in climate science and climate prediction into climate-related decision frameworks and decision tools.

Still, these programs are small in comparison with current federally funded science focused initiatives and are only just beginning to make inroads into the vast arena of effectively increasing the use and utility of climate and hydrologic data and forecast products.

data and forecast products that support hydrologic and water supply forecasting efforts in the United States. In Section 2.4, we provide a more detailed discussion of pathways for improving the skill and utility in hydrologic and climate forecasts and data products.

Section 2.5 contains a brief review of operational considerations and efforts to improve the utility of forecast and data products through efforts to improve the forecast evaluation and development process. These efforts include cases in which forecast providers and users have been engaged in sustained interactions to improve the use and utility of forecast and data products, and have led to many improvements and innovations in the data and forecast products generated by national centers. In recent years, a small number of water resource agencies have also developed end-to-end forecasting systems (*i.e.* forecasting systems that integrate observations and forecast models with decision-support tools) that utilize climate forecasts to directly inform hydrologic and water resources forecasts.

2.2 HYDROLOGIC AND WATER RESOURCES: MONITORING AND PREDICTION

The uses of hydrologic monitoring and prediction products, and specifically those that are relevant for water, hazard and energy management, vary depending on the forecast lead time (Figure 2.1). The shortest climate and hydrologic lead-time forecasts, from minutes to hours, are applied to such uses as warnings for floods and extreme weather, wind power scheduling, aviation, recreation, and wild fire response management. In contrast, at lead times of years to decades, predictions are used for strategic planning purposes rather than operational management of resources. At SI lead times, climate and hydrologic forecast applications span a wide range that includes the management of water, fisheries, hydropower and agricultural production, navigation and recreation. Table 2.2 lists aspects of forecast products at these time scales that are relevant to decision makers.

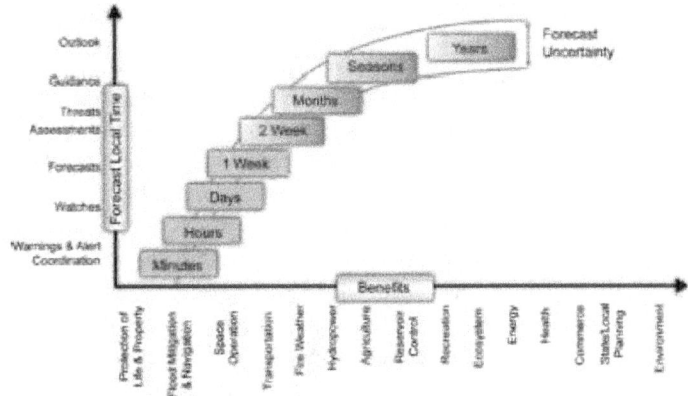

Figure 2.1 The correspondence of climate and hydrologic forecast lead time to user sectors in which forecast benefits are realized (from National Weather Service Hydrology Research Laboratory). The focus of this Product is on climate and hydrologic forecasts with lead times greater than two weeks and up to approximately one year.

2.2.1 Prediction Approaches

The primary climate and hydrologic prediction approaches used by operational and research centers fall into four categories: statistical, dynamical, statistical-dynamical hybrid, and consensus. The first three approaches are objective in the sense that the inputs and methods are formalized, outputs are not modified on an *ad hoc* basis, and the resulting forecasts are potentially reproducible by an independent forecaster using the same inputs and methods. The fourth major category of approach, which might also be termed blended knowledge, requires subjective weighting of results from the other approaches. These types of approaches are discussed in Box 2.2.

Other aspects of dynamical prediction schemes related to model physical and computational structure are important in distinguishing one model or model version from another. These aspects are primary indicators of the sophistication of an evolving model, relative to other models, but are not of much interest to the forecast user community. Examples include the degree of coupling of model components, model vertical resolution, cloud microphysics package, nature of data assimilation approaches and of the data assimilated, and the ensemble generation scheme, among many other forecast system features.

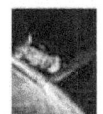

Climate and hydrologic lead-time forecasts range from minutes to years.

2.2.2 Forecast Producers and Products
Federal, regional, state, and local agencies, as well as private sector companies, such as utilities, produce hydrologic forecasts. In contrast to climate forecasts, hydrologic forecast products more directly target end use sectors—e.g., water, energy, natural resource or hazard management—and are often region-specific. Prediction methods and forecast products vary from region to region and are governed by many factors, but depend in no small measure on the hydroclimatology, institutional traditions and sectoral concerns in each region. A representative sampling of typical forecast producers and products is given in Appendix A.1. Forecasting activities at the federal, state, regional, and local scales are discussed in the following subsections.

2.2.2.1 FEDERAL
The primary federal streamflow forecasting agencies at SI lead times are the NOAA, National Weather Service (NWS) and the U.S. Department of Agriculture (USDA) National

Resource Conservation Service (NRCS) National Water and Climate Center (NWCC). The NWCC's four forecasters produce statistical forecasts of summer runoff volume in the western United States using multiple linear regression to estimate future streamflow from current observed snow water equivalent, accumulated water year precipitation, streamflow, and in some locations, using ENSO indicators such as the Niño 3.4 index (Garen, 1992; Pagano and Garen, 2005). Snowmelt runoff is critical for a wide variety of uses (water supply, irrigation, navigation, recreation, hydropower, environmental flows) in the relatively dry summer season. The regression approach has been central to the NRCS since the mid-1930s, before which similar snow-survey based forecasting was conducted by a number of smaller groups. Forecasts are available to users both in the form of tabular summaries (Figure 2.2) that convey the central tendency of the forecasts and estimates of uncertainty, and maps showing the median forecast anomaly for each river basin area for which the forecasts are operational

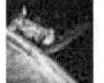

Table 2.2 Aspects of forecast products that are relevant to users.

Forecast Product Aspect	Description / Example
Forecast product variables	Precipitation, temperature, humidity, wind speed, and atmospheric pressure
Forecast product spatial resolution	Grid cell longitude by latitude, climate division
Domain	Watershed, river basin, regional, national, and global
Product time step (temporal resolution)	Hourly, sub-daily, daily, monthly, and seasonal
Range of product lead times	1 to 15 days, 1 to 13 months
Frequency of forecast product update	Every 12 hours, every month
Lag of forecast product update	The length of time from the forecast initialization time before forecast products are available: e.g., two hours for a medium range forecast, one day for a monthly to seasonal forecast.
Existence of historical climatology	Many users require a historical climatology showing forecast model performance to use in bias-correction, downscaling, and/or verification.
Deterministic or probabilistic	Deterministic forecasts have a single prediction for each future lead time. Probabilistic forecasts frame predicted values within a range of uncertainty, and consist either of an ensemble of forecast sequences spanning all lead times, or of a distinct forecast distribution for each future lead time.
Availability of skill/accuracy information	Published or otherwise available information about the performance of forecasts is not always available, particularly for forecasts that are steadily evolving. In principle, the spread of probabilistic forecasts contains such information about the median of the forecast; but the skill characteristics pertaining to the spread of the forecast are not usually available.

BOX 2.2: Forecast Approaches

Dynamical: Computer models designed to represent the physical features of the oceans, atmosphere and land surface, at least to the extent possible given computational constraints, form the basis for dynamical predictions. These models have, at their core, a set of physical relationships describing the interactions of the Earth's energy and moisture states. Inputs to the models include estimates of the current moisture and energy conditions needed to initialize the state variables of the model (such as the moisture content of an atmospheric or soil layer), and of any physical characteristics (called parameters—one example is the elevation of the land surface) that must be known to implement the relationships in the model's physical core. In theory, the main advantage of dynamical models is that influence of any one model variable on another is guided by the laws of nature as we understand them. As a result, the model will correctly simulate the behavior of the earth system even under conditions that may not have occurred in the period during which the model is verified, calibrated and validated. The primary disadvantages of dynamical models, however, are that their high computational and data input demands require them to approximate characteristics of the Earth system in ways that may compromise their realism and therefore performance. For example, the finest computational grid resolution that can be practically achieved in most atmospheric models (on the order of 100 to 200 kilometers per cell) is still too coarse to support a realistic representation of orographic effects on surface temperature and precipitation. Dynamical hydrologic models can be implemented at much finer resolutions (down to ten meters per cell, for catchment-scale models) because they are typically applied to much smaller geographic domains than are atmospheric models. While there are many aspects that distinguish one model from another, only a subset of those (listed in Table 1.1) is appreciated by the forecast user, as opposed to the climate modeler, and is relevant in describing the dynamical forecast products.

Statistical: Statistical forecast models use mathematical models to relate observations of an earth system variable that is to be predicted to observations of one or more other variables (and/or of the same variable at a prior time) that serve as predictors. The variables may describe conditions at a point location (e.g., flow along one reach of a river) or over a large domain, such as sea surface temperatures along the equator. The mathematical models are commonly linear relationships between the predictors and the predictand, but also may be formulated as more complex non-linear systems.

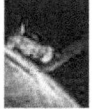

Statistical models are often preferred for their computational ease relative to dynamical models. In many cases, statistical models can give equal or better performance to dynamical models due in part to the inability of dynamical models to represent fully the physics of the system (often as a result of scale or data limitations), and in part to the dependence of predictability in many systems on predominantly linear dynamics (Penland and Magorian, 1993; van den Dool, 2007). The oft-cited shortcomings of statistical models, on the other hand, include their lack of representation of physical causes and effects, which, in theory, compromise their ability to respond to unprecedented events in a fashion that is consistent with the physical constraints of the system. In addition, statistical models may require a longer observational record for "training" than dynamical models, which are helped by their physical structure.

Objective hybrids: Statistical and dynamical tools can be combined using objective approaches. A primary example is a weighted merging of the tools' separate predictions into a single prediction (termed an objective consolidation; van den Dool, 2007). A second example is a tool that has dynamical and statistical subcomponents, such as a climate prediction model that links a dynamical ocean submodel to a statistical atmospheric model. A distinguishing feature of these hybrid approaches is that an objective method exists for linking the statistical and dynamical schemes so as to produce a set of outputs that are regarded as "optimal" relative to the prediction goals. This objectivity is not preserved in the next consensus approach.

Blended Knowledge or Subjective consensus: Some forecast centers release operational predictions, in which expert judgment is subjectively applied to modify or combine outputs from prediction approaches of one or more of the first three types, thereby correcting for perceived errors in the objective approaches to form a prediction that has skill superior to what can be achieved by objective methods alone. The process by which the NOAA Climate Predication Center (CPC) and International Research Institute for Climate and Society (IRI) constructs their monthly and seasonal outlooks for example, includes subjective weighting of the guidance provided by different climate forecast tools. The weighting is often highly sensitive to recent evolution and current state of the tropical ENSO, but other factors, like decadal trends in precipitation and surface temperature, also have the potential to influence the final official climate forecasts.

Streamflow Forecasts as of June 1, 2008

Stream and Station	Forecast Period	Most Probable		Reasonable		30 Year '71-'00 Average Runoff
		kaf	%avg	Max %avg	Min %avg	kaf
Arkansas River Basin						
Arkansas River						
Granite at, CO	Apr-Sep	250	124	177	118	210
Salida at, CO	Apr-Sep	450	145	177	118	310
Canon City at, CO	Apr-Sep	540	136	172	111	397
Pueblo abv, CO	Apr-Sep	650	134	157	105	485
Grape Creek West-cliffe nr, CO	Apr-Sep	33.0	168	245	107	19.6
Cucharas River						
La Veta nr, CO	Apr-Sep	11.1	85	108	58	13.0
Purgatoire River-Trinidad at, CO	Apr-Sep	32.0	73	107	48	44
Huerfano River						
Redwing nr, CO	Apr-Sep	12.8	83	103	65	15.5
Chalk Creek						
Nathrop nr, CO	Apr-Sep	43.0	159	211	115	27
Vermejo River						
Dawson nr, NM	Mar-Jun	6.20	89	113	73	7.0
Eagle Nest Reservoir Reservoir Inflow, NM	Mar-Jun	14.70	126	143	118	11.7
Cimarron River						
Cimarron nr, NM	Mar-Jun	18.60	117	138	106	15.9
Ponil Creek						
Cimarron nr, NM	Mar-Jun	6.10	91	109	81	6.7
Rayado Creek						
Sauble Ranch, NM	Mar-Jun	5.90	83	101	73	7.1

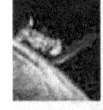

Figure 2.2 Example of NRCS tabular summer runoff (streamflow) volume forecast summary, showing median ("most probable") forecasts and probabilistic confidence intervals, as well as climatological flow averages. Flow units are thousand-acre-feet (KAF), a runoff volume for the forecast period. This table was downloaded from <http://www.wcc.nrcs.usda.gov/wsf/wsf.html>.

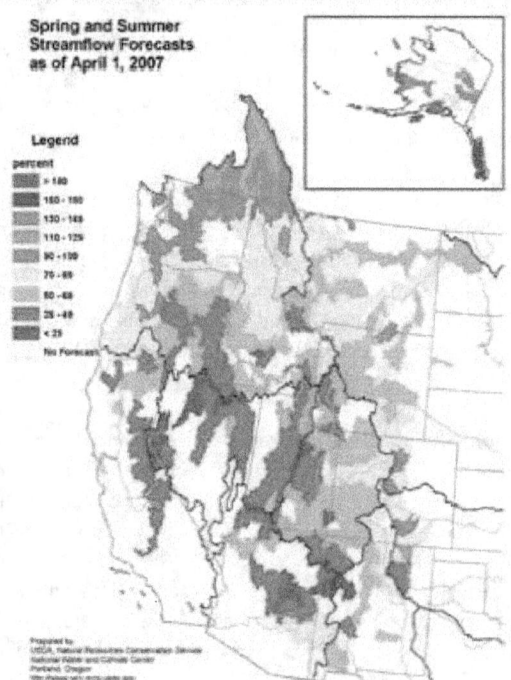

Spring and Summer
Streamflow Forecasts
as of April 1, 2007

Legend
percent

Figure 2.3 Example of NRCS spatial summer runoff (April-September streamflow) volume forecast summary, showing median runoff forecasts as an anomaly (percent of average).

(Figure 2.3). Until 2006, the NWCC's forecasts were released near the first of each month, for summer flow periods such as April through July or April through September. In 2006, the NWCC began to develop automated daily updates to these forecasts, and the daily product is likely to become more prevalent as development and testing matures. The NWCC has also just begun to explore the use of physically-based hydrologic models as a basis for forecasting.

NWCC water supply forecasts are coordinated subjectively with a parallel set of forecasts produced by the western U.S. NWS River Forecast Centers (RFCs), and with forecasts from Environment Canada's BC Hydro. The NRCS-NWS joint, official forecasts are of the subjective consensus type described earlier, so the final forecast products are subjective combinations of information from different sources, in this case, objective statistical tools (*i.e.*, regression models informed by observed snow water equivalent, accumulated water year precipitation, and streamflow) and model based forecast results from the RFCs.

The NWS surface water supply forecast program began in the 1940s in the Colorado Basin. It has since expanded to include seasonal forecasts (of volume runoff during the spring to summer snow melt period) for most of the snowmelt-dominated basins important to water management in the western United States. These forecasts rely on two primary tools: Statistical Water Supply (SWS), based on multiple-linear regression, and Ensemble Streamflow Prediction (ESP), a technique based on hydrologic modeling (Schaake, 1978; Day, 1985). Results from both approaches are augmented by forecaster experience and the coordination process with other forecasting entities. In contrast to the western RFCs, RFCs in the eastern United States are more centrally concerned with short to medium-range flood risk and drought-related water availability out to about a three month lead time. At some eastern RFC websites, the seasonal forecast is linked only to the CPC Drought Outlook rather than an RFC-generated product (Box 2.3).

The streamflow prediction services of the RFCs have a national presence, and, as such, are able to leverage a number of common technological elements, including models, databases and software for handling meteorological and hydrological data, and for making, assessing and disseminating forecasts (*i.e.*, website structure). Nonetheless, the RFCs themselves are regional entities with regional concerns.

The NWS's ESP approach warrants further discussion. In the mid 1970s, the NWS developed the hydrologic modeling, forecasting and analysis system—NWS River Forecast System (NWSRFS)—the core of which is the Sacramento soil moisture accounting scheme coupled to the Snow-17 temperature index snow model, for ESP-based prediction (Anderson, 1972, 1973; Burnash *et al.*, 1973). The ESP approach uses a deterministic simulation of the hydrologic state during a model spin-up (initialization) period, leading up to the forecast start date to estimate current hydrologic conditions, and then uses an ensemble of historical meteorological sequences as model inputs (*e.g.*, temperature and precipitation) to simulate hydrology in the future (or forecast period). Until several years ago, the RFC dissemination of ESP-based forecasts for streamflows at SI lead times was rare, and the statistical forecasts were the accepted standard. Now, as part of the NWS Advanced Hydrologic Prediction Service (AHPS) initiative, ESP forecasts are being aggressively implemented for basins across the United States (Figure 2.4) at lead times from hours to SI (McEnery *et al.*, 2005).

At the seasonal lead times, several western RFCs use graphical forecast products for the summer period streamflow forecasts that convey the probabilistic uncertainty of the forecasts. A unified web based suite of applications that became operational in 2008 provides forecast users with a number of avenues for exploring the RFC water supply forecasts. For example, Figure 2.5 shows (in clockwise order from top left) (a) a western United States depiction of the median water supply outlook for the RFC forecast basins, (b) a progression of forecasts (median and bounds) during the water year together with flow normals and observed flows; (c) monthly forecast distributions, with the option to display individual forecast ensemble members (*i.e.*, single past years) and also select ENSO-based categorical forecasts (ESP subsets); and (d) various skill measures,

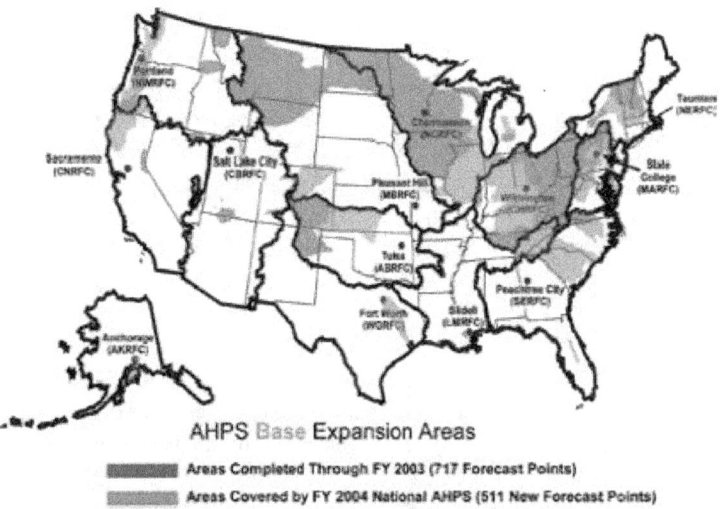

AHPS Base Expansion Areas

▓▓ Areas Completed Through FY 2003 (717 Forecast Points)
░░ Areas Covered by FY 2004 National AHPS (511 New Forecast Points)

Figure 2.4 Areas covered by the NWS Advanced Hydrologic Prediction Service (AHPS) initiative (McEnery *et al.*, 2005).

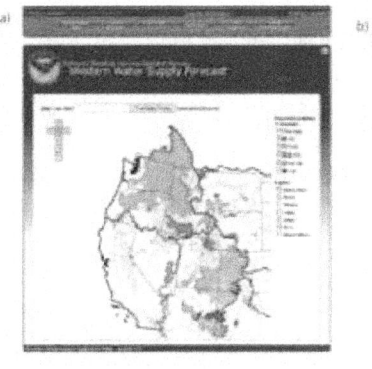
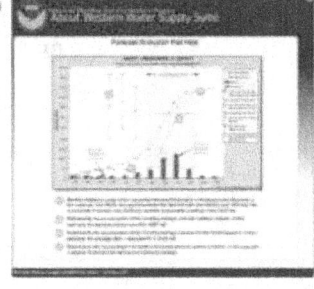
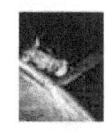
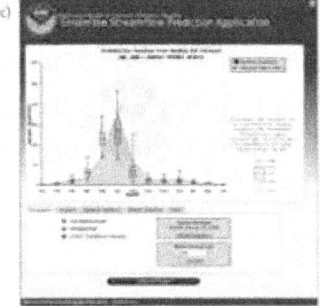
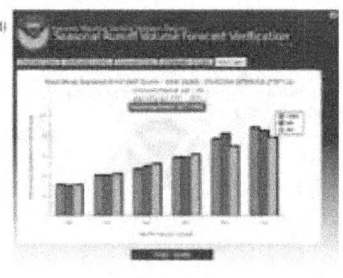

Figure 2.5 A graphical forecast product from the NWS River Forecast Centers, showing a forecast of summer (April through July) period streamflow on the Colorado River, Colorado to Arizona. These figures were obtained from <http://www.nwrfc.noaa.gov/westernwater>.

such as mean absolute error, for the forecasts based on hindcast performance. Access to raw ensemble member data is also provided from the same website.

The provision of a service that assists hydrologic forecast users in either customizing a selection of ESP possibilities to reflect, perhaps, the users' interest in data from past years that they perceive as analogues to the current year, or the current ENSO state, is a notable advance from the use of "climatological" ESP (*i.e.*, using all traces from a historical period) in the prior ESP-related seasonal forecast products. Some western RFCs have also experimented with using the CPC seasonal climate outlooks as a basis for adjusting the precipitation and temperature inputs used in climatological ESP, but it was found that the CPC outlook anomalies were generally too small to produce a distinct forecast from the climatological ESP (Hartmann *et al.*, 2002). In some RFCs, NWS statistical water supply forecasts have also provided perspective (albeit more limited) on the effect of future climate assumptions on future runoff by including results from projecting 50, 75, 100, 125 and 150 percent of normal precipitation in the remaining water year. At times, the official NWS statistical forecasts have adopted such assumptions, *e.g.*, that the first month following the forecast date would contain other than 100 percent of expected precipitation, based on forecaster judgment and consideration of a range of factors, including ENSO state and CPC climate predictions.

Figure 2.6 shows the performance of summer streamflow volume forecasts from both the NWS and NRCS over a recent ten-year period; this example is also part of the suite of forecast products that the western RFC designed to improve the communication of forecast

performance and provide verification information. Despite recent literature (Welles *et al.*, 2007) that has underscored a general scarcity of such information from hydrologic forecast providers, the NWS has recently codified verification approaches and developed verification tools, and is in the process of disbursing them throughout the RFC organization (NWS, 2006). The existence in digitized form of the retrospective archive of seasonal forecasts is critical for the verification of forecast skill. The ten-year record shown in Figure 2.6, which is longer than the record available (internally or to the public) for many public agency forecast variables, is of inadequate length for some types of statistical assessment, but is an undeniable advance in forecast communication relative to the services that were previously available. Future development priorities include a climate change scenario application, which would leverage climate change scenarios from IPCC or similar to produce inputs for future water supply planning exercises. In addition, forecast calibration procedures (*e.g.*, Seo *et al.*, 2006; Wood and Schaake, 2008) are being developed for the ensemble forecasts to remove forecast biases. The current NOAA/NWS web service Internet web address is: <http://www.nwrfc.noaa.gov/westernwater>

A contrast to these probabilistic forecasts is the deterministic five-week forecast of lake water level in Lake Lanier, GA, produced by the U.S. Army Corps of Engineers (USACE) based on probabilistic inflow forecasts from the NWS southeastern RFC. Given that the lake is a managed system and the forecast has

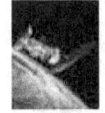

The existence in digitized form of the retrospective archive of seasonal forecasts is critical for the verification of forecast skill.

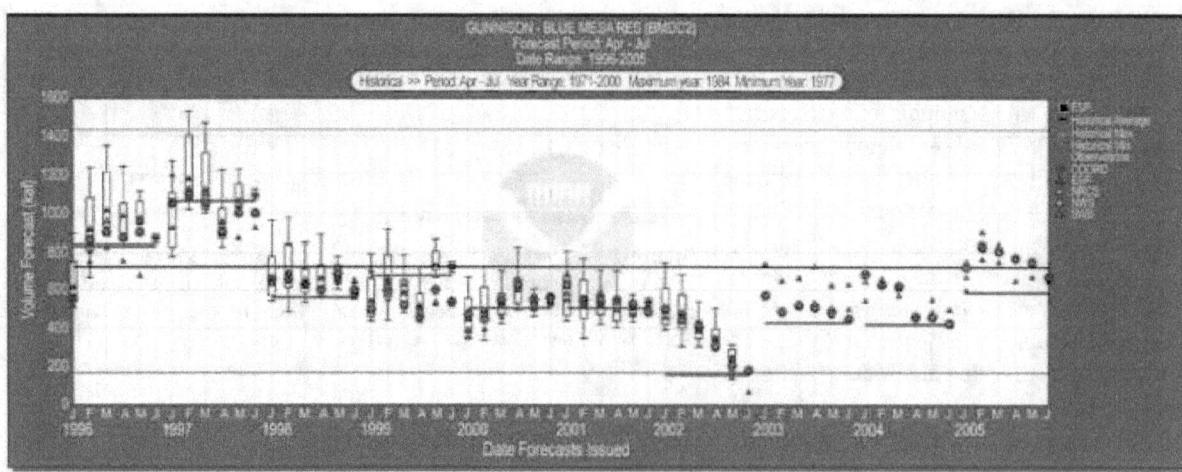

Figure 2.6 Comparing ESP and statistical forecasts from the NRCS and NWS for a recent 10-year period. The forecasts are for summer (April through July) period streamflow on the Gunnison River, Colorado.

a sub-seasonal lead time, the single-valued outlook may be justified by the planned management strategy. In such a case, the lake level is a constraint that requires transferring uncertainty in lake inflows to a different variable in the reservoir system, such as lake outflow. Alternatively, the deterministic depiction may result from an effort to simplify probabilistic information in the communication of the lake outlook to the public.

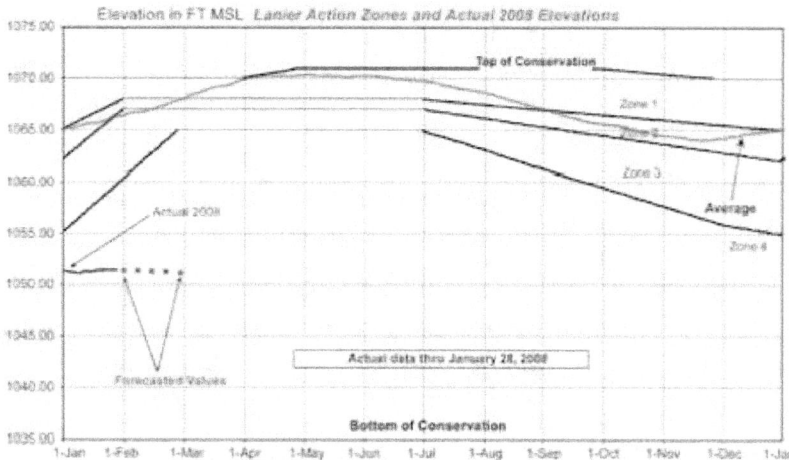

Figure 2.7 A deterministic five-week forecast of reservoir levels in Lake Lanier, Georgia, produced by USACE <http://water.sam.usace.army.mil/lanfc.htm>.

2.2.2.2 STATE AND REGIONAL

Regionally-focused agencies such as the U.S. Bureau of Reclamation (USBR), the Bonneville Power Administration (BPA), the Tennessee Valley Authority (TVA), and the Great Lakes Environmental Research Laboratory (GLERL) also produce forecasts targeting specific sectors within their priority areas. Figure 2.8 shows an example of an SI lead forecast of lake levels produced by GLERL. GLERL was among the first major public agencies to incorporate climate forecast information into operational forecasts using hydrologic and water management variables. Forecasters use coarse-scale climate forecast information to adjust climatological probability distribution functions (PDFs) of precipitation and temperature that are the basis for generating synthetic ensemble inputs to hydrologic and water management models, the outputs of which include lake level as shown in the figure. In this case, the climate forecast information is from the CPC seasonal outlooks (method described in Croley, 1996).

The Bonneville Power Administration (BPA), which helps manage and market power from the Columbia River reservoir system, is both a consumer and producer of hydrologic forecast products. The BPA generates their own ENSO-state conditioned ESP forecasts of reservoir system inflows as input to management decisions, a practice supported by research into the benefits of ENSO information for water management (Hamlet and Lettenmaier, 1999).

A number of state agencies responsible for releasing hydrologic and water resources forecasts also make use of climate forecasts in

the process of producing their own hydrologic forecasts. The South Florida Water Management District (SFWMD) predicts lake (e.g., Lake Okeechobee) and canal stages, and makes drought assessments, using a decision tree in which the CPC seasonal outlooks play a role. SFWMD follows GLERL's lead in using the Croley (1996) method for translating the CPC seasonal outlooks to variables of interest for their system.

2.2.2.3 LOCAL

At an even smaller scale, some local agencies and private utilities may also produce forecasts or at least derive applications-targeted forecasts from the more general climate or hydrology forecasts generated at larger agencies or centers.

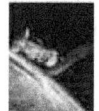

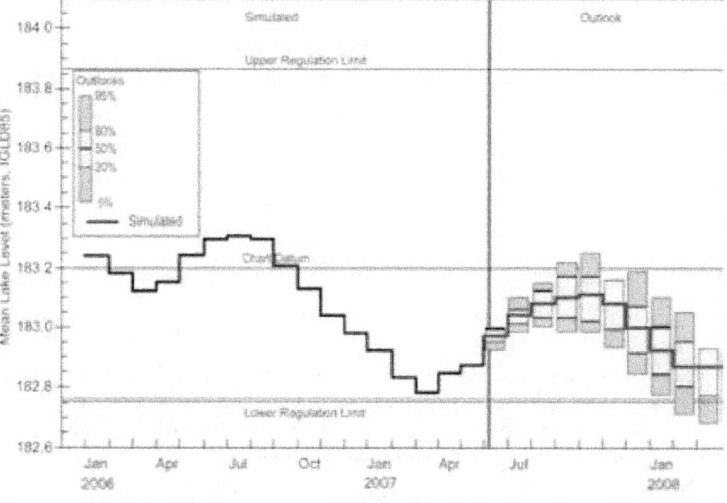

Figure 2.8 Probabilistic forecasts of future lake levels disseminated by GLERL. From: <http://www.glerl.noaa.gov/wr/ahps/curfcst/>.

Seattle Public Utilities (SPU; see Experiment 4, Section 4.2.1), for example, operates a number of reservoirs for use primarily in municipal water supply. SPU makes SI reservoir inflow forecasts using statistical methods based on observed conditions in their watersheds (*i.e.*, snow and accumulated precipitation), and on the current ENSO state, in addition to consulting the Northwest River Forecast Center (NWRFC) volume runoff forecasts. The SPU forecasts are made and used internally rather than disseminated to the public.

2.2.2.4 RESEARCH

Research institutions such as universities also produce hydrologic forecasts of a more experimental nature. A prime example is the Integrated Forecast and Reservoir Management (INFORM) project housed at the Hydrologic Research Center (HRC), which produces not only streamflow forecasts in the State of California, but also reservoir system forecasts. This project is discussed at greater length in Chapter 4 (Georgakakos *et al.*, 2005). Approximately five years ago, researchers at the University of Washington and Princeton University launched an effort to produce operational hydrologic and streamflow predictions using distributed land surface models that were developed by an interagency effort called the Land Data Assimilation System (LDAS) project (Mitchell *et al.*, 2004). In addition to generating SI streamflow forecasts in the western and eastern United States, the project also generates real-time forecasts for land surface variables such as runoff, soil moisture, and snow water equivalent (Wood and Lettenmaier, 2006; Luo and Wood, 2008), some of which are used in federal drought monitoring and prediction activities (Wood, 2008; Luo and Wood, 2007). Figure 2.9 shows an example (a runoff forecast) from this body of work that is based on the use of the Climate Forecast System (CFS) and CPC climate outlooks. Similar to the NWS ESP predictions, these hydrologic and streamflow forecasts are physically-based, dynamical and objective. The effort is supported primarily by NOAA, and like the INFORM project collaborates with public forecast agencies in developing research-level prediction products. The federal funding is provided with the intent of migrating operational forecasting advances that arise in the course of these efforts into the public agencies, a topic discussed briefly in Section 2.1.

2.2.3 Skill in Seasonal-to-Interannual Hydrologic and Water Resource Forecasts

This Section focuses on the skill of hydrologic forecasts; Section 2.5 includes a discussion of forecast utility. Forecasts are statements about events expected to occur at specific times and places in the future. They can be either deterministic, single-valued predictions about specific outcomes, or probabilistic descriptions of likely outcomes that typically take the form of ensembles, distributions, or weighted scenarios.

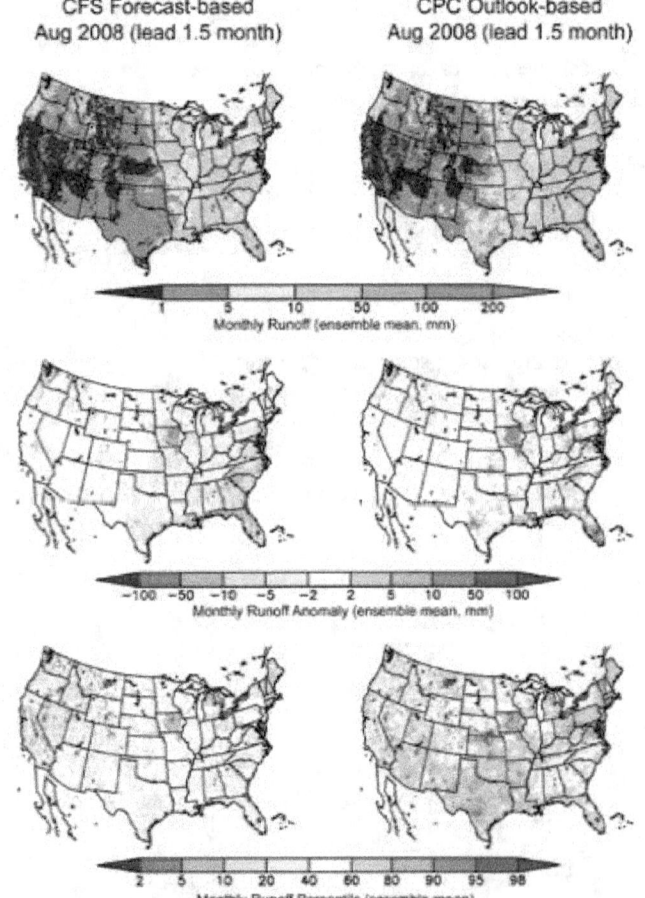

Figure 2.9 Ensemble mean forecasts of monthly runoff at lead 1.5 months created using an LDAS hydrologic model driven by CFS and CPS climate outlooks. The hydrologic prediction techniques were developed at the University of Washington and Princeton University as part of a real-time streamflow forecasting project sponsored by NOAA. Other variables, not shown, include soil moisture, snow water equivalent, and streamflow. This map is based on those available from <http://hydrology.princeton.edu/~luo/research/FORECAST/forecast.php>.

The hydrologic and water resources forecasts made for water resources management reflect three components of predictability: the seasonality of the hydrologic cycle, the predictability associated with large-scale climate teleconnections, and the persistence of anomalies in hydrologic initial conditions. Evapotranspiration, runoff (e.g., Pagano et al., 2004) and ground-water recharge (e.g., Earman et al., 2006) all depend on soil moisture and (where relevant) snowpack conditions one or two seasons prior to the forecast windows, so that these moisture conditions, directly or indirectly, are key predictors to many hydrologic forecasts with lead times up to six months. Although hydrologic initial conditions impart only a few months of predictability to hydrologic systems, during their peak months of predictability, the skill that they contribute is often paramount. This is particularly true in the western United States, where much of the year's precipitation falls during the cool season, as snow, and then accumulates in relatively easily observed form, as snowpack, until it predictably melts and runs off in the warm season months later. Information about large-scale climatic influences, like the current and projected state of ENSO, are valued because some of the predictability that they confer on water resources has influence even before snow begins to accumulate or soil-recharging fall storms arrive. ENSO, in particular, is strongly synchronized with the annual cycle so that, in many instances, the first signs of an impending warm (El Niño) or cold (La Niña) ENSO event may be discerned toward the end of the summer before the fluctuation reaches its maturity and peak of influence on the United States climate in winter. This advance warning for important aspects of water year climate allows forecasters in some locations to incorporate the expected ENSO influences into hydrologic forecasts before or near the beginning of the water year (e.g., Hamlet and Lettenmaier, 1999).

These large-scale climatic influences, however, rarely provide the high level of skill that can commonly be derived later in the water year from estimates of land surface moisture state, i.e., from precipitation accumulated during the water year, snow water equivalent or soil moisture, as estimated indirectly from streamflow. Finally, the unpredictable, random component of variability remains to limit the skill of all

real-world forecasts. The unpredictable component reflects a mix of uncertainties and errors in the observations used to initialize forecast models, errors in the models, and the chaotic complexities in forecast model dynamics and in the real world.

Many studies have shown that the single greatest source of forecast error is unknown precipitation after the forecast issue date. Schaake and Peck (1985) estimate that for the 1947 to 1984 forecasts for inflow to Lake Powell, almost 80 percent of the January 1st forecast error is due to unknown future precipitation; by April 1st, Schaake and Peck find that future precipitation still accounts for 50 percent of the forecast error. Forecasts for a specific area can perform poorly during years with abnormally high spring precipitation or they can perform poorly if the spring precipitation in that region is normally a significant component of the annual cycle. For example, in California, the bulk of the moisture falls from January to March and it rarely rains in spring (April to June), meaning that snowpack-based April 1st forecasts of spring-summer streamflow are generally very accurate. In comparison (see Figure 2.10), in eastern Wyoming and the Front Range of Colorado, April through June is the wettest time of year and, by April 1st, the forecaster can only guess at future precipitation events because of an inability to skillfully forecast springtime precipitation in this region one season in advance.

Pagano et al. (2004) determined that the second greatest factor influencing forecasting skill is how much influence snowmelt has on the hydrology of the basin and how warm the basin is during the winter. For example, in basins high in the mountains of Colorado, the temperature remains below freezing for most of the winter. Streamflow is generally low through April until temperatures rise and the snow starts to melt. The stream then receives a major pulse of snowmelt over the course of several weeks. Spring precipitation may supplement the streamflow, but any snow that falls in January is likely to remain in the basin until April when the forecast target season starts. In comparison, in western Oregon, warm rain-producing storms can be interspersed with snow-producing winter storms. Most of the runoff occurs during the winter and it is possible for a large snowpack in Febru-

Forecasts made for water resources management reflect three components of predictability: the seasonality of the hydrologic cycle, the predictability associated with large-scale climate teleconnections, and the persistence of anomalies in hydrologic initial conditions.

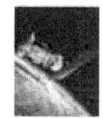

ary to be melted and washed away by March rains. For the forecaster, predicting April-to-July streamflow is difficult, particularly in anticipating the quantity of water that is going to "escape" before the target season begins. Additional forecast errors in snowmelt river basins can arise from the inability to accurately predict the sublimation of snow (sublimation occurs when ice or snow converts directly into atmospheric water vapor without first passing through the liquid state), a complex process that is influenced by cloudiness, sequences of meteorological conditions (wind, relative humidity as well as temperature) affecting crust, internal snow dynamics, and vegetation.

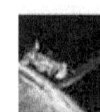

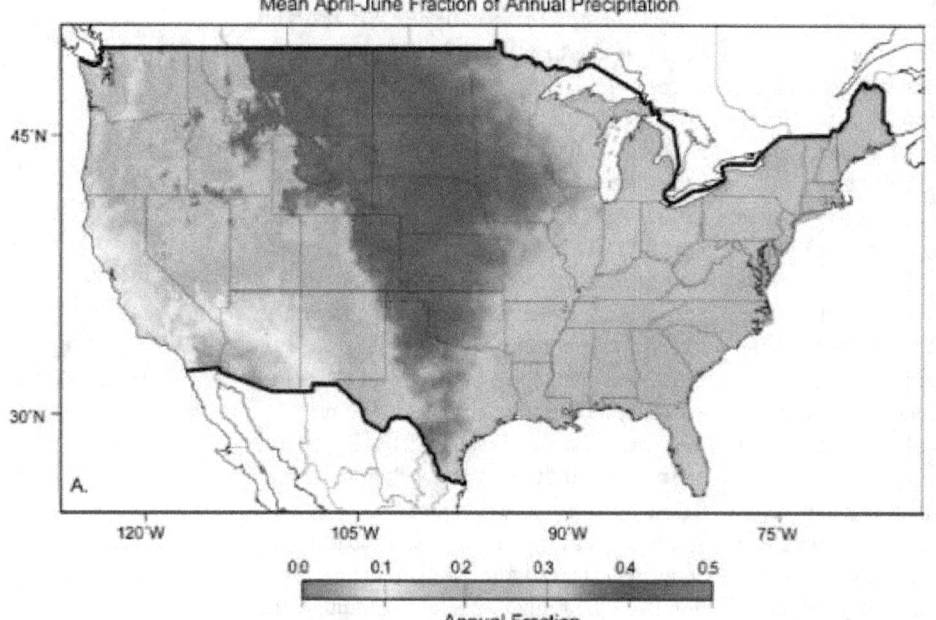

Figure 2.10 Mean percentages of annual precipitation that fell from April through June, 1971 to 2000 (based on 4-km PRISM climatologies). This figure was obtained from <http://www.prism.oregonstate.edu/>.

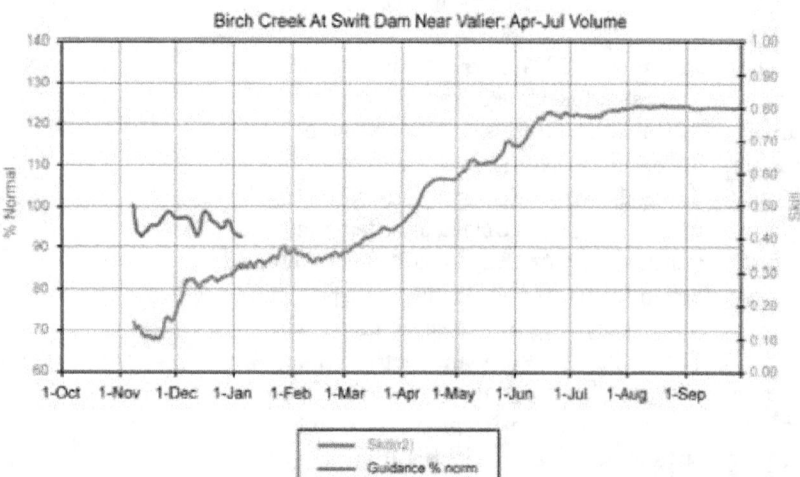

Figure 2.11 Recent operational National Water and Climate Center (NWCC) forecasts of April-July 2007 streamflow volume in Birch Creek at Swift Dam near Valier, Montana, showing daily median-forecast values of percentages of long-term average streamflow total for summer 2007 (blue) and the long-term estimates of correlation-based forecast skill corresponding to each day of the year. Figure obtained from the NWCC <http://www.wcc.nrcs.usda.gov/>.

Some element of forecast accuracy depends on the variability of the river itself. It would be easy to incur a 100 percent forecast error on, for example, the San Francisco River in Arizona, whose observations vary between 17 percent to more than 750 percent of average. It would be much more difficult to incur such a high error on a river such as the Stehekin River in Washington, where the streamflow ranges only between 60 percent and 150 percent of average. A user may be interested in this aspect of accuracy (e.g., percent of normal error), but most forecasters use skill scores (e.g., correlation) that would normalize for this effect and make the results from these two basins more comparable. As noted by Hartmann *et al.* (2002), consumers of forecast information may be more interested in measures of forecast skill other than correlations.

2.2.3.1 SKILL OF CURRENT SEASONAL HYDROLOGIC AND WATER-SUPPLY FORECASTS

As previously indicated, hydrologic and streamflow forecasts that extend to a nine-month lead time are made for western United States rivers, primarily during the winter and spring, whereas in other parts of the United States, where seasonality of precipitation is less pronounced, the forecasts link to CPC drought products, or are qualitative (the NWS Southeastern RFC, for instance, provides water supply related briefings from their website), or are in other regards less amenable to skill evaluation. For this reason, the following discussion of water supply forecast skill focuses mostly on western United States streamflow forecasting, and in particular water supply (*i.e.*, runoff volume) forecasts, for which most published material relating to SI forecasts exists.

In the western United States, the skill of operational forecasts generally improves progressively during the winter and spring months leading up to the period being forecasted, as increasing information about the year's land surface water budget are observable (*i.e.*, reflected in snowpack, soil moisture, streamflow and the like). An example of the long-term average seasonal evolution of NWCC operational forecast skill at a particular stream gage in Montana is shown in Figure 2.11. The flow rates that are judged to have a 50 percent chance of not being exceeded (*i.e.*, the 50th percentile or median) are shown by the blue curve for the early part of 2007. The red curve shows that, early in the water year, the April to July forecast has little skill, measured by the regression coefficient of determination (r^2, or correlation squared), with only about ten percent of historical variance captured by the forecast equations. By about April 1st, the forecast equations predict about 45 percent of the historical variance, and at the end of the season, the variance explained is about 80 percent. This measure of skill does not reach 100 percent because the observations available for use as predictors do not fully explain the observed hydrologic variation.

Comparisons of "hindcasts"—seasonal flow estimates generated by applying the operational forecast equations to a few decades (lengths of records differ from site to site) of historical input variables at each location with observed flows provide estimates of the expected skill of current operational forecasts. The actual skill of the forecast equations that are operationally used at as many as 226 western stream gages are illustrated in Figure 2.12, in which skill is measured by correlation of hindcast median with observed values.

The symbols in the various panels of Figure 2.12 become larger and bluer in hue as the hindcast dates approach the start of the April to July seasons being forecasted. They begin with largely unskillful beginnings each year in the January 1st forecast; by April 1st the forecasts are highly skillful by the correlation measures (predicting as much as 80 percent of the year-to-year fluctuations) for most of the California, Nevada, and Idaho rivers, and many stations in Utah and Colorado.

The general increases in skill and thus in numbers of stations with high (correlation) skill scores as the April 1st start of the forecast period approaches is shown in Figure 2.13.

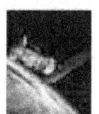

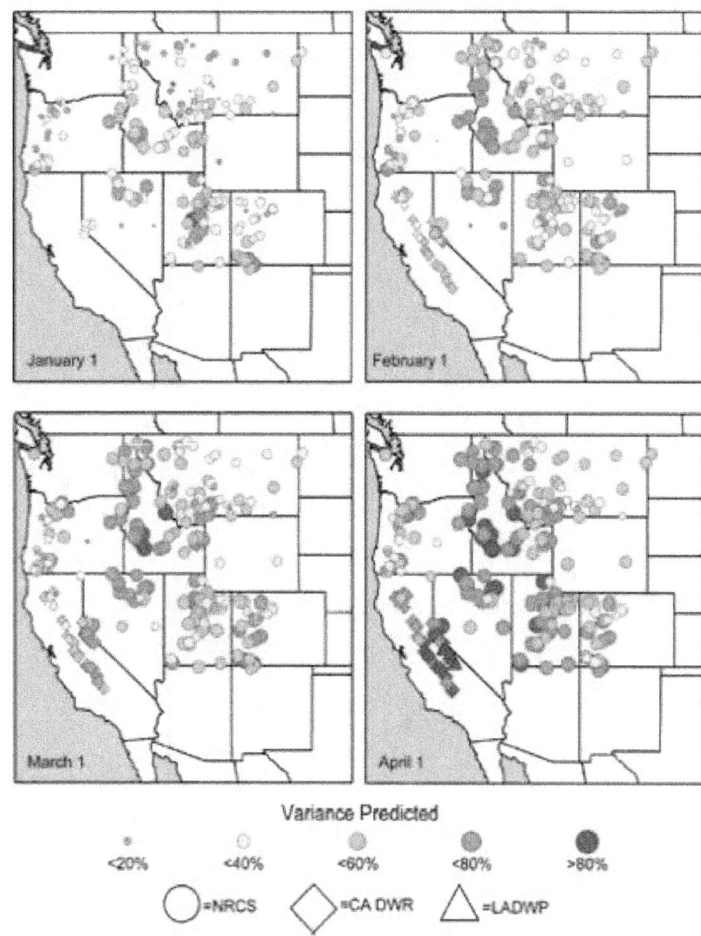

Figure 2.12 Skills of forecast equations used operationally by NRCS, California Department of Water Resources, and Los Angeles Department of Water and Power, for predicting April to July water supplies (streamflow volumes) on selected western rivers, as measured by correlations between observed and hindcasted flow totals over each station's period of forecast records. Figure provided by Tom Pagano, USDA NRCS.

A question not addressed in this Product relates to the probabilistic skill of the forecasts: How reliable are the confidence limits around the median forecasts that are provided by the published forecast quantiles (10th and 90th

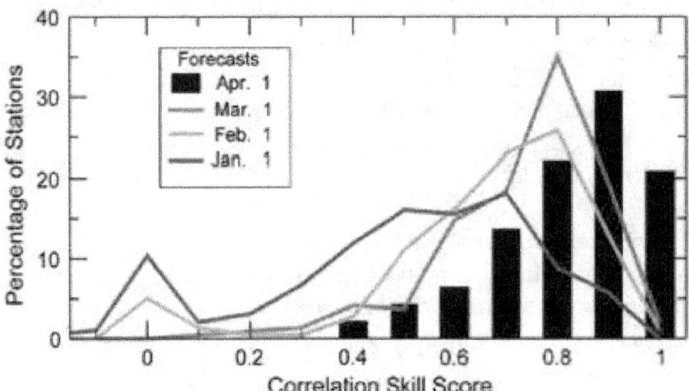

Figure 2.13 Percentages of stations with various correlation skill scores in the various panels (forecast dates) of Figure 2.12.

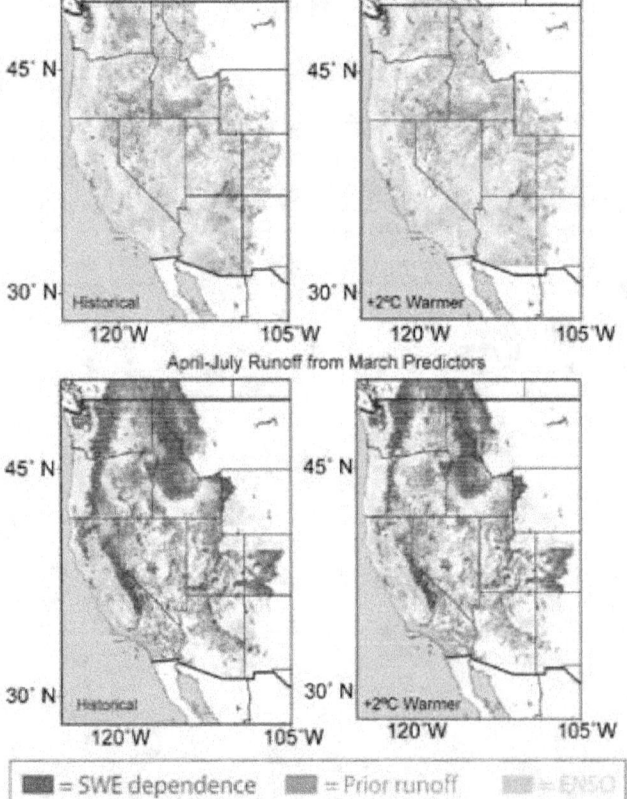

Figure 2.14 Potential contributions of antecedent snowpack conditions, runoff, and Niño 3.4 sea-surface temperatures to seasonal forecast skills in hydrologic simulations under historical, 1950 to 1999, meteorological conditions (left panels) and under those same conditions but with a 2°C uniform warming imposed (Dettinger, 2007).

percentiles, for example)? In a reliable forecast, the frequencies with which the observations fall between various sets of confidence bounds matches the probability interval set by those bounds. That is, 80 percent of the time, the observed values fall between the 10th and 90th percentiles of the forecast. Among the few analyses that have been published focusing on the probabilistic performance of United States operational streamflow forecasts, Franz *et al.* (2003) evaluated Colorado River basin ESP forecasts using a number of probabilistic measures and found reliability deficiencies for many of the streamflow locations considered.

2.2.3.2 THE IMPLICATIONS OF DECADAL VARIABILITY AND LONG TERM CHANGE IN CLIMATE FOR SEASONAL HYDROLOGIC PREDICTION SKILL

In the earlier discussion of sources of water-supply forecast skill, we highlighted the amounts and sources of skill provided by snow, soil moisture, and antecedent runoff influences. IPCC projections of global and regional warming, with its expected strong effects on western United States snowpack (Stewart *et al.*, 2004; Barnett *et al.*, 2008), raises the concern that prediction methods, such as regression, that depend on a consistent relationship between these predictors, and future runoff may not perform as expected if the current climate system is being altered in ways that then alters these hydro-climatic relationships. Decadal climate variability, particularly in precipitation (*e.g.*, Mantua *et al.*, 1997; McCabe and Dettinger, 1999), may also represent a challenge to such methods, although some researchers suggest that knowledge of decadal variability can be beneficial for streamflow forecasting (*e.g.*, Hamlet and Lettenmaier, 1999). One view (*e.g.*, Wood and Lettenmaier, 2006) is that hydrologic model-based forecasting may be more robust to the effects of climate change and variability due to the physical constraints of the land surface models, but this thesis has not been comprehensively explored.

The maps shown in Figure 2.14 are based on hydrologic simulations of a physically-based hydrologic model, called the Variable Infiltration Capacity (VIC) model (Liang *et al.*, 1994), in which historical temperatures are uniformly increased by 2°C. These figures show that the

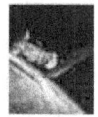

losses of snowpack and the tendencies for more precipitation to fall as rain rather than snow in a warmer world reduce overall forecast skill, shrinking the areas where snowpack contributes strong predictability and also making antecedent runoff a less reliable predictor. Thus, many areas where warm-season runoff volumes are accurately predicted historically are likely to lose some forecast skill along with their snowpack. Overall, the average skill declines by about 2 percent (out of a historical average of 35 percent) for the January to March volumes and by about 4 percent (out of a historical average of 53 percent) for April to July. More importantly, though, are the declines in skill at grid cells where historical skills are greatest, nearly halving the occurrence of high-end (>0.8) January-to-March skills and reducing high-end April-to-July skills by about 15 percent (Figure 2.15).

This enhanced loss among the most skillful grid cells reflects the strong reliance of those grid cells on historical snowpacks for the greater part of their skill, snowpacks which decline under the imposed 2°C warmer conditions. Overall, skills associated with antecedent runoff are more strongly reduced for the April-to-July runoff volumes, with reductions from an average contribution of 24 percent of variance predicted (by antecedent runoff) historically to 21 under the 2°C warm conditions; for the January to March volumes, skill contributed by antecedent runoff only declines from 18.6 percent to 18.2 percent under the imposed warmer conditions. The relative declines in the contributions from snowpack and antecedent runoff make antecedent runoff (or, more directly, soil moisture, for which antecedent runoff is serving as a proxy here) a more important predictor to monitor in the future (for a more detailed discussion, see Section 2.4.2).

It is worth noting that the changes in skill contributions illustrated in Figure 2.14 are best-case scenarios. The skills shown are skills that would be provided by a complete recalibration of forecast equations to the new (imposed) warmer conditions, based on 50 years of runoff

history. In reality, the runoff and forecast conditions are projected to gradually and continually trend towards increasingly warm conditions, and fitting new, appropriate forecast equations (and models) will always be limited by having only a brief reservoir of experience with each new degree of warming. Consequently, we must expect that regression-based forecast equations will tend to be increasingly and perennially out of date in a world with strong warming trends. This problem with the statistics of forecast skill in a changing world suggests development and deployment of more physically based, less statistically based forecast models should be a priority in the foreseeable future (Herrmann, 1999; Gleick, 2000; Milly *et al.*, 2008).

2.2.3.3 SKILL OF CLIMATE FORECAST-DRIVEN HYDROLOGIC FORECASTS

The extent to which the ability to forecast U.S. precipitation and temperature seasons in advance can be translated into long-lead hydrologic forecasting has been evaluated by Wood *et al.* (2005). That evaluation compared hydrologic variables in the major river basins of the western conterminous United States as simulated by the VIC hydrologic model (Liang *et al.*, 1994), forced by two different sources of temperature and precipitation data: (1) observed historical meteorology (1979 to 1999); and (2) by hindcast climate-model-derived six-month-lead climate forecasts.

The Wood *et al.* (2005) assessment quantified and reinforced an important aspect of the hydro-

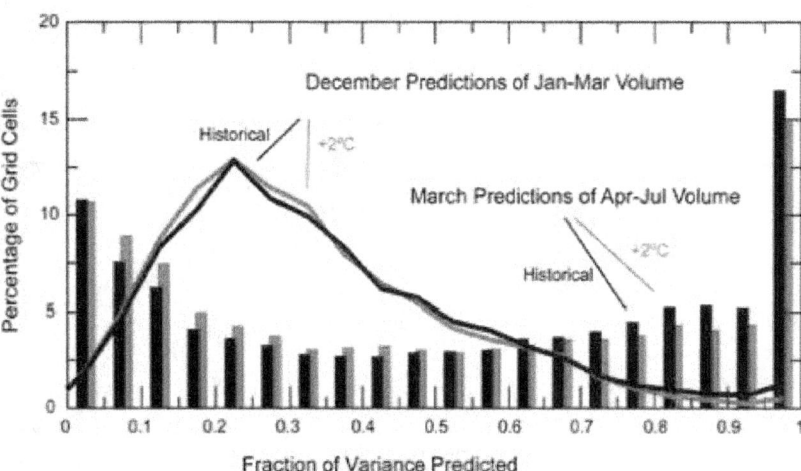

Figure 2.15 Distributions of overall fractions of variance predicted, in Figure 2.13, of January to March (curves) and April to July (histograms) runoff volumes under historical (black) and +2°C warmer conditions (Dettinger, 2007).

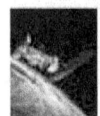

logic forecasting community's intuition about the current levels of hydrologic forecast skill using long-lead climate forecasts generated from various sources. The analysis first underscored the conclusions that, depending on the season, knowledge of initial hydrologic conditions conveys substantial forecast skill. A second finding was that the additional skill available from incorporating current (at the time) long-lead climate model forecasts into hydrologic prediction is limited when all years are considered, but can improve streamflow forecasts relative to climatological ESP forecasts in extreme ENSO years. If performance in all years is considered, the skill of current climate forecasts (particularly of precipitation) is inadequate to provide readily extracted hydrologic-forecast skill at monthly to seasonal lead times. This result is consistent with findings for North American climate predictability (Saha *et al.*, 2006). During El Niño years, however, the climate forecasts have

adequate skill for temperatures, and mixed skill for precipitation, so that hydrologic forecasts for some seasons and some basins (especially California, the Pacific Northwest and the Great Basin) provide measurable improvements over the ESP alternative.

The authors of the Wood *et al.* (2005) assessment concluded that "climate model forecasts presently suffer from a general lack of skill, [but] there may be locations, times of year and conditions (*e.g.*, during El Niño or La Niña) for which they improve hydrologic forecasts relative to ESP". However, their conclusion was that improvements to hydrologic forecasts based on other forms of climate forecasts, *e.g.*, statistical or hybrid methods that are not completely reliant on a single climate model, may prove more useful in the near term in situations where alternative approaches yield better forecast skill than that which currently exists in climate models.

2.3 CLIMATE DATA AND FORECAST PRODUCTS

2.3.1 A Sampling of Seasonal-to-Interannual Climate Forecast Products of Interest to Water Resource Managers

At SI lead times, a wide array of dynamical prediction products exist. A representative sample of SI climate forecast products is listed in Appendix A.1. The current dynamical prediction scheme used by NCEP, for example, is a system of models comprising individual models of the oceans, global atmosphere and continental land surfaces. These models were developed and originally run for operational forecast purposes in an uncoupled, sequential mode, an example of which is the so-called "Tier 2" framework in which the ocean model runs first, producing ocean surface boundary conditions that are prescribed as inputs for subsequent atmospheric model runs. Since 2004, a "Tier 1" scheme was introduced in which the models, together called the Coupled Forecast System (CFS) (Saha *et al.*, 2006), were fully coupled to allow dynamic exchanges of moisture and energy across the interfaces of the model components.

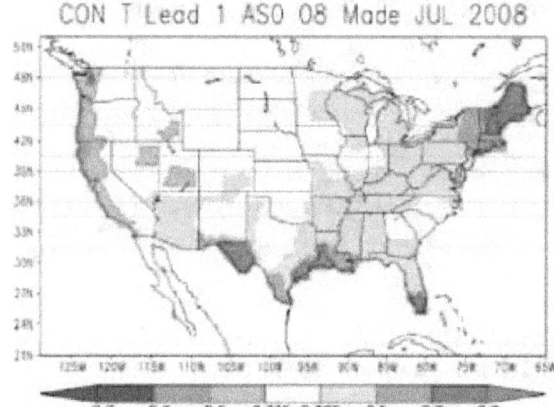

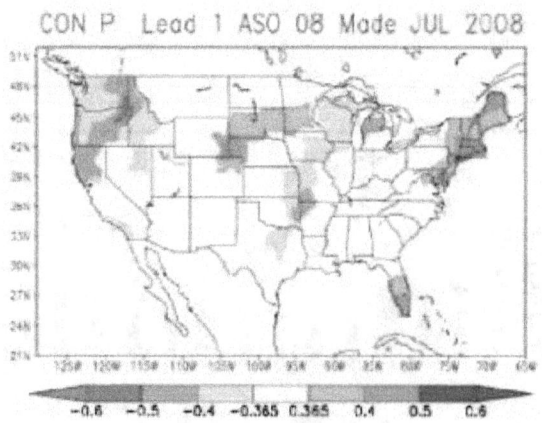

Figure 2.16 CPC objective consolidation forecast made in June 2007 (lead 1 month) for precipitation and temperature for the three month period Aug-Sep-Oct 2007. Figure obtained from <http://www.cpc.ncep.noaa.gov>.

At NCEP, the dynamical tool, CFS, is complemented by a number of statistical forecast tools, three of which, Screening Multiple Linear Regression (SMLR), Optimal Climate Normals (OCN), and Canonical Correlation Analysis (CCA), are merged with the CFS to form an objective consolidation forecast product (Figure 2.16). While the consolidated forecast exceeds the skill of the individual tools, the official seasonal forecast from CPC involves a subjective merging of it with forecast and nowcast information sources from a number of different sources, all accessible to the public at CPC's monthly briefing. The briefing materials comprise 40 different inputs regarding the past, present and expected future state of the land, oceans and atmosphere from sources both internal and external to CPC. These materials are posted online at: <http://www.cpc.ncep.noaa.gov/products/predictions/90day/tools/briefing/>.

The resulting official forecast briefing has been the CPC's primary presentation of climate forecast information each month. Forecast products are accessible directly from CPC's root level home page in the form of maps of the probability anomalies for precipitation and temperature in three categories, or "terciles", representing below-normal, normal and above-normal values; a two-category scheme (above and below normal) is also available. This framework is used for the longer lead outlooks (Figure 2.17). The seasonal forecasts are also available in the form of maps of climate anomalies in degrees Celsius for temperature and inches for precipitation (Figure 2.18). The forecasts are released monthly, have a time-step of three months, and have a spatial unit of the climate division (Figure 2.19). For users desiring more information about the probabilistic forecast than is given in the map products, a "probability of exceedence" (POE) plot, with associated parametric information, is also available for each climate division (Figure 2.20). The POE plot shows the shift of the forecast probability distribution from the climatological distribution for each lead-time of the forecast.

In addition to NCEP, a few other centers, (e.g., the International Research Institute for Climate and Society [IRI]) produce similar consensus forecasts and use a similar map-based, tercile-

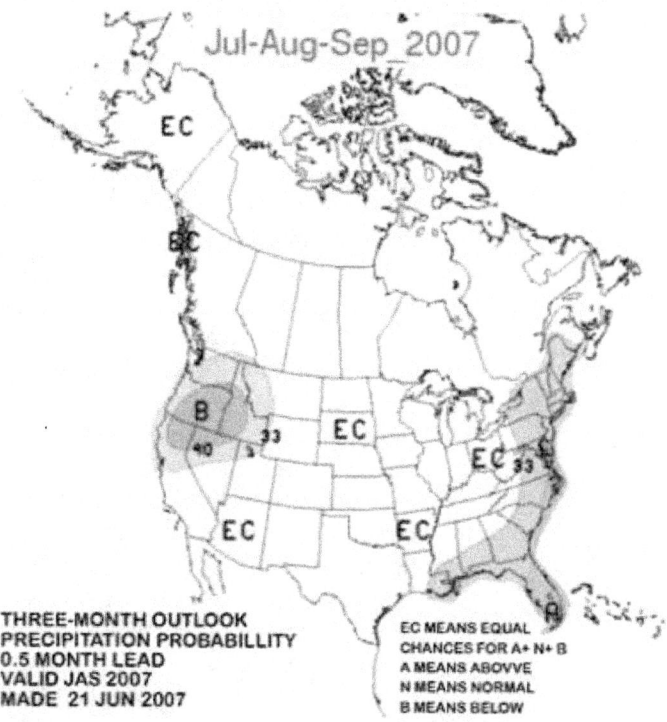

Figure 2.17 The National Center for Environmental Predictions CPC seasonal outlook for precipitation also shown as a tercile probability map. Tan/brown (green) shading indicates regions where the forecast indicates an increased probability for precipitation to be in the dry (wet) tercile, and the degree of shift is indicated by the contour labels. EC means the forecast predicts equal chances for precipitation to be in the A (above normal), B (below normal), or N (normal) terciles. Figure obtained from <http://www.cpc.ncep.noaa.gov/products/predictions/multi_season/13_seasonal_outlooks/color/page2.gif>.

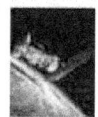

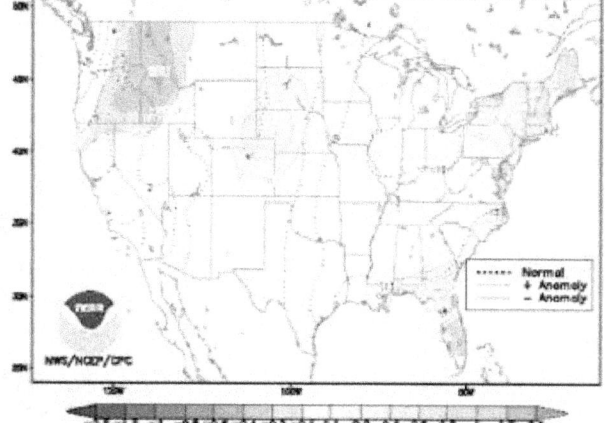

Figure 2.18 The National Center for Environmental Predictions CPC seasonal outlook for precipitation shown as inches above or below the total normal precipitation amounts for the 3-month target period (compare with the probability of exceedence forecast product shown in Figure 2.20). Figure obtained from <http://www.cpc.ncep.noaa.gov/products/predictions/long_range/poe_index.php?lead=3&var=p>.

focused framework for exhibiting their results. A larger number of centers run dynamical forecast tools, and the NOAA Climate Diagnostics Center, which produces monthly climate outlooks internally using statistical tools, also provides summaries of climate forecasts from a number of major sources, both in terms of probabilities or anomalies, for selected surface and atmospheric variables. Using dynamical models, the Experimental Climate Prediction Center (ECPC) at Scripps Institute provides monthly and seasonal time step forecasts of both climate and land surface variables at a national and global scale. Using these model outputs, ECPC also generates forecasts for derived variables that target wildfire management—e.g., soil moisture and the Fireweather Index (see Chapter 4 for a more detailed description of Water Resource Issues in Fire-Prone U.S. Forests and the use of this index). The CPC has made similar efforts in the form of the Hazards Assessment, a short- to medium-range map summary of hazards related to extreme weather (such as flooding and wildfires), and the CPC Drought Outlook (Box 2.3), a subjective consensus product focusing on the evolution of large-scale droughts that is released once a month, conveying expectations for a three-month outlook period.

The foregoing is a brief survey of climate forecast products from major centers in the United States, and, as such, is far from a comprehen-

> Seasonal-to-interannual forecast products are national to global in scale.

sive presentation of the available sources. It does, however, provide examples from which the following observations about the general nature of climate prediction in the United Sates may be drawn. First, that operational SI climate forecasting is conducted at a relatively small number of federally-funded centers, and the resulting forecast products are national to global in scale. These products tend to have a coarse resolution in space and time, and are typically for basic earth system variables (e.g., temperature, precipitation, atmospheric pressure) that are of general interest to many sectors. Forecasts are nearly always probabilistic, and the major products attempt to convey the inherent uncertainty via maps or data detailing forecast probabilities, although deterministic reductions (such as forecast variable anomalies) are also available.

2.3.2 Sources of Climate-Forecast Skill for North America

Much as with hydrologic forecasts, the skill of forecasts of climate variables (notably, temperature and precipitation) is not straightforward as it varies from region to region as well as with the forecast season and lead time; it is also limited by the chaotic and uncertain character of the climate system and derives from a variety of sources. While initial conditions are an important source for skill in SI hydrologic forecasts, the initial conditions of an atmospheric forecast are of little use after about 8 to 10 days as other forecast errors and/or disturbances rapidly grow, and therefore have no influence on SI climate forecast skill (Molteni et al., 1996). SI forecasts are actually forecasts of those variations of the climate system that reflect predictable changes in boundary conditions, like sea surface temperatures (SSTs), or in external 'forcings,' disturbances in the radiative energy budget of the Earth's climate system. At time scales of decades-to-centuries, potential skill rests in predictions for slowly varying components of the climate system, like the atmospheric concentrations of carbon dioxide that influence the greenhouse effect, or slowly

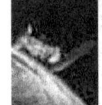

Figure 2.19 The CPC climate division spatial unit upon which the official seasonal forecasts are based. Figure obtained from <http://www.cpc.ncep.noaa.gov/products/predictions/long_range/poe_index.php?lead=3&var=p>.

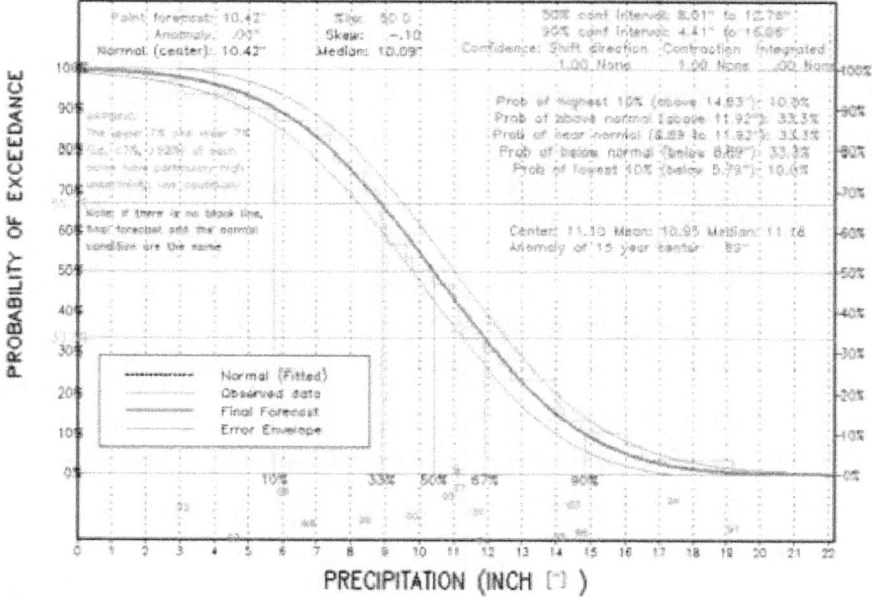

Figure 2.20 The NCEP CPC seasonal outlook for precipitation in the Seattle Region Climate Division (Division 75 in Figure 2.19) shown as the probability of exceedence for total precipitation for the three-month target period <http://www.cpc.ncep.noaa.gov/products/predictions/long_range/poe_graph_index.php?lead=3&climdiv=75&var=p.>.

evolving changes in ocean circulation that can alter SSTs and thereby change the boundary conditions for the atmosphere. Not all possible sources of SI climate-forecast skill have been identified or exploited, but contributors that have been proposed and pursued include a variety of large-scale air-sea connections (e.g., Redmond and Koch, 1991; Cayan and Webb, 1992; Mantua *et al.*, 1997; Enfield *et al.*, 2001; Hoerling and Kumar, 2003), snow and sea-ice patterns (e.g., Cohen and Entekhabi, 1999; Clark and Serreze, 2000; Lo and Clark, 2002; Liu *et al.*, 2004), and soil moisture and vegetation regimes (e.g., Koster and Suarez, 1995, 2001; Ni-Meister *et al.*, 2005).

In operational practice, however, most of the forecast skill provided by current forecast systems (especially including climate models) derives from our ability to predict the evolution of ENSO events on time scales of 6 to 12 months, coupled with the teleconnections from the events in the tropical Pacific to many areas of the globe. Barnston *et al.* (1999), in their explanation of the advent of the first operational long-lead forecasts from the NOAA Climate Prediction Center, stated that "while

some extratropical processes probably develop independently of the Tropics... much of the skill of the forecasts for the extratropics comes from anomalies of ENSO-related tropical sea surface temperatures". Except for the changes associated with diurnal cycles, seasonal cycles, and possibly the (30 to 60 day) Madden-Julian Oscillation of the tropical ocean-atmosphere system, "ENSO is the most predictable climate fluctuation on the planet" (McPhaden *et al.*, 2006). Diurnal cycles and seasonal cycles are predictable on time scales of hours-to-days and months-to-years, respectively, whereas ENSO mostly provides predictability on SI time scales. Figure 2.21a shows that temperatures over the tropical oceans and lands and extratropical oceans are more correlated from season to season than the extratropical continents. To the extent that they can anticipate the slow evolution of the tropical oceans, indicated by these correlations, SCFs in the extratropics that derive their skill from an ability to forecast conditions in the tropical oceans are provided a basis for prediction skill. To the extent that the multi-seasonal long-term potential predictability of the ENSO episodes (Figure 2.21b) can be drawn upon in certain regions at certain times of year,

Most of the skill
provided by current
forecast systems
derives from our
ability to predict the
evolution of El Niño–
Southern Oscillation
events on time scales
of 6 to 12 months.

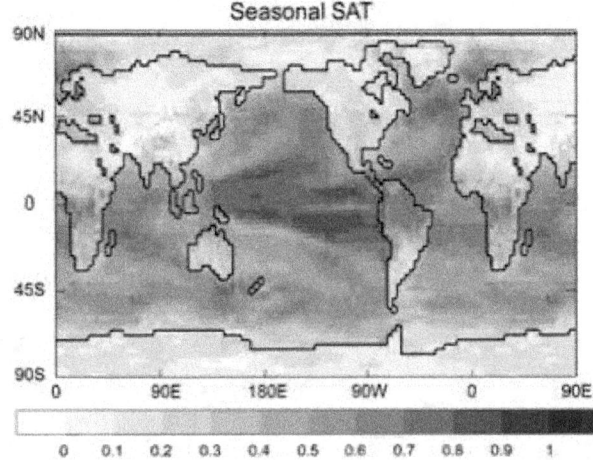

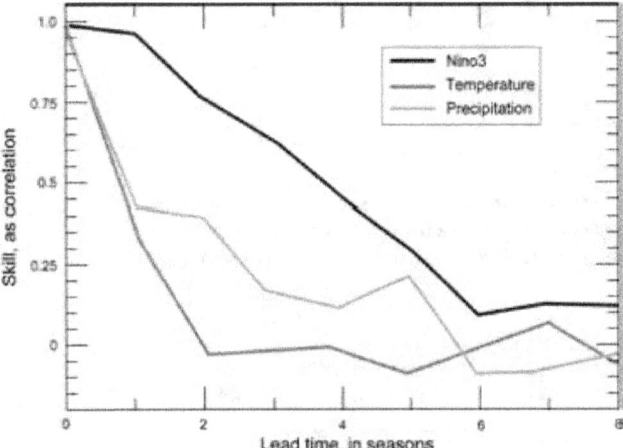

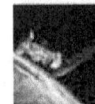

Figure 2.21 (a, top) Map of correlations between surface-air temperatures in each season and the following season in 600 years of historical climate simulation by the HadCM3 model (Collins 2002); (b, bottom) Potential predictability of a common ENSO index (Niño3 SST, the average of SSTs between 150°W and 90W, 5°S, and 5°N), average temperatures over the United States and Canada, and average precipitation over the United States and Canada, with skill measured by anomaly correlations and plotted against the forecast lead times; results extracted from Collins (2002), who estimated these skills from the reproducibility among multiple simulations of 30 years of climate by the HadCM3 coupled ocean-atmosphere model. Correlations below about 0.3 are not statistically significant at the 95 percent level.

the relatively meager predictabilities of North American temperatures and precipitation can be extended.

The scattered times between ENSO events drastically limits skillful prediction of events until, at least, the first faltering steps towards the initiation of an ENSO event have been observed. ENSO events, however, are frequently (but not always) phase-locked (synchronized) with aspects of the seasonal cycle (Neelin *et al.*, 2000), so that (a) forecasters know when to

look most diligently for those "first faltering steps" and (b) the first signs of the initiation of an event are often witnessed 6 to 9 months prior to ENSO's largest expressions in the tropics and Northern Hemisphere (*e.g.*, Penland and Sardeshmukh, 1995). Thus, ENSO influences, however irregular and unpredictable they are on multiyear time scales, regularly provide the basis for SI climate forecasts over North America. ENSO events generally begin their evolution sometime in late (northern) spring or early summer, growing and maturing until they most often reach full strength (measured by either their SST expressions in the tropical Pacific or by their influences on the Northern Hemisphere) by about December – March (*e.g.*, Chen and van den Dool 1997). An ENSO event's evolution in the tropical ocean and atmosphere during the interim period is reproducible enough that relatively simple climate indices that track ENSO-related SST and atmospheric pressure patterns in the tropical Pacific provide predictability for North American precipitation patterns as much as two seasons in advance. Late summer values of the Southern Oscillation Index (SOI), for instance, are significantly correlated with a north-south see-saw pattern of wintertime precipitation variability in western North America (Redmond and Koch, 1991).

2.4 IMPROVING WATER RESOURCES FORECAST SKILL AND PRODUCTS

Although forecast skill is only one measure of the value that forecasts provide to water resources managers and the public, it is an important measure, and current forecasts are generally understood to fall short of the maximum possible skill on SI time scales (*e.g.*, <http://www.clivar.org/organization/wgsip/spw/spw_position.php>). Schaake *et al.* (2007) describe the SI hydrologic prediction process for model-based prediction in terms of several components: (1) development, calibration and/or downscaling of SI climate forecasts; (2) estimation of hydrologic initial conditions, with or without data assimilation; (3) SI hydrologic forecasting models and methods; and (4) calibration of the resulting forecasts. Notable opportunities for forecast skill improvement in each area are discussed here.

2.4.1 Improving Seasonal-to-Interannual Climate Forecast Use for Hydrologic Prediction

SI climate forecast skill is a function of the skill of climate system models, the efficacy of model combination strategies if multiple models are used, the accuracy of climate system conditions from which the forecasts are initiated, and the performance of post-processing approaches applied to correct systematic errors in numerical model outputs. Improvements are sought in all of these areas.

2.4.1.1 CLIMATE FORECAST USE

Many researchers have found that SI climate forecasts must be downscaled, disaggregated and statistically calibrated to be suitable as inputs for applied purposes (*e.g.*, hydrologic prediction, as in Wood *et al.*, 2002). Downscaling is the process of bridging the spatial scale gap between the climate forecast resolution and the application's climate input resolution, if they are not the same. If the climate forecasts are from climate models, for instance, they are likely to be at a grid resolution of several hundred kilometers, whereas the application may require climate information at a point (*e.g.*, station location). Disaggregation is similar to downscaling, but in the temporal dimension—for exapmple, seasonal climate forecasts may need to be translated into daily or sub-daily temperature and precipitation inputs for a given application. Forecast calibration is a process by which the statistical properties (such as bias and spread errors) of a probabilistic forecast are corrected to match their observed error statistics (*e.g.*, Atger, 2003; Hamill *et al.*, 2006). These procedures may be distinct from each other, or they may be inherent parts of a single approach (such as the analogue techniques of Hamill *et al.*, 2006). These steps do not necessarily improve the signal to noise ratio of the climate forecast, but done properly, they do correct bias and reliability problems that would otherwise render impossible their use in applications. For shorter lead predictions, corrections to forecast outputs have long been made based on (past) model output statistics (MOS; Glahn and Lowry, 1972). MOS are sets of statistical relations (*e.g.*, multiple linear regression) that effectively convert numerical model outputs into unbiased, best climate predictions for selected areas or stations, where "best" relates to

past performance of the model in reproducing observations. MOS corrections are widely used in weather prediction (Dallavalle and Glahn, 2005). Corrections may be as simple as removal of mean biases indicated by historical runs of the model, with the resulting forecasted anomalies superimposed on station climatology. More complex methods specifically address spatial patterns in climate forecasts based on specific inadequacies of the models in reproducing key teleconnection patterns or topographic features (*e.g.*, Landman and Goddard, 2002; Tippett *et al.*, 2003).

A primary limitation on calibrating SI forecasts is the relatively small number of retrospective forecasts available for identifying biases. Weather predictions are made every day, so even a few years of forecasts provide a large number of examples from which to learn. SI forecasts, in contrast, are comparatively infrequent and even the number of forecasts made over several decades may not provide an adequate resource with which to develop model-output corrections (Kumar, 2007). This limitation is exacerbated when the predictability and biases themselves vary between years and states of the global climate system. Thus, there is a clear need to expand current "reforecast" practices for fixed SI climate models over long historical periods to provide both for quantification (and verification) of the evolution of SI climate forecast skills and for post-processing calibrations to those forecasts.

2.4.1.2 DEVELOPMENT OF OBJECTIVE MULTI-MODEL ENSEMBLE APPROACHES

The accuracy of SI climate forecasts has been shown to increase when forecasts from groups of models are combined into multi-model ensembles (*e.g.*, Krishnamurti *et al.*, 2000; Palmer *et al.*, 2004; Tippett *et al.*, 2007). Multi-model forecast ensembles yield greater overall skill than do any of the individual forecasts included, in principle, as a result of cancellation of errors between ensemble members. Best results thus appear to accrue when the individual models are of similar skill and when they exhibit errors and biases that differ from model to model. In part, these requirements reflect the current uncertainties about the best strategies for choosing among models for inclusion in the ensembles used and, especially for weighting

> Seasonal-to-interannual climate forecasts must be downscaled, disaggregated and statistically calibrated to be suitable as inputs for applied purposes.

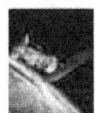

and combining the model forecasts within the ensembles. Many methods have been proposed and implemented (e.g., Rajagopalan *et al.*, 2002; Yun *et al.*, 2005), but strategies for weighting and combining ensemble members are still an area of active research (e.g., Doblas-Reyes *et al.*, 2005; Coelho *et al.*, 2004). Multi-model ensemble forecast programs are underway in Europe (DEMETER, Palmer *et al.*, 2004) and in Korea (APEC; e.g., Kang and Park, 2007). In the United States, IRI forms an experimental multi-model ensemble forecast, updating monthly, from seasonal forecast ensembles run separately at seven centers, a "simple multi-model" approach that compares well with centrally organized efforts such as DEMETER (Doblas-Reyes *et al.*, 2005). The NOAA Climate Test Bed Science Plan also envisions such a capability for NOAA (Higgins *et al.*, 2006).

2.4.1.3 IMPROVING CLIMATE MODELS, INITIAL CONDITIONS, AND ATTRIBUTIONS

Improvements to climate models used in SI forecasting efforts should be a high priority. Several groups of climate forecasters have identified the lack of key aspects of the climate system in current forecast models as important weaknesses, including underrepresented linkages between the stratosphere and troposphere (Baldwin and Dunkerton, 1999), limited processes and initial conditions at land surfaces (Beljaars *et al.*, 1996; Dirmeyer *et al.*, 2006; Ferranti and Viterbo, 2006), and lack of key biogeochemical cycles like carbon dioxide.

Because climate prediction is, by most definitions, a problem determined by boundary condition rather than an initial condition, specification of atmospheric initial conditions is not the problem for SI forecasts that it is for weather forecasts. However, SI climate forecast skill for most regions comes from knowledge of current SSTs or predictions of future SSTs, especially those in the tropics (Shukla *et al.*, 2000; Goddard and Dilley, 2005; Rosati *et al.*, 1997). Indeed, forecast skill over land (worldwide) increases directly with the strength of an ENSO event (Goddard and Dilley, 2005). Thus, an important determinant of recent improvements in SI forecast skill has been the quality and placement of tropical ocean observations, like the TOGA-TAO (Tropical Atmosphere Ocean project) network of buoys that monitors

the conditions that lead up to and culminate in El Niño and La Niña events (Trenberth *et al.*, 1998; McPhaden *et al.*, 1998; Morss and Battisti, 2004). More improvements in all of the world's oceans are expected from the broader Array for Real-time Geostrophic Oceanography (ARGO) upper-ocean monitoring arrays and Global Ocean Observing System (GOOS) programs (Nowlin *et al.*, 2001). In many cases, and especially with the new widespread ARGO ocean observations, ocean data assimilation has improved forecast skill (e.g., Zheng *et al.*, 2006). Data assimilation into coupled ocean-atmosphere-land models is a difficult and unresolved problem that is an area of active research (e.g., Ploshay and Anderson, 2002; Zheng *et al.*, 2006). Land-surface and cryospheric conditions also can influence the seasonal-scale dynamics that lend predictability to SI climate forecasting, but incorporation of these initial boundary conditions into SI climate forecasts is in an early stage of development (Koster and Suarez, 2001; Lu and Mitchell, 2004; Mitchell *et al.*, 2004). Both improved observations and improved avenues for including these conditions into SI climate models, especially with coupled ocean-atmosphere-land models, are needed. Additionally, education and expertise deficiencies contribute to unresolved problems in data assimilation for geophysical modeling. The Office of the Federal Coordinator for Meteorology (2007) documents that there is a need for more students (either undergraduate or graduate) who have sufficient mathematics and computer science skills to engage in data assimilation work in the research and/or operational environment.

Finally, a long-standing but little explored approach to improving the value of SI climate forecasts is the attribution of the causes of

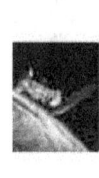

Seasonal-to-interannual climate forecast skill for most regions comes from knowledge of current sea surface temperatures or predictions of future sea surface temperatures, especially those in the tropics.

climate variations. The rationale for an attribution effort is that forecasts have greater value if we know why the forecasted event happened, either before or after the event, and why a forecast succeeded or failed, after the event. The need to distinguish natural from human-caused trends, and trends from fluctuations, is likely to become more and more important as climate change progresses. SI forecasts are likely to fail from time to time or to realize less probable ranges of probabilistic forecasts. Knowing that forecasters understand the failures (in hindsight) and have learned from them will help to build increasing confidence through time among users. Attempts to attribute causes to important climate events began as long ago as the requests from Congress to explain the 1930s Dust Bowl. Recently NOAA has initiated a Climate Attribution Service (see: <http://www.cdc.noaa.gov/CSI/>) that will combine historical records, climatic observations, and many climate model simulations to infer the principal causes of important climate events of the past and present. Forecasters can benefit from knowledge of causes and effects of specific climatic events as well as improved feedbacks as to what parts of their forecasts succeed or fail. Users will also benefit from knowing the reasons for prediction successes and failures.

2.4.2 Improving Initial Hydrologic Conditions for Hydrologic and Water Resource Forecasts

Operational hydrologic and water resource forecasts at SI time scales derive much of their skill from hydrologic initial conditions, with the particular sources of skill depending on seasons and locations. Better estimation of hydrologic initial conditions will, in some seasons, lead to improvements in SI hydrologic and consequently, water resources forecast skill. The four main avenues for progress in this area are: (1) augmentation of climate and hydrologic observing networks; (2) improvements in hydrologic models (*i.e.*, physics and resolution); (3) improvements in hydrologic model calibration approaches; and (4) data assimilation.

2.4.2.1 HYDROLOGIC OBSERVING NETWORKS

As discussed previously (in Section 2.2), hydrologic and hydroclimatic monitoring networks provide crucial inputs to hydrologic and water resource forecasting models at SI time scales. Continuous or regular measurements of streamflow, precipitation and snow water contents provide important indications of the amount of water that entered and left river basins prior to the forecasts and thus directly or indirectly provide the initial conditions for model forecasts.

Observed snow water contents are particularly important sources of predictability in most of the western half of the United States, and have been measured regularly at networks of snow courses since the 1920s and continually at SNOTELs (automated and telemetered snow instrumentation sites) since the 1950s. Snow measurements can contribute as much as three-fourths of the skill achieved by warm-season water supply forecasts in the West (Dettinger, 2007). However, recent studies have shown that measurements made at most SNOTELs are not representative of overall basin water budgets, so that their value is primarily as indices of water availability rather than as true monitors of the overall water budgets (Molotch and Bales, 2005). The discrepancy arises because most SNOTELs are located in clearings, on flat terrain, and at moderate altitudes, rather than the more representative snow courses that historically sampled snow conditions throughout the complex terrains and micrometeorological conditions found in most river basins. The discrepancies limit some of the usefulness of SNOTEL measurements as the field of hydrologic forecasting moves more and more towards physically-based, rather than empirical-statisti-

The need to distinguish natural from human-caused trends, and trends from fluctuations, is likely to become more and more important as climate change progresses.

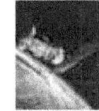

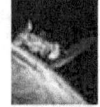

Groundwater level networks already are contributing to drought monitors and response plans in many states.

cal models. To remedy this situation, and to provide more diverse and more widespread inputs as required by most physically-based models, combinations of remotely sensed snow conditions (to provide complete areal coverage) and extensions of at least some SNOTELs to include more types of measurements and measurements at more nearby locations will likely be required (Bales *et al.*, 2006).

Networks of ground-water level measurements are also important because: (1) these data support operations and research, and (2) the networks' data may be critical to some aspects of future hydrologic forecast programs. Groundwater level measurements are made at thousands of locations around the United States, but they have only recently been made available for widespread use in near-real time (see: <http://ogw01.er.usgs.gov/USGSGWNetworks. asp>). Few operational surface water resource forecasts have been designed to use ground-water measurements. Similarly climate-driven SI groundwater resource forecasts are rare, if made at all. However, surface water and groundwater are interlinked in nearly all cases and, in truth, constitute a single resource (Winter *et al.*, 1998). With the growing availability of real-time groundwater data dissemination, opportunities for improving water resource forecasts by better integration and use of surface- and groundwater data resources may develop. Groundwater level networks already are contributing to drought monitors and response plans in many states.

Similarly, long-term soil-moisture measurements have been relatively uncommon until recently, yet are of potentially high value for many land management activities including range management, agriculture, and drought forecasting. Soil moisture is an important

control on the partitioning of water between evapotranspiration, groundwater recharge, and runoff, and plays an important (but largely unaddressed) role in the quantities addressed by water resource forecasts. Soil moisture varies rapidly from place to place (Vinnikov *et al.*, 1996; Western *et al.*, 2004) so that networks that will provide representative measurements have always been difficult to design (Wilson *et al.*, 2004). Nonetheless, the Illinois State Water Survey has monitored soil moisture at about 20 sites in Illinois for many years (see: <http://www.sws.uiuc.edu/warm/soilmoist/ ISWSSoilMoistureSummary.pdf>), but was alone in monitoring soil moisture at the state scale for most of that time. As the technologies for monitoring soil moisture have become less troublesome, more reliable, and less expensive in recent years, more agencies are beginning to install soil-moisture monitoring stations (e.g., the NRCS is augmenting many of its SNOTELs with soil-moisture monitors and has established a national Soil Climate Analysis Network (SCAN; <http://www.wcc.nrcs.usda.gov/scan/ SCAN-brochure.pdf>); Oklahoma's Mesonet micrometeorological network includes soil-moisture measurements at its sites; California is on the verge of implementing a state-scale network at both high and low altitudes). With the advent of regular remote sensing of soil-moisture conditions (Wagner *et al.*, 2007), many of these *in situ* networks will be provided context so that their geographic representativeness can be assessed and calibrated (Famligietti *et al.*, 1999). As with groundwater, soil moisture has not often been an input to water resource forecasts on the SI time scale. Instead, if anything, it is being simulated, rather than measured, where values are required. Increased monitoring of soil moisture, both remotely and *in situ*, will provide important checks on the models of soil-moisture reservoirs that underlie nearly all of our water resources and water resource forecasts, making hydrological model improvements possible.

Augmentation of real-time stream gauging networks is also a priority, a subject discussed in the Synthesis and Assessment Product 4.3 (CCSP, 2008).

2.4.2.2 IMPROVEMENTS IN HYDROLOGIC MODELING TECHNIQUES

Efforts to improve hydrologic simulation techniques have been pursued in many areas since the inception of hydrologic modeling in the 1960s and 1970s when the Stanford Watershed Model (Crawford and Linsley, 1966), the Sacramento Model (Burnash *et al.*, 1973) and others were created. More recently, physically-based, distributed and semi-distributed hydrologic models have been developed, both at the watershed scale (*e.g.*, Wigmosta *et al.*, 1994; Boyle *et al.*, 2000) to account for terrain and climate inhomogeneity, and at the regional scale (Liang *et al.*, 1994 among others). Macroscale models (like the Sacramento Model and the Stanford Watershed Model) were partly motivated by the need to improve land surface representation in climate system modeling approaches (Mitchell *et al.*, 2004), but these models have also been found useful for hydrologic applications related to water management (*e.g.*, Hamlet and Lettenmaier, 1999; Maurer and Lettenmaier, 2004; Wood and Lettenmaier, 2006). The NOAA North American Land Data Assimilation Project (Mitchell *et al.*, 2004) and NASA Land Information System (Kumar *et al.*, 2006) projects are leading agency-sponsored research efforts that are focused on advancing the development and operational deployments of the regional/physically based models. These efforts include research to improve the estimation of observed parameters (*e.g.*, use of satellite remote sensing for vegetation properties and distribution), the accuracy of meteorological forcings, model algorithms and computational approaches. Progress in these areas has the potential to improve the ability of hydrologic models to characterize land surface conditions for forecast initialization, and to translate future meteorology and climate into future hydrologic response.

Aside from improving hydrologic models and inputs, strategies for hydrologic model implementation are also important. Model calibration—, the identification of optimal parameter sets for simulating particular types of hydrologic output (single or multiple)—has arguably been the most extensive area of research toward improving hydrologic modeling techniques (*e.g.*, Wagener and Gupta, 2005, among others). This body of work has yielded advances in the understanding of the model calibration problem from both practical and theoretical perspectives. The work has been conducted using models at the watershed scale to a greater extent than the regional scale, and the potential for applying these techniques to the regional scale models has not been explored in depth.

Data assimilation is another area of active research (*e.g.*, Andreadis and Lettenmaier 2006; Reichle *et al.*, 2002; Vrugt *et al.*, 2005; Seo *et al.*, 2006). It is a process in which verifying observations of model state or output variables are used to adjust the model variables as the model is running, thereby correcting simulation errors on the fly. The primary types of observations that can be assimilated include snow water equivalent and snow covered area, land surface skin temperature, remotely sensed or *in situ* soil moisture, and streamflow. NWS-RFS has the capability to do objective data assimilation. In practice, NWS (and other agencies) perform a qualitative data assimilation, in which forecaster judgment is used to adjust model states and inputs to reproduce variables such as streamflow, snow line elevation and snow water equivalent prior to initializing an ensemble forecast.

2.4.3 Calibration of Hydrologic Model Forecasts

Even the best real-world hydrologic models have biases and errors when applied to specific gages or locations. Statistical models often are tuned well enough so that their biases are relatively small, but physically-based models often exhibit significant biases. In either case, further improvements in forecast skill can be obtained, in principle, by post-processing model forecasts to remove or reduce any remaining systematic errors, as detected in the performance of the models in hindcasts. Very little research has been performed on the best methods for such post-processing (Schaake *et al.*, 2007), which is closely related to the calibration corrections regularly made to weather forecasts. Seo *et al.* (2006), however, describe an effort being undertaken by the National Weather Service for short lead hydrologic forecasts, a practice that is more common than for longer lead hydrologic forecasts. Other examples include work by Hashino *et al.* (2007) and Krzysztofowicz (1999). At least one example of an application

Efforts to improve hydrologic simulation techniques have been pursued in many areas since the inception of hydrologic modeling in the 1960s and 1970s.

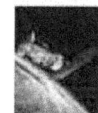

for SI hydrologic forecasts is given in Wood and Schaake (2008); but as noted earlier, a major limitation for such approaches is the limited sample sizes available for developing statistical corrections.

2.5 IMPROVING PRODUCTS: FORECAST AND RELATED INFORMATION PACKAGING AND DELIVERY

There is wide support for a comparative and relative "now *versus* normal *versus* last year" form of characterizing hydrologic and climate forecasts.

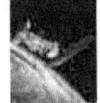

The value of SI forecasts can depend on more than their forecast skill. The context that is provided for understanding or using forecasts can contribute as much or more to their value to forecast users. Several avenues for re-packaging and providing context for SI forecasts are discussed in the following paragraphs.

Probabilistic hydrologic forecasts typically represent summaries of collections of forecasts, forecasts that differ from each other due to various representations of the uncertainties at the time of forecast or likely levels of climate variation after the forecast is made, or both (Schaake *et al.*, 2007). For example, the "ensemble streamflow prediction" methodology begins its forecasts (generally) from a single best estimate of the initial conditions from which the forecasted quantity will evolve, driven by copies of the historical meteorological variations from each year in the past (Franz *et al.*, 2003). This provides ensembles of as many forecasts as there are past years of appropriate meteorological records, with the ensemble scatter representing likely ranges of weather variations during the forecast season. Sometimes deterministic forecasts are extended to represent ranges of possibilities by directly adding various measures of past hydrologic or climatic variability. More modern probabilistic methods are based on multiple climate forecasts, multiple initial conditions or multiple parameterizations (including multiple downscalings) (Clark *et al.*, 2004; Schaake *et al.*, 2007). However accomplished, having made numerous forecasts that represent ranges of uncertainty or variability, the probabilistic forecaster summarizes the results in terms of statistics of the forecast ensemble and presents the probabilistic forecast in terms of selected statistics, like probabilities of being more or less than normal.

In most applications, it is up to the forecast user to interpret these statistical descriptions in terms of their own particular data needs, which frequently entails (1) application of various corrections to make them more representative of their local setting and (2), in some applications, essentially a deconvolution of the reported probabilities into plausible examples that might arise during the future described by those probabilities. Forecast users in some cases may be better served by provision of historical analogs that closely resemble the forecasted conditions, so that they can analyze their own histories of the results during the analogous (historical) weather conditions. For example, Wiener *et al.* (2000) report that there is wide support for a comparative and relative "now *versus* normal *versus* last year" form of characterizing hydrologic and climate forecasts. Such qualitative characterizations would require careful and explicit caveats, but still have value as reference to historical conditions in which most current managers learned their craft and in which operations were institutionalized or codified. While "normal" is increasingly problematic, "last year" may be the best and most accessible analogue for the wide variety of relevant market conditions in which agricultural water users (and their competitors), for example, operate.

Alternatively, some forecast users may find that elements from the original ensembles of forecasts would provide useful examples that could be analyzed or modeled in order to more clearly represent the probabilistic forecast in concrete terms. The original forecast ensemble members are the primary source of the probabilistic forecasts and can offer clear and definite examples of what the forecasted future *could* look like (but not specifically what it *will* look like). Thus, along with the finished forecasts, which should remain the primary forecast products, other representations of what the forecasts are and how they would appear in the real world could be useful and more accessible complements for some users, and would be a desirable addition to the current array of forecast products.

Another approach to providing context (and, potentially, examples) for the SI water resource forecasts involves placing the SI forecasts in the context of paleoclimate reconstructions for the prior several centuries. The twentieth century

BOX 2.3: The CPC Seasonal Drought Outlook

The CPC Drought Outlook (DO) is a categorical prediction of drought evolution for the three months forward from the forecast date. The product, which is updated once per month, comprises a map that is accompanied by a text discussion of the rationale for the categories depicted on the map.

The starting conditions for the DO are given by the current Drought Monitor (DM) (a United States map that is updated weekly showing the status of drought nationwide located: <http://www.drought.unl.edu/DM/monitor.html>), and the DO shows likely changes in and adjacent to the current DM drought areas. The DO is a subjective consensus forecast that is assembled each month by a single author (rotating between CPC and the National Drought Mitigation Center [NDMC]) with feedback from a panel of geographically distributed agency and academic experts. The basis for estimating future drought evolution includes a myriad of operational climate forecast products: from short- and medium-range weather forecasts to seasonal predictions from the CPC climate outlooks and the NCEP CFS outputs; consideration of climate tendencies for current El Nino–Southern Oscillation state; regional hydroclimatology; and medium-range to seasonal soil moisture and runoff forecasts from a variety of sources.

The DO makes use of the most advanced objective climate and hydrologic prediction products currently available, including not only operational, but experimental products, although the merging of the different inputs is based on expert judgment rather than an objective system. The DO is verified by comparing the DM drought assessments at the start and end of the DO forecast period; verification skill scores have been tracked for the last seven years. The DO is the primary drought-related agency forecast produced in the United States, and is widely used by the drought management and response community from local to regional scales.

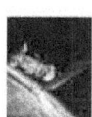

The DO was developed in the context of new drought assessment partnerships between the CPC, U.S. Department of Agriculture and the NDMC following the passage of the National Drought Policy Act of 1998. The DM was released as an official product in August, 1999, with the expectation that a weekly or seasonal drought forecast capacity would be added in the future. A drought on the Eastern Seaboard in the fall of 1999 required briefings for the press and the Clinton Administration; internal discussions between DM participants at the CPC led to the formation of the first version of the DO (maps and text) for these briefings. These were released informally to local, state and federal agency personnel throughout the winter of 1999 to 2000, and received positive feedback.

The CPC decided to make the products official, provided public statements and developed product specifications, and made the product operational in March 2000. The initial development process was informal and lasted about six months. In November 2000, the first Drought Monitor Forum was held, at which producers and users (agency, state, private, academic) came together to evaluate the DM in its first year and plan for its second, providing, in addition, a venue for discussion of the DO. This forum still meets bi-annually, focusing on both DM- and DO-relevant issues. Developmental efforts for the DO are internal at CPC or within NCEP, and the primary avenues for feedback are the website and at presentations by DO authors at workshops and conferences. The DO authors also interact with research efforts funded by the National Oceanic and Atmospheric Administration (NOAA) Climate Program Office and other agency funding sources, and with NOAA research group efforts (such as at NCEP), as part of the ongoing development effort. URL: <http://www.cpc.noaa.gov/products/expert_assessment/drought_assessment.shtml>.

has, by and large, been climatically benign in much of the nation, compared to previous centuries (Hughes and Brown, 1992; Cook *et al.*, 1999). As a consequence, the true likelihood of various forecasted, naturally-occurring climate and water resource anomalies may best be understood in the context of longer records, which paleoclimatic reconstructions can provide. At present, approaches to incorporating paleoclimatic information into responses to SI forecasts are uncommon and only beginning to develop, but eventually they may provide a clearer framework for understanding and perfecting probabilistic SI water resource forecasts. One approach being investigated is the statistical synthesis of examples (scenarios) that reflect both the long-term climate variability identified in paleo-records and time-series-based deterministic long-lead forecasts (Kwon *et al.*, 2007).

2.6 THE EVOLUTION OF PROTOTYPES TO PRODUCTS AND THE ROLE OF EVALUATION IN PRODUCT DEVELOPMENT

Studies of what makes forecasts useful have identified a number of common characteristics in the process by which forecasts are generated, developed, and taught to and disseminated among users (Cash and Buizer, 2005). These characteristics include: ensuring that the problems that forecasters address are themselves driven by forecast users; making certain that knowledge-to-action networks (the process of interaction between scientists and users which produces forecasts) are end-to-end inclusive; employing "boundary organizations" (groups or other entities that bridge the communication void between experts and users) to perform translation and mediation functions between the producers and consumers of forecasts; fostering a social learning environment between producers and users (*i.e.*, emphasizing adaptation); and providing stable funding and other support to keep networks of users and scientists working together.

This Section begins by providing a review of recent processes used to take a prototype into an operational product, with specific examples from the NWS. Some examples of interactions between forecast producers and users that have

lead to new forecast products are then reviewed, and finally a vision of how user-centric forecast evaluation could play a role in setting priorities for improving data and forecast products in the future is described.

2.6.1 Transitioning Prototypes to Products
During testimony for this Product, heads of federal operational forecast groups all painted a relatively consistent picture of how most in-house innovations currently begin and evolve. Although formal and quantitative innovation planning methodologies exist (see Appendix A.3: Transitioning NWS Research into Operations and How the Weather Service Prioritizes the Development of Improved Hydrologic Forecasts), for the most part, the operational practice is often relatively *ad hoc* and unstructured except for the larger and longer-term projects. The Seasonal Drought Outlook is an example of a product that was developed under a less formal process than that used by the NWS (Box 2.3).

Climate and water resource forecasters are often aware of small adjustments or "tweaks" to forecasts that would make their jobs easier; these are often referred to as "forecasts of opportunity". A forecaster may be aware of a new dataset or method or product that he/she believes could be useful. Based on past experience, production of the forecast may seem feasible and it could be potentially skillful. In climate forecasting in particular, where there is very high uncertainty in the forecasts themselves and there is marginal user adoption of existing products, the operational community often focuses more on potential forecast skill than likely current use. The belief is that if a product is skillful, a user base could be cultivated. If there is no skill, even if user demand exists, forecasting would be futile.

Attractive projects may also develop when a new method comes into use by a colleague of the forecaster (someone from another agency, alumni, friend or prior collaborator on other projects). For example, Redmond and Koch (1991) published the first major study of the impacts of ENSO on streamflow in the western United States. At the time the study was being done, a NRCS operational forecaster was one of Koch's graduate students. The student put

The true likelihood of various forecasted, naturally-occurring climate and water resource anomalies may best be understood in the context of longer records, which paleoclimatic reconstructions can provide.

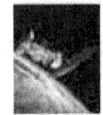

Koch's research to operational practice at the NRCS after realizing that forecast skill could be improved.

Efficiency is also often the inspiration for an innovation. A forecaster may be looking for a way to streamline or otherwise automate an existing process. For example, users frequently call the forecaster with a particular question; if it is possible to automate answering that question with a new Internet-based product, the forecaster may be freed up to work on other tasks. While most forecasters can readily list several bottlenecks in the production process, this knowledge often comes more from personal experience than any kind of structured system review.

At this stage, many ideas exist for possible innovations, although only some small subset of them will be pursued. The winnowing process continues with the forecaster and/or peers evaluating the feasibility of the innovation: Is the method scientifically defensible? Are the data reliably available to support the product? Are the computers powerful enough to complete the process in a reasonable time? Can this be done with existing resources, would it free up more resources than it consumes, or is the added value worth the added operational expense? In other words, is the total value of the advance worth the effort? Is it achievable and compatible with legacy systems or better than the total worth of the technology, installed base and complementary products?

If it is expected to be valuable, some additional questions may be raised by the forecaster or by management about the appropriateness of the solution. Would it conflict with or detract from another product, especially the official suite (*i.e.*, destroy competency)? Would it violate an agency policy? For example, a potential product may be technically feasible but not allowed to exist because the agency's webpage does not permit interactivity because of increasingly stringent congressionally-mandated cyber-security regulations. In this case, to the agency as a whole, the cost of reduced security is greater than the benefit of increased interactivity. It is important to note that if security and interactivity in general are not at odds, the issue may be that a particular form of interactivity is not compatible with the existing security architecture.

If a different security architecture is adopted or a different form of interactivity used (*e.g.*, written in a different computer language), then both may function together, assuming one has the flexibility and ability to change.

Additionally, an agency policy issue can sometimes be of broader, multi-organizational scope and would require policy decisions to settle. For example, no agency currently produces water quality forecasts. Which federal agency should be responsible for this: the U.S. Department of Agriculture, Environmental Protection Agency, U.S. Geological Survey or National Weather Service? What of soil moisture forecasts? Should it be the first agency to develop the technical proficiency to make such forecasts? Or should it be established by a more deliberative process to prevent "mission creep?" Agencies are also concerned about whether innovations interfere with the services provided by the private sector.

If appropriate, the forecaster may then move to implement the solution on a limited test basis, iteratively developing and adapting to any unforeseen challenges. After a successful functional prototype is developed, it is tested in-house using field personnel and/or an inner circle of sophisticated customers and gradually made more public as confidence in the product increases. In these early stages, many of the "kinks" of the process are smoothed out, developing the product format, look and feel; and adapting to initial feedback (*e.g.*, "please make the map labels larger") but, for the most part, keeping the initial vision intact.

There is no consistent formal procedure across agencies for certifying a new method or making a new product official. A product may be run and labeled "experimental" for one to two years in an evaluation period. The objectives and duration of the evaluation period are sometimes not formalized and one must just assume that if a product has been running for an extended period of time with no obvious problems, then it succeeds and the experimental label removed. Creating documentation of the product and process is often part of the transition from experimental to official, either in the form of an internal technical memo, conference

No agency currently produces water quality forecasts.

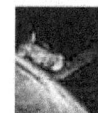

proceedings or peer-reviewed journal article, if appropriate.

If the innovation involves using a tool or technique that supplements the standard suite of tools, some of the evaluation may involve running both tools in parallel and comparing their performance. Presumably, ease of use and low demand on resources are criteria for success (although the task of running models in parallel can, by itself, be a heavy demand on resources). Sometimes an agency may temporarily stretch its resources to accommodate the product for the evaluation period and if additional resources are not acquired by the end of the evaluation (for one of a number of reasons, some of which may not be related to the product but, rather, are due to variability in budgets), the product may be discontinued.

Sometimes skill is used to judge success, but this can be a very inefficient measure. This is because seasonal forecast skill varies greatly from year to year, primarily due to the variability of nature. Likewise, individual tools may perform better than other tools in some years but not others. In the one to two years of an evaluation period the new tool may be lucky (or unlucky) and artificially appear better (or worse) than the existing practice.

If the agency recognizes that a tool has not had a fair evaluation, more emphasis is placed on "hindcasting", using the new tool to objectively and retrospectively generate realistic "forecasts" for the last 20 to 30 years and comparing the results to hindcasts of the existing system and/or official published forecasts. The comparison is much more realistic and effective, although hindcasting has its own challenges. It can be operationally demanding to produce the actual forecasts each month (e.g., the agency may have to compete for the use of several hours of an extremely powerful computer to run a model), much less do the equivalent of 30 years worth at once. These hindcast datasets, however, have their own uses and have proven to be very valuable (e.g., Hamill et al., 2006 for medium range weather forecasting and Franz et al., 2003 for seasonal hydrologic forecasting). Oftentimes, testbeds are better suited for operationally realistic hindcasting experiments (Box 2.4).

During the evaluation period, the agency may also attempt to increasingly "institutionalize" a process by identifying and fixing aspects of a product or process that do not conform to agency guidelines. For example, if a forecasting model is demonstrated as promising but the operating system or the computer language it is written in does not match the language chosen by the agency, a team of contract programmers may rewrite the model and otherwise develop interfaces that make the product more user-friendly for operational work. A team of agency personnel may also be assembled to help transfer the research idea to full operations, from prototype to project. For large projects, many people may be involved, including external researchers from several other agencies.

During this process of institutionalization, the original innovation may change in character. There may be uncertainty at the outset and the development team may consciously postpone certain decisions until more information is available. Similarly, certain aspects of the original design may not be feasible and an alternative solution must be found. Occasionally, poor communication between the inventor and the developers may cause the final product to be different than the original vision. Davidson et al. (2002) found success in developing a hydrologic database using structured, iterative development involving close communication between users and developers throughout the life of the project. This model is in direct contrast to that of the inventor generating a ponderous requirements document at the outset, which is then passed on to a separate team of developers who execute the plan in isolation until completion.

2.6.2 Evaluation of Forecast Utility

As mentioned in Section 2.1, there are many ways to assess the usefulness of forecasts, one of which is forecast skill. While there are inherent limitations to skill (due to the chaotic nature of the atmosphere), existing operational systems also fall short of their potential maximum skill for a variety of reasons. Section 2.4 highlighted ways to improve operational skill, such as by having better models of the natural system or denser and more detailed climate and hydrologic monitoring networks. Other factors, such as improved forecaster training or better

There is no consistent formal procedure across agencies for certifying a new method or making a new product official.

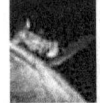

BOX 2.4: What Role Can a "Testbed" Play in Innovation?

For an innovation to be deemed valuable, it must be able to stand on its own and be better than the entire existing system, or marginally better than the existing technology, if it is compatible with the rest of the framework of the existing system. If the innovation is not proven or believed likely to succeed, its adoption is less likely to be attempted. However, who conducts the experiments to measure this value? And who has the resources to ensure backwards-compatibility of the new tools in an old system?

This model lacks any direct communication between user and producer and leaves out the necessary support structure to help users make the most of the product (Cash et al., 2006). Similarly, testbeds are designed as an alternative to the "Loading Dock Model" of transferring research to operations. A loading dock model is one in which scientists prepare models, products, forecasts or other types of information for general dissemination, in somewhat of a vacuum, without consulting with and/or understanding the needs of the people who will be using that information, with the anticipation that others will find these outputs useful.

Previously, a researcher might get a short-term grant to develop a methodology, and conduct an idealized, focused study of marginal operational realism. The results might be presented at research conferences or published in the scientific literature. While a researcher's career may have a unifying theme, for the most part, this specific project may be finished when publication is accomplished and the grant finishes. Meanwhile, the operational forecaster is expected to seek out the methodology and attempt to implement it, although, often, the forecaster does not have the time, resources or expertise to use the results. Indeed, the forecaster may not be convinced of the incremental advantage of the technique over existing practices if it has not endured a realistic operational test and been compared to the results of the official system.

Testbeds are intermediate activities, a hybrid mix of research and operations, serving as a conduit between the operational, academic and research communities. A testbed activity may have its own resources to develop a realistic operational environment. However, the testbed would not have real-time operational responsibilities and instead, would be focused on introducing new ideas and data to the existing system and analyzing the results through experimentation and demonstration. The old and new system may be run in parallel and the differences quantified. The operational system may even be deconstructed to identify the greatest sources of error and use that as the motivation to drive new research to find solutions to operations-relevant problems. The solutions are designed to be directly integrated into the mock-operational system and therefore should be much easier to directly transfer to actual production.

NOAA has many testbeds currently in operation: Hydrometeorological (floods), Hazardous Weather (thunderstorms and tornadoes), Aviation Weather (turbulence and icing for airplanes), Climate (ENSO, seasonal precipitation and temperature), and Hurricanes. The Joint Center for Satellite Data Assimilation is also designed to facilitate the operational use of new satellite data. A testbed for seasonal streamflow forecasting does not exist. Generally, satisfaction with testbeds has been high, rewarding for operational and research participants alike.

visualization tools, also play a role. This Section addresses the role of forecast evaluation in driving the technology development agenda.

Understanding the current skill of forecast products is a key component to ensuring the effectiveness of programs to improve the skill of these products. There are several motivations for verifying forecasts including administrative, scientific and economic (Brier and Allen, 1951). Evaluation of very recent forecasts can also play a role in helping operational forecasters make mid-course adjustments to different components of the forecast system before issuing an official product.

Of particular interest to forecasting agencies is administrative evaluation because of its ability to describe the overall skill and efficiency of the forecast service in order to inform and guide decisions about resource allocation, research directions and implementation strategies (Welles, 2005). For example, the development of numerical weather prediction (NWP) forecasting models is conducted by numerous, unaffiliated groups following different approaches, with the results compared through objective measures of performance. In other words, the forecasts are verified, and the research is driven, not by *ad hoc* opinions postulated by subject matter experts, but by the actual performance of the forecasts as determined with objective measures (Welles *et al.*, 2007). The most important sources of error are identified quantitatively and systematically, and are paired with objective measures of the likely improvement resulting from an innovation in the system.

Recently, the NWS adopted a broad national-scale administrative initiative of hydrologic forecast evaluation. This program defines a standard set of evaluation measures, establishes a formal framework for forecast archival and builds flexible tools for access to results. It is designed to provide feedback to local forecasters and users on the performance of the regional results, but also to provide an end-to-end assessment of the elements of the entire system (HVSRT, 2006). Welles *et al.* (2007) add that these activities would be best served by cultivating a new discipline of "hydrologic forecast science" that engages the research community to focus on operational-forecast-specific issues.

While administrative evaluation is an important tool for directing agency resources, innovation should ultimately be guided by the anticipated benefit to forecast users. Some hydrologists would prefer not to issue a forecast that they suspect the user could not use or would misinterpret (Pielke, Jr., 1999). Additionally, evaluations of forecasts should be available and understandable to users. For instance, it might be valuable for some users to know that hydrologic variables in particular regions of interest lack predictability. Uncertainty about the accuracy of forecasts precludes users from making more effective use of them (Hartmann *et al.*, 2002). Users want to know how good the forecasts are so they know how much confidence to place in them. Agencies want to focus on the aspects of the forecast that are most important to users. Forecast evaluation should be more broadly defined than skill alone; it should also include measures of communication and understandability, as well as relevance. In determining these critical aspects, agencies must make a determination of the key priorities to address given the number and varied interest of potential forecast users. The agencies can not fully

> Forecast evaluation should be more broadly defined than skill alone; it should also include measures of communication and understandability, as well as relevance.

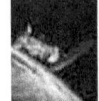

BOX 2.5: The Advanced Hydrologic Prediction Service

Short- to medium-range forecasts (those with lead times of hours to days) of floods are a critical component of National Weather Service hydrological operations, and these services generate nearly $2 billion of benefits annually (NHWC, 2002). In 1997 the NWS Office of Hydrologic Development began the Advanced Hydrologic Prediction Service (AHPS) program to advance technology for hydrologic products and forecasts. This 16-year multi-million dollar program seeks to enhance the agency's ability to issue and deliver specific, timely, and accurate flood forecasts. One of its main foci is the delivery of probabilistic and visual information through an Internet-based interface. One of its seven stated goals is also to "Expand outreach and engage partners and customers in all aspects of hydrologic product development" (NRC, 2006).

Starting in 2004, the National Research Council reviewed the AHPS program and also analyzed the extent that users were actually playing in the development of products and setting of the research agenda (NRC, 2006). The study found that AHPS had largely a top-down structure with technology being developed at a national center to be delivered to regional and local offices. Although there was a wide range of awareness, understanding and acceptance of AHPS products inside and outside the NWS, little to no research was being done in early 2004 on effective communication of information, and some of the needs of primary customers were not being addressed. From the time the NRC team carried out its interviews, the NWS started acting on the perceived deficiencies, so that, by the time the report was issued in late 2006, the NWS had already made some measurable progress. This progress included a rigorous survey process in the form of focus groups, but also a more engaged suite of outreach, training, and educational activities that have included presentations at the national floodplain and hydrologic manager's conferences, the development of closer partnerships with key users, committing personnel to education activities, conducting local training workshops, and awarding a research grant to social scientists to determine the most effective way to communicate probabilistic forecasts to emergency and floodplain managers.

satisfy all users. The Advanced Hydrologic Prediction System (AHPS) of the NWS provides a nice case study of product development and refinement in response to user-driven feedback (Box 2.5).

There is another component to forecast skill beyond the assessment of how the forecast quantities are better (or worse) than a reference forecast. Thinking of forecast assessment more broadly, the forecasts should be evaluated for their "skill" at communicating their information content in ways that can be correctly interpreted both easily and reliably—*i.e.*, no matter what the quantity (*e.g.*, wet, dry, or neutral tercile) of the forecast, the user can still correctly interpret it (Hartmann *et al.*, 2002).

Finally, it seems important to stress that agencies should provide for user-centric forecast assessment as part of the process for moving prototypes to official products. This would include access to user tools for assessing forecast skill (*i.e.*, the Forecast Evaluation Tool, which is linked to by the NWS Local 3-month Temperature Outlook [Box 2.6]), and field testing of the

BOX 2.6: National Weather Service Local 3-Month Outlooks for Temperature and Precipitation

In January 2007, the National Weather Service made operational the first component of a new set of climate forecast products called Local 3-Month Outlooks (L3MO). Accessible from the NWS Weather Forecast Offices (WFO), River Forecast Centers (RFC), and other NWS offices, the Local 3-Month Temperature Outlook (L3MTO) is designed to clarify and downscale the national-scale CPC Climate Outlook temperature forecast product. The corresponding local product for precipitation is still in development as of the writing of this Product. The local outlooks were motivated by ongoing National Oceanic and Atmospheric Administration, NWS activities focusing on establishing a dialog with NWS climate product users <http://www.nws.noaa.gov/directives/>. In particular, a 2004 NWS climate product survey (conducted by Claes Fornell International for the NOAA Climate Services Division) found that a lack of climate product clarity lowered customer satisfaction with NWS CPC climate outlook products; and presentations and interactions at the annual Climate Prediction Application Science Workshop (CPASW) highlighted the need for localized CPC climate outlooks in numerous and diverse applications.

In response to these user-identified issues, CSD collaborated with the NWS Western Region Headquarters, CPC, and the National Climatic Data Center (NCDC) to develop localized outlook products. The collaboration between the four groups, which linked several line offices of NOAA (e.g., NCDC, NWS), took place in the context of an effort that began in 2003 to build a climate services infrastructure within NOAA. The organizations together embarked on a structured process that began with a prototype development stage, which included identifying resources, identifying and testing methodologies, and defining the product delivery method. To downscale the CPC climate outlooks (which are at the climate division scale) to local stations, the CSD, and WR development team assessed and built on internal, prior experimentation at CPC that focused on a limited number of stations. To increase product clarity, the team added interpretation, background information, and a variety of forecast displays providing different levels of data density. A NWS products and services team made product mockups that were reviewed by all 102 WFOs, CPC and CSD representatives and a small number of non-agency reviewers. After product adjustments based on the reviews, CSD moved toward an experimental production stage, providing NWS staff with training and guidelines, releasing a public statement about the product and writing product description documentation. Feedback was solicited via the experimental product website beginning in August 2006, and the products were again adjusted. Finally, the products were finalized, the product directive was drafted and the product moved to an operational stage with official release. User feedback continues via links on the official product website <http://www.weather. gov/climate/l3mto.php>.

In general, the L3MO development process exhibited a number of strengths. Several avenues existed for user needs to reach developers, and user-specified needs determined the objectives of the product development effort. The development team, spanning several parts of the agency, then drew on internal expertise and resources to propose and to demonstrate tentative products responding to those needs. The first review stage of the process gave mostly internal (i.e., agency) reviewers an early opportunity for feedback, but this was followed by an opportunity for a larger group of users in the experimental stage, leading to the final product. An avenue for continued review is built into the product dissemination approach.

communication effectiveness of the prototype products. Just as new types of forecasts should show (at least) no degradation in predictive skill, they should also show no degradation in their communication effectiveness.

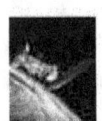

CHAPTER 3

Decision-Support Experiments Within the Water Resource Management Sector

Convening Lead Authors: David L. Feldman, Univ. of California, Irvine; Katharine L. Jacobs, Arizona Water Institute

Lead Authors: Gregg Garfin, Univ. of Arizona; Aris Georgakakos, Georgia Inst. of Tech.; Barbara Morehouse, Univ. of Arizona; Robin Webb, NOAA; Brent Yarnal, Penn. State Univ.

Contributing Authors: John Kochendorfer, Riverside Technology, Inc. & NOAA; Cynthia Rosenzweig, NASA; Michael Sale, ORNL; Brad Udall, NOAA; Connie Woodhouse, Univ. of Arizona

KEY FINDINGS

Decision-support experiments that test the utility of seasonal-to-interannual (SI) information for use by water resource decision makers have resulted in a growing set of successful applications. However, there is significant opportunity for expansion of applications of climate-related data and decision-support tools, and for developing more regional and local tools that support management decisions within watersheds. Among the constraints that limit tool use are:

- The range and complexity of water resources decisions: This is compounded by the numerous organizations responsible for making these decisions, and the shared responsibility for implementing them. These organizations include water utility companies, irrigation management districts and other entities, and government agencies.
- Inflexible policies and organizational rules that inhibit innovation: Large institutions historically have been reluctant to change practices in part because of value differences; risk aversion; fragmentation; the primacy accorded water rights, which often vary from region to region, and among various users; and sharing of authority. This conservatism impacts how decisions are made as well as whether to use newer, scientifically generated information, including SI forecasts and observational data.
- Different spatial and temporal frames for decisions: Spatial scales for decision making range from local, state, and national levels to international. Temporal scales range from hours to multiple decades impacting policy, operational planning, operational management, and near real-time operational decisions. Resource managers often make multidimensional decisions spanning various spatial and temporal frames.
- Lack of appreciation of the magnitude of potential vulnerability to climate impacts: Communication of the risks differs among scientific, political, and mass media elites, each systematically selecting aspects of these issues that are most salient to their conception of risk, and thus, socially constructing and communicating its aspects most salient to a particular perspective.

Decision-support systems are not often well integrated into planning and management activities, making it difficult to realize the full benefits of these tools. Because use of many climate products requires special training or access to data that are not easily available, decision-support products may not equitably reach all audiences. Moreover, over-specialization and narrow disciplinary perspectives make it difficult for information providers, decision makers, and the public to communicate with one another. Three lessons stem from this:

- Decision makers need to understand the types of predictions that can be made, and the trade-offs between longer-term predictions of information at the local or regional scale on the one hand, and potential decreases in accuracy resulting from transition to smaller spatial scales on the other.
- Decision makers and scientists need to work together in formulating research questions relevant to the spatial and temporal scale of problems the former manage that can be supported by current understandings of physical conditions.
- Scientists should aim to generate findings that are accessible and viewed as useful, accurate and trustworthy by stakeholders by working to enhance transparency of the scientific process.

3.1 INTRODUCTION

Over the past century, the United States has built a vast and complex infrastructure to provide clean water for drinking and for industry, dispose of wastes, facilitate transportation, generate electricity, irrigate crops, and reduce the risks of floods and droughts. To the average citizen, the nation's dams, aqueducts, reservoirs, treatment plants, and pipes are taken for granted. Yet they help insulate us from wet and dry years and moderate other aspects of our naturally variable climate. Indeed they have permitted us to almost forget about our complex dependences on climate. We can no longer ignore these close connections (Gleick, 2000).

This Chapter synthesizes and distills lessons for the water resources management sector from efforts to apply decision-support experiments and evaluations using SI forecasts and observational climate data. Its thesis is that,

while there is a growing, theoretically-grounded body of knowledge on how and why resource decision makers use information, there is little research on barriers to use of decision-support products in the water management sector. Much of what we know about these barriers comes from case studies on the application of SI forecast information and by efforts to span organizational boundaries dividing scientists and users. Research is needed on factors that can be generalized beyond these single cases in order to develop a strong, theoretically-grounded understanding of the processes that facilitate information dissemination, communication, use, and evaluation, and to predict effective methods of boundary spanning between decision makers and information generators.

Decision support is a three-fold process that encompasses: (1) the generation of climate science products; (2) the translation of those products into forms useful for decision makers (*i.e.*, user-centric information); and, (3) the processes that facilitate the dissemination, communication, and use of climate science products, information, and tools (NRC, 2007). As shall be seen, because users include many private and small users, as well as public and large users serving multiple jurisdictions and entities, effective decision support is difficult to achieve.

Section 3.2 describes the range of major decisions water users make, their decision-support needs, and the role decision-support systems can play in meeting them. We examine the attributes of water resource decisions, their spatial and temporal characteristics, and the implications of complexity, political fragmentation, and shared responsibility on forecast use. We also

discuss impediments to forecast information
use by decision makers, including mistrust,
uncertainty, and lack of agency coordination,
and discuss four cases whose problem foci
range from severe drought to flooding, where
efforts to address these impediments are being
undertaken with mixed results.

Section 3.3 examines challenges in fostering
closer collaboration between scientists and
decision makers in order to communicate,
translate, and operationalize climate forecasts
and hydrology information into integrated
water management decisions. We review what
the social and decision sciences have learned
about barriers in interpreting, deciphering,
and explaining climate forecasts and other
meteorological and hydrological models and
forecasts to decision makers, including issues
of relevance, accessibility, organizational constraints on decision makers, and compatibility
with users' values and interests. Case studies
reveal how these issues manifest themselves in
decision-support applications. Chapter 4, which
is a continuation of these themes in the context
of how to surmount these problems, examines
how impediments to effectively implementing
decision-support systems can be overcome in
order to make them more useful, useable, and
responsive to decision-maker needs.

3.2 WHAT DECISIONS DO WATER USERS MAKE, WHAT ARE THEIR DECISION-SUPPORT NEEDS, AND WHAT ROLES CAN DECISION-SUPPORT SYSTEMS PLAY IN MEETING THESE NEEDS?

This section reviews the range and attributes of
water resource decisions, including complexity,
political fragmentation, shared decision making, and varying spatial scale. We also discuss
the needs of water resource managers for climate variability forecast information, and the
multi-temporal and multi-spatial dimensions of
these needs. Finally, we examine how climatic
variability affects water supply and quality.
Embedded in this examination is discussion of
the risks, hazards, and vulnerability of water
resources (and human activities dependent on
them) from climatic variability.

3.2.1 Range and Attributes of Water Resource Decisions

As discussed in Chapter 1, and as illustrated in
Table 1.1, decisions regarding water resources
in the United States are many and varied, and
involve public and private sector decision makers such as farmers, ranchers, electric power
utilities, and eminent domain landowners who
use a large percentage of the country's water.
Spatial scales for decision making range from
local, state, and national levels to international
political jurisdictions, the latter with some say
in the way United States water resources are
managed (Hutson *et al.*, 2004; Sarewitz and
Pielke, 2007; Gunaji, 1995; Wagner, 1995).
These characteristics dictate that information
must be tailored to the particular roles, responsibilities, and concerns of different decision
makers to be useful. Chapter 1 also suggested
that the way water issues are framed—a process
determined partly by organizational commit-
ments and perceptions, and in part by changing demands imposed by external events and
actors—determines how information must be
tailored to optimally impact various decision-
making constituencies and how it will likely
be used once tailored. In Chapter 3, we focus
on the implications of this multiple-actor,
multi-jurisdictional environment for delivery
of climate variability information.

3.2.1.1 INSTITUTIONAL COMPLEXITY, POLITICAL FRAGMENTATION, AND SHARED DECISION MAKING: IMPACTS ON INFORMATION USE

The range and complexity of water resource
decisions, the numerous organizations responsible for making these decisions, and the shared
responsibility for implementing them affect
how water resource decision makers use climate
variability information in five ways:

1. a tendency toward institutional conservatism by water agencies;
2. a decision-making climate that discourages innovation;
3. a lack of national-scale coordination of
   decisions
4. difficulties in providing support for decisions at varying spatial and temporal scales
   due to vast variability in "target audiences"
   for products; and

Decisions regarding
water resources in
the United States
are many and varied,
and involve public
and private sector
decision makers such
as farmers, ranchers,
electric power utilities,
and eminent domain
landowners who use
a large percentage of
the country's water.

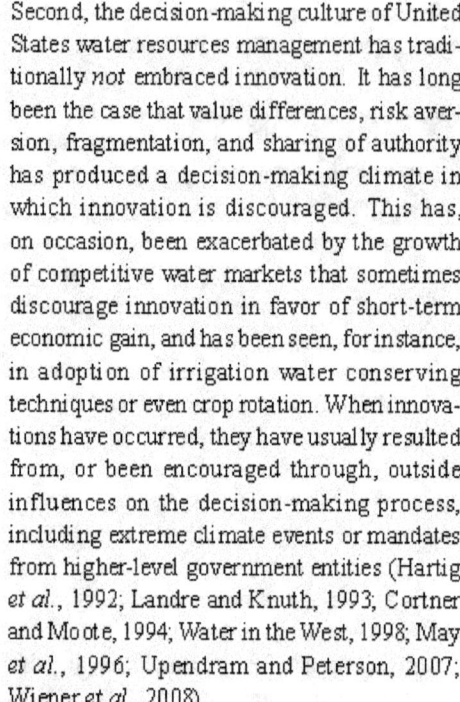

5. growing recognition that rational choice models that attempt to explain information use as a function of decision-maker needs for "efficiency" are overly simplistic.

These are discussed in turn in this Section and the following two Sections.

First, institutions that make water resource decisions, particularly government agencies, operate in domains where they are beholden to powerful constituencies. These constituencies have historically wanted public works projects for flood control, hydropower, water supply, navigation, and irrigation. They also have worked hard to maximize their benefits within current institutional structures, and are often reluctant to change practices that appear antiquated or inefficient to observers.

There have been various efforts to seek greater synchronization of decisions at the national level, in part, to better respond to environmental protection, economic development, water supply, and other goals.

The success of these constituencies in leveraging federal resources for river and harbor improvements, dams, and water delivery systems is in part due to mobilizing regional development interests. Such interests commonly resist change and place a premium on engineering predictability and reliability (Feldman, 1995, 2007; Ingram and Fraser, 2006; Merritt, 1979; Holmes, 1979). This conservatism not only affects how these agencies and organizations make decisions, it also impacts how they employ, or do not employ, scientifically generated information, including information that related to SI climate variability. Information that conflicts with their mandates, traditions, or roles may not be warmly received, as surveys of water resource managers have shown (e.g., O'Connor et al., 1999 and 2005; Yarnal et al., 2006; Dow et al., 2007).

Second, the decision-making culture of United States water resources management has traditionally not embraced innovation. It has long been the case that value differences, risk aversion, fragmentation, and sharing of authority has produced a decision-making climate in which innovation is discouraged. This has, on occasion, been exacerbated by the growth of competitive water markets that sometimes discourage innovation in favor of short-term economic gain, and has been seen, for instance, in adoption of irrigation water conserving techniques or even crop rotation. When innovations have occurred, they have usually resulted from, or been encouraged through, outside influences on the decision-making process, including extreme climate events or mandates from higher-level government entities (Hartig et al., 1992; Landre and Knuth, 1993; Cortner and Moote, 1994; Water in the West, 1998; May et al., 1996; Upendram and Peterson, 2007; Wiener et al., 2008).

Third, throughout the history of United States water resources management there have been various efforts to seek greater synchronization of decisions at the national level, in part, to better respond to environmental protection, economic development, water supply, and other goals. These efforts hold many lessons for understanding the role of climate change information and its use by decision makers, as well as how to bring about communication between decision makers and climate information producers. While there has been significant investment of federal resources to provide for water infrastructure improvements, there has been little national-scale coordination over decisions, or over the use of information employed in making them (Kundell et al., 2001). The system does not encourage connectivity between the benefits of the federal investments and those who actually pay for them, which leaves little incentive for improvements in efficiency and does not reward innovation (see Wahl, 1989).

3.2.1.2 IMPLICATIONS OF THE FEDERAL ROLE IN WATER MANAGEMENT

In partial recognition of the need to coordinate across state boundaries to manage interstate rivers, in the 1960s, groups of northeastern states formed the Delaware River Basin Commission (DRBC) and the Susquehanna River Basin

Commission (SRBC) to pave the way for conflict resolution. These early federal interstate commissions functioned as boundary organizations that mediated communication between supply and demand functions for water and climate information (Sarewitz and Pielke, 2007). They relied on frequent, intensive, face-to-face negotiations; coordination among politically-neutral technical staffs; sharing of study findings among partners; willingness to sacrifice institutional independence when necessary, and commission authority to implement decisions so as to transcend short-term pressures to act expediently (Cairo, 1997; Weston, 1995)[1].

An ambitious effort to coordinate federal water policy occurred in 1965 when Congress established the Water Resources Council (WRC), under the Water Resources Planning Act, to coordinate federal programs. Due to objections to federal intervention in water rights issues by some states, and the absence of vocal defenders for the WRC, Congress de-funded WRC in 1981 (Feldman, 1995). Its demise points out the continued frustration in creating a national framework to coordinate water management, especially for optimal management in the context of climate variability. Since termination of the WRC, coordination of federal programs, when it has occurred, has come variously from the Office of Management and Budget, White House Council on Environmental Quality, and *ad hoc* bodies (*e.g.*, Task Force on Floodplain Management)[2]. A lesson in all of this is that innovation in promoting the use of information requires a concerted effort across agencies and

political jurisdictions. Sometimes this may best be facilitated by local collaboration encouraged by federal government incentives; at other times, federal coordination of information may be needed, as shown by a number of case studies noted in Chapter 4.

Fourth, the physical and economic challenge in providing decision support due to the range of "target audiences" (*e.g.*, Naim, 2003) and the controversial role of the federal government in such arenas is illustrated by efforts to improve the use of SI climate change information for managing water resources along the United States—Mexico border, as well as the United States—Canada border. International cross-boundary water issues in North America bring multiple additional layers of complexity, in part because the federal governments of Canada, Mexico and the United States often are ill-equipped to respond to local water and wastewater issues. Bringing the U.S. State Department into discussions over management of treatment plants, for example, may not be an effective way to resolve technical water treatment or supply problems.

In the last decade, climate-related issues that have arisen between Mexico and the United States regarding water revolve around disagreements among decision makers on how to define extraordinary drought, allocate shortages, and cooperatively prepare for climate extremes. These issues have led to renewed efforts to better consider the need for predictive information and ways to use it to equitably distribute water under drought conditions. Continuous monitoring of meteorological data, consumptive water uses, calculation of drought severity, and detection of longer-term climate trends could, under the conditions of these agreements, prompt improved management of the cross-boundary systems (Gunaji, 1995; Mumme, 2003, 1995; Higgins *et al.*, 1999). The 1906 Rio Grande Convention and 1944 Treaty between the United States and Mexico, the latter established the *International Boundary Water Commission*, contain specific clauses related to "extraordinary droughts". These clauses prescribe that the United States government apprise Mexico of the onset of drought conditions as they develop, and adjust water deliveries to both United States and Mexican customers accordingly (Gunaji,

Innovation in promoting the use of information requires a concerted effort across agencies and political jurisdictions.

---

[1] Compact entities were empowered to allocate interstate waters (including groundwater and interbasin diversions), regulate water quality, and manage interstate bridges and ports. DRBC includes numerous federal partners such as the Department of Interior and Army Corps of Engineers officials (DRBC, 1998; DRBC, 1961; Weston, 1995; Cairo, 1997). One of the forces giving rise to DRBC was periodic drought that helped exacerbate conflict between New York City and other political entities in the basin. This led to DRBC's empowerment, as the nation's first federal interstate water commission, in all matters relating to the water resources of its basin, ranging from flooding to fisheries to water quality.

[2] Today the need for policy coordination, according to one source, "stems from the ... environmental and social crises affecting the nation's rivers" (Water In the West, 1998: xxvii). In nearly every basin in the West, federal agencies are responding to tribal water rights, growing urban demands, endangered species listings, and Clean Water Act lawsuits. Climate change is expected to exacerbate these problems.

1995). However, there is reluctance to engage in conversations that could result in permanent reduced water allocations or reallocations of existing water rights.

For the United States and Canada, a legal regime similar to that between the United States and Mexico has existed since the early 1900s. The anchor of this regime is the 1909 Boundary Waters Treaty that established an *International Joint Commission* with jurisdiction over threats to water quality, anticipated diversions, and protection of instream flow and water supply inflow to the Great Lakes. Climate change-related concerns have continued to grow in the Great Lakes region in recent years due, especially, to questions arising over calls to treat its water resources as a marketable commodity, as well as concerns over what criteria to use to resolve disputes over these and other questions (Wagner, 1995; International Joint Commission, 2000).

3.2.1.3 INSTITUTIONS AND DECISION MAKING

Fifth, there is growing recognition of the limits of so-called *rational choice models* of information use, which assume that decision makers deliberately focus on optimizing organizational performance when they use climate variability or other water resource information. This recognition is shaping our understanding of the impacts of institutional complexity on the use of climate information. An implicit assumption in much of the research on probabilistic forecasting of SI variation in climate is that decision makers on all levels will value and use improved climate predictions, monitoring data, and forecast tools that can predict changes to conditions affecting water resources (e.g., Nelson and Winter, 1960). *Rational choice* models of decision making are predicated on the assumption that decision makers seek to make optimal decisions (and perceive that they have the flexibility and resources to implement them).

A widely-cited study of four water management agencies in three locations—the Columbia River system in the Pacific Northwest, the Metropolitan Water District of Southern California, and the Potomac River Basin and Chesapeake Bay in the greater Washington, D.C. area— examined the various ways water agencies at different spatial scales use probabilistic climate forecast information. The study found that not only the multiple geographic scales at which these agencies operate but also the complexity of their decision-making systems dramatically influence how, and to what extent, they use probabilistic climate forecast information. An important lesson is that the complexity of these systems' sources of supply and infrastructure, and the stakeholders they serve are important influences on their capacity to use climate information. Decision systems may rely on multiple sources of data, support the operation of various infrastructure components, straddle political (and hydrological) boundaries, and serve stakeholders with vastly different management objectives (Rayner *et al.*, 2005). Thus, science is only one of an array of potential elements influencing decisions.

The cumulative result of these factors is that water system managers and operations personnel charged with making day-to-day decisions tend toward an overall institutional conservatism when it comes to using complex meteorological information for short- to medium-term decisions. Resistance to using new sources of information is affected by the complexity of the institutional setting within which managers work, dependency on craft skills and local knowledge, and a hierarchy of values and processes designed to ensure their political invisibility. Their goal is to smooth out fluctuations in operations and keep operational issues out of the public view (Rayner *et al.*, 2005).

In sum, the use of climate change information by decision makers is constrained by a politically-fragmented environment, a regional economic development tradition that has inhibited, at least until recently, the use of innovative information (e.g., conservation, integrated resource planning), and multiple spatial and temporal frames for decisions. All this makes the target audience for climate information products vast and complex.

The interplay of these factors, particularly the specific needs of target audiences and the inherently conservative nature of water management, is shown in the case of how Georgia has come to use drought information to improve long-term water supply planning. As shall be

The use of climate change information by decision makers is constrained by a politically-fragmented environment, a regional economic development tradition that has inhibited, at least until recently, the use of innovative information, and multiple spatial and temporal frames for decisions.

BOX 3.1: Georgia Drought

Background

Two apparent physical causes of the 2007/2008 Southeast drought include a lack of tropical storms and hurricanes, which usually can be counted on to replenish declining reservoirs and soil moisture, and the development of a La Niña episode in the tropical Pacific, which continues to steer storms to the north of the region (Box Figure 3.1). Drought risk is frequently modeled as a function of hazard (e.g., lack of precipitation) and vulnerability (i.e., susceptibility of society to the hazard) using a multiplicative formula, risk = hazard × vulnerability (Hayes *et al.*, 2004). In 2007, Atlanta, Georgia received only 62 percent of its average annual precipitation, the second driest calendar year on record; moreover, streamflows were among the lowest recorded levels on several streams. By June 2007, the National Climatic Data Center reported that December through May precipitation totals for the Southeast were at new lows. Spring wildfires spread throughout southeastern Georgia which also recorded its worst pasture conditions in 12 years. Georgia's Governor Purdue extended a state of emergency through June 30; however, the state's worst drought classification, accompanied by a ban on outdoor water use, was not declared until late September.

While progressive state drought plans, such as Georgia's (which was adopted in March, 2003), emphasize drought preparedness and mitigation of impacts through mandatory restrictions in some water use sectors, they do not commonly factor in the effect of population growth on water supplies. Moreover, conservation measures in a single state cannot address water allocation factors affecting large, multi-state watersheds, such as the Apalachicola–Chattahoochee–Flint (ACF), which encompasses parts of Georgia, Alabama, and Florida.

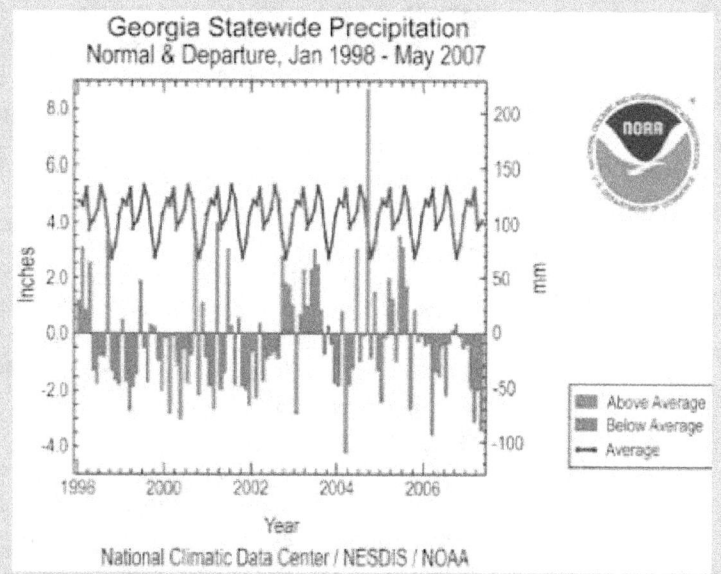

Figure Box 3.1 Georgia statewide precipitation: 1998 to 2007

Institutional barriers and problems

The source of water woes in this Southeastern watershed dates back to a 1987 decision by the Army Corps of Engineers to reallocate 20 percent of power generation flow on the Chattahoochee River to municipal supply for Atlanta, which sits near the headwaters of the river. Alabama and Florida soon demanded an assessment of the environmental and economic effects of that decision, which set off a series of on-again, off-again disputes and negotiations between the three states, known as the "Tri-State Water Wars", that have not been resolved (as of June, 2008). At the heart of the disputes is a classic upstream-downstream water use and water rights dispute, pitting municipal water use for the rapidly expanding Atlanta metropolitan region against navigation, agriculture, fishing, and environmental uses downstream in Alabama and Georgia. The situation is further complicated by water quality concerns, as downstream users suffer degraded water quality, due to polluted urban runoff and agricultural waste, pesticide, and fertilizer leaching. Despite the efforts of the three states and Congress to create water compacts, by engaging in joint water planning and developing and sharing common data bases, the compacts have never been implemented as a result of disagreements over what constitutes equitable water allocation formulae (Feldman, 2007).

Political and sectoral disputes continue to exacerbate lack of coordination on water-use priorities, and there is a continuing need to include climate forecast information in these activities, as underscored by continuing drought in the Southeast. The result is that water management decision making is constrained, and there are few opportunities to insert effective decision-support tools, aside from the kinds of multi-stakeholder shared-vision modeling processes developed by the U.S. Army Corps of Engineers Institute for Water Resources.

seen in Section 3.3.1, while the good news in this case is that information is beginning to be used by policymakers, the downside is that *some* information use is being inhibited by institutional impediments, namely, interstate political conflicts over water.

Spatial scale of decisions
In addition to the challenges created by institutional complexity, the spatial scale of decisions made by water management organizations ranges from small community water systems to large, multi-purpose metropolitan water service and regional water delivery systems (Rayner *et al.*, 2005). Differences in spatial scale of management also affect information needed—an issue discussed in Chapter 4 when we analyze Regional Integrated Science Assessment (RISA) experiences. These problems of diverse spatial scale are further compounded by the fact that most water agency boundaries do not conform to hydrological units. While some entities manage water resources in ways that conform to hydrological constraints (*i.e.*, watershed, river basin, aquifer or other drainage basin, Kenney and Lord, 1994; Cairo, 1997), basin-scale management is not the most common United States management approach. Because most hydrologic tools focus on watershed boundaries, there is a disconnect between the available data and the decision context.

Decision makers often share authority for decisions across local, state, and national jurisdictions. In fact, the label "decision maker" embraces a vast assortment of elected and appointed local, state, and national agency officials, as well as public and private sector managers with policy-making responsibilities in various water management areas (Sarewitz and Pielke, 2007). Because most officials have different management objectives while sharing authority for decisions, it is likely that their specific SI climate variability information needs will vary not only according to spatial scale, but also according to institutional responsibilities and agency or organization goals.

Identifying who the decision makers are is equally challenging. The Colorado River basin illustrates the typical array of decision makers on major U.S. streams. A recent study in Arizona identified an array of potential decision

makers affected by water shortages during drought, including conservation groups, irrigation districts, power providers, municipal water contractors, state water agencies, several federal agencies, two regional water project operators (the Central Arizona and Salt River projects), tribal representatives, land use jurisdictions, and individual communities (Garrick *et al.*, 2008). This layering of agencies with water management authority is also found at the national level.

There is no universally agreed-upon classification system for defining *water users*. Taking as one point of departure the notion that water users occupy various "sectors" (*i.e.*, activity areas distinguished by particular water uses), the U.S. Geological Survey (USGS) monitors and assesses water use for eight user categories: public supply, domestic use, irrigation, livestock, aquaculture, industrial, mining, and thermo-electric power. These user categories share freshwater supplies withdrawn from streams and/or aquifers and, occasionally, from saline water sources as well (Hutson *et al.*, 2004). However, the definitions of these classes of users vary from state to state.

One limitation in this user-driven classification scheme in regards to identifying information needs for SI climate forecasts is that it inadvertently excludes in-stream water users, those who do not remove water from streams or aquifers. Instream uses are extremely important, as they affect aquatic ecosystem health, recreation, navigation, and public health (Gillilan and Brown, 1997; Trush and McBain, 2000; Rosenberg *et al.*, 2000; Annear *et al.*, 2002). Moreover, instream uses and wetland habitats have been found to be among the most vulnerable to impacts of climate variability and change (NAST, 2001)[3].

Finally, decision makers' information needs are also influenced by the time frame for decisions, and to a greater degree than scientists' needs.

---

[3] In general, federal law protects instream uses only when an endangered species is affected. Protection at the state level varies, but extinction of aquatic species suggests the relatively low priority given to protecting flow and habitat. Organizations with interests in the management of instream flows are diverse, ranging from federal land management agencies to state natural resource agencies and private conservation groups, and their climate information needs widely vary (Pringle, 2000; Restoring the Waters, 1997).

Decision makers often share authority for decisions across local, state, and national jurisdictions.

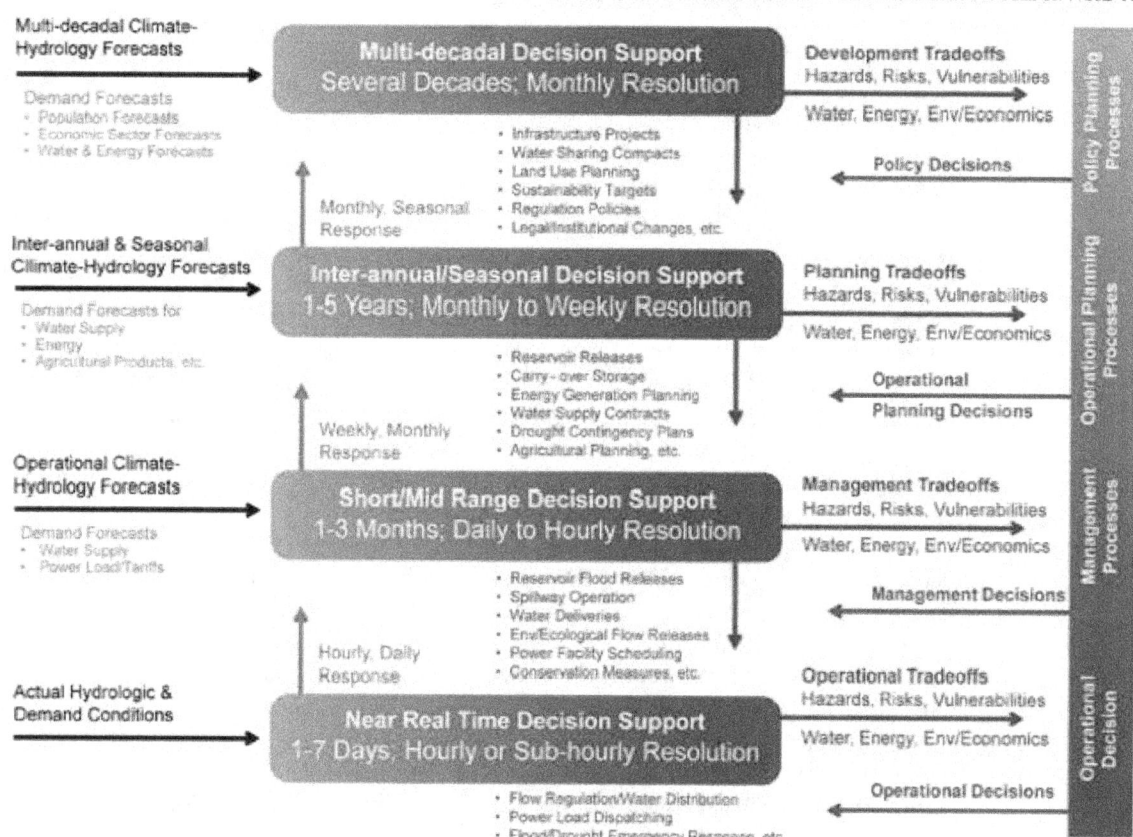

Figure 3.1 Water resources decisions: range and attributes.

For example, while NOAA researchers commonly distinguish between weather prediction information, produced on an hours-to-weeks time frame, and climate predictions, which may be on a SI time frame, many managers make decisions based on annual operating requirements or on shorter time frames that may not match the products currently produced.

Two important points stem from this. First, as longer-term predictions gain skill, use of longer-term climate information is likely to expand, particularly in areas with economic applications. Second, short-term decisions may have long-term consequences. Thus, identifying the information needed to make better decisions in all time frames is important, especially since it can be difficult to get political support for research that focuses on long-term, incremental increases in knowledge that are the key to significant policy changes (Kirby, 2000). This poses a challenge for decision makers concerned about adaptation to global change. Multi-decadal climate-hydrology forecasts and demand forecasts (including population and economic sector forecasts and forecasts of water

and energy demand) are key inputs for policy decisions. Changes in climate that affect these hydrology and water demand forecasts are particularly important for policy decisions, as they may alter the anticipated streams of benefits and impacts of a proposal. Information provided to the policy planning process is best provided in the form of tradeoffs assessing the relative implications, hazards, risks, and vulnerabilities associated with each policy option[4].

3.2.2 Decision-Support Needs of Water Managers for Climate Information
As we have noted, the decision-support needs of water resource decision makers for information on climate variability depend upon the temporal and spatial scale of the decisions that they make. The complexity of the decision process

Two important points stem from this. First, as longer-term predictions gain skill, use of longer-term climate information is likely to expand, particularly in areas with economic applications. Second, short-term decisions may have long-term consequences.

is graphically illustrated in Figure 3.1 (Georgakakos, 2006; HRC-GWRI, 2006). This figure includes four temporal scales ranging from multiple decades to hours. The first decision level includes *policy decisions pertaining to multi-decadal time scales and involving infrastructure changes* (e.g., storage projects, levee systems, energy generation facilities, waste water treatment facilities, inter-basin transfer works, sewer/drainage systems, well fields, and monitoring networks), as well as water sharing compacts, land use planning, agricultural investments, environmental sustainability requirements and targets, regulations, and other legal and institutional requirements (see Wiener *et al.*, 2000). Policy decisions may also encompass many political entities. Decisions pertaining to trans-boundary water resources are particularly challenging, as noted in Section 3.2.1.1, because they aim to reconcile benefits and impacts measured and interpreted by different standards, generated and accrued by stakeholders of different nations, and regulated under different legal and institutional regimes (Naim, 2003; Mumme, 2003,1995; Higgins *et al.*, 1999).

The second decision level involves *operational planning decisions pertaining to inter-annual and seasonal time scales*. These and other lower-level decisions are made within the context set by the policy decisions and pertain to inter-annual and seasonal reservoir releases, carry-over storage, hydro-thermal energy generation plans, agreements on tentative or final water supply and energy contracts, implementation of drought contingency plans, and agricultural planning decisions, among others. The relevant spatial scales for operational planning decisions may be as large as those of the policy decisions, but are usually associated with individual river basins as opposed to political jurisdictions. Interannual and seasonal hydro-climatic and demand forecasts (for water supply, energy, and agricultural products) are critical inputs for this decision level.

The third decision level pertains to *operational management decisions associated with short- and mid-range time scales of one to three months*. Typical decisions include reservoir releases during flood season; spillway operations; water deliveries to urban, industrial, or

agricultural areas; releases to meet environmental and ecological flow requirements; power facility operation; and drought conservation measures. The benefits and impacts of these decisions are associated with daily and hourly system response (high resolution). This decision level requires operational hydro-climatic forecasts and forecasts of water and power demand and pricing. The decision process is similar to those of the upper decision layers, although, as a practical matter, general stakeholder participation is usually limited, with decisions taken by the responsible operational authorities. This is an issue relevant to several cases discussed in Chapter 4.

The final decision level pertains to *near real time operations associated with hydrologic and demand conditions*. Typical decisions include regulation of flow control structures, water distribution to cities, industries, and farms, operation of power generation units, and implementation of flood and drought emergency response measures. Data from real time monitoring systems are important inputs for daily to weekly operational decisions. Because such decisions are made frequently, stakeholder participation may be impractical, and decisions may be limited to government agencies or public sector utilities according to established operational principles and guidelines.

While the above illustration addresses water resources complexity (*i.e.*, multiple temporal and spatial scales, multiple water uses, multiple decision makers), it cannot be functionally effective (*i.e.*, create the highest possible value) unless it exhibits consistency and adaptiveness. *Consistency* across the decision levels can be achieved by ensuring that (1) lower level forecasts, decision support systems, and stakeholder processes operate within the limits established by upper levels (as represented by the downward pointing feedback links in Figure 3.1, and (2) upper decision levels capture the benefits and impacts associated with the high resolution system response (as represented by the upward pointing feedback links in Figure 3.1). *Adaptiveness*, as a number of studies indicate, requires that decisions are continually revisited as system conditions change and new information becomes available, or as institutional

The decision process includes policy decisions, operational planning decisions, operational management decisions, and near real-time operations.

frameworks for decision making are amended (Holling, 1978; Walters, 1986; Lee, 1993).

3.2.3 How Does Climate Variability Affect Water Management?

Water availability is essential for human health, economic activity, ecosystem function, and geophysical processes. Climate variability can have dramatic seasonal and interannual effects on precipitation, drought, snow-pack, runoff, seasonal vegetation, water quality, groundwater, and other variables. Much recent research on climate variability impacts on water resources is linked to studies of long-term climate change, necessitating some discussion of the latter. In fact, there is a relative paucity of information on the potential influence of climate change on the underlying patterns of climate variability (e.g., CCSP, 2007). At the close of this Section, we explore one case—that of drought in the Colorado River basin—exemplifying several dimensions of this problem, including adaptive capacity, risk perception, and communication of hazard.

According to the Intergovernmental Panel on Climate Change (IPCC), while total annual precipitation is increasing in the northern latitudes, and average precipitation over the continental United States has increased, the southwestern United States (and other semitropical areas worldwide) appear to be tending towards reduced precipitation, which in the context of higher temperatures, results in lower soil moisture and a substantial effect on runoff in rivers (IPCC, 2007b). The observed trends are expected to worsen due to continued warming over the next century. Observed impacts on water resources from changes that are thought to have already occurred include increased surface temperatures and evaporation rates, increased global precipitation, an

increased proportion of precipitation received as rain rather than snow, reduced snowpack, earlier and shorter runoff seasons, increased water temperatures and decreased water quality (IPCC, 2007a, b).

Additional effects on water resources result from sea-level rise of approximately 10 to 20 centimeters since the 1890s (IPCC, 2007a)[5], an unprecedented rate of mountain glacier melting, seasonal vegetation emerging earlier in the spring and a longer period of photosynthesis, and decreasing snow and ice cover with earlier melting. Climate change is also likely to produce increases in intensity of extreme precipitation events (e.g., floods, droughts, heat waves, violent storms) that could "exhaust the social buffers that underpin" various economic systems such as farming; foster dynamic and interdependent consequences upon other resource systems (e.g., fisheries, forests); and generate "synergistic" outcomes due to simultaneous multiple human impacts on environmental systems (i.e., an agricultural region may be simultaneously stressed by degraded soil and changes in precipitation caused by climate change) (Rubenstein, 1986; Smith and Reeves, 1988; Atwood et al., 1988; Homer-Dixon, 1999).

Studies have concluded that changes to runoff and stream flow would have considerable regional-scale consequences for economies as well as ecosystems, while effects on the latter are likely to be more severe (Milly et al., 2005). If elevated aridity in the western United States is a natural response to climate warming, then any trend toward warmer temperatures in the future could lead to serious long-term increase in droughts, highlighting both the extreme vulnerability of the semi-arid West to anticipated precipitation deficits caused by global warming, and the need to better understand long-term drought variability and its causes (Cook et al., 2004).

The impacts of climate variability are largely regional, making the spatial and temporal scale of information needs of decision makers likewise regional. This is why we focus (Section 3.2.3.1) on specific regional hazards, risks, and

> The impacts of climate variability are largely regional, making the spatial and temporal scale of information needs of decision makers likewise regional.

⁵ According to the IPCC 2007 Fourth Assessment Report, sea level has risen an average of 1.8 mm per year over the period 1961 to 2003 (IPCC, 2007a)

vulnerabilities of climate variability on water resources. TOGA and RISA studies focus on the regional scale consequences of changes to runoff and stream flow on economies as well as ecosystems (Milly *et al.*, 2005).

3.2.3.1 HAZARDS, RISKS, AND VULNERABILITIES OF CLIMATE VARIABILITY

A major purpose of decision-support tools is to reduce the risks, hazards, and vulnerabilities to water resources from SI climate variation, as well as to related resource systems, by generating climate science products and *translating* these products into forms useful to water resource managers (NRC, 2008). In general, what water managers need help in translating is *how* changes resulting from weather and SI climate variation can affect the functioning of the systems they manage. Numerous activities are subject to risk, hazard, and vulnerability, including fires, navigation, flooding, preservation of threatened or endangered species, and urban infrastructure. At the end of this Section, we focus on three less visible but nonetheless important challenges: water quality, groundwater depletion, and energy production.

Despite their importance, hazard, risk, and vulnerability can be confusing concepts. A *hazard* is an event that is potentially damaging to people or to things they value. Floods and droughts are two common examples of hazards that affect water resources. *Risk* indicates the probability of a particular hazardous event occurring. Hence, while the hazard of drought is a concern to all water managers, drought risk varies considerably with physical geography, management context, infrastructure type and condition, and many other factors so that some

> Water managers need help in translating how changes resulting from weather and Seasonal to Interannual climate variation can affect the functioning of the systems they manage.

water resource systems are more at-risk than others (Stoltman *et al.*, 2004; NRC, 1996; Wilhite, 2004).

A related concept, *vulnerability*, is more complex and can cause further confusion[6]. Although experts dispute precisely what the term means, most agree that vulnerability considers the likelihood of harm to people or things they value and it entails physical as well as social dimension (*e.g.*, Blaikie *et al.*, 1994; Cutter 1996; Hewitt, 1997; Schröter *et al.*, 2005; Handmer, 2004). Physical vulnerability relates to exposure to harmful events, while social vulnerability entails the factors affecting a system's sensitivity and capacity to respond to exposure. Moreover, experts accept some descriptions of vulnerability more readily than others. One commonly accepted description considers vulnerability to be a function of exposure, sensitivity, and adaptive capacity (Schneider and Sarukhan, 2001). Exposure is the degree to which people and the places or things they value, such as their water supply, are likely to be impacted by a hazardous event, such as a flood. The "things they value" include not only economic value and wealth but also cultural, spiritual, and personal values. This concept also refers to physical infrastructure (*e.g.*, water pipelines and dams) and social infrastructure (*e.g.*, water management associations). Valued components include intrinsic values like water quality and other outcomes of water supply availability such as economic vitality.

Sensitivity is the degree to which people and the things they value can be harmed by exposure. Some water resource systems, for example, are more sensitive than others when exposed to the same hazardous event. All other factors being equal, a water system with old infrastructure will be more sensitive to a flood or drought than one with new state-of-the-art infrastructure; in a century, the newer infrastructure will be considerably more sensitive to a hazardous event than it is today because of aging.

---

[6] Much of this discussion on vulnerability is modified from Yarnal (2007). See also Polsky *et al.* (2007), and Dow *et al.* (2007) for definitions of vulnerability, especially in relation to water resource management.

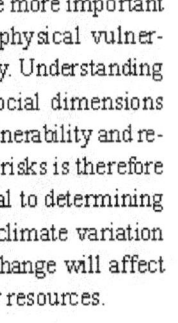

Adaptive capacity is the least explored and most controversial aspect of vulnerability. The understanding of adaptive capacity favored by the climate change research community is the degree to which people can mitigate the potential for harm—that is, reduce vulnerability—by taking action to reduce exposure or sensitivity, both before and after the hazardous event. The physical, social, economic, spiritual, and other resources they possess, including such resources as educational level and access to technology, determine the capacity to adapt. For instance, all things being equal, a community water system that has trained managers and operators with up-to-date computer technology will be less vulnerable than a neighboring system with untrained volunteer operators and limited access to computer technology[7].

Some people or things they value can be highly vulnerable to low-impact events because of high sensitivity or low adaptive capacity. Others may be less vulnerable to high-impact events because of low sensitivity or high adaptive capacity. A hazardous event can result in a patchwork pattern of harm due to variation in vulnerability over short distances (Rygel *et al.*, 2006). Such variation means that preparing for or recovering from flood or drought may require different preparation and recovery efforts from system to system.

3.2.3.2 PERCEPTIONS OF RISK AND VULNERABILITY—ISSUE FRAMES AND RISK COMMUNICATION

Much of the research on vulnerability of water resources to climate variability has focused on *physical vulnerability* (*i.e.*, the exposure of water resources and water resource systems to harmful events). Cutter *et al.* (2003) and many others have noted, however, that *social vulnerability*—the social factors that affect a system's sensitivity to exposure, and that influence its capacity to respond and adapt in order to lessen its exposure or sensitivity—can of-

ten be more important than physical vulnerability. Understanding the social dimensions of vulnerability and related risks is therefore crucial to determining how climate variation and change will affect water resources.

The perception of risk
is perhaps the most-studied of the social factors relating to climate information and the management of water resources. At least three barriers stemming from their risk perceptions prevent managers from incorporating weather and climate information in their planning; each barrier has important implications for communicating climate information to resource managers and other stakeholders (Yarnal *et al.*, 2005). A fourth barrier relates to the underlying public perceptions of the severity of climate variability and change and thus, implicit public support for policies and other actions that might impel managers to incorporate climate variability into decisions.

The first conceptual problem is that managers who find climate forecasts and projections to be reliable appear in some cases no more likely to use them than managers who find them to be unreliable (O'Connor *et al.*, 1999, 2005)[8]. Managers most likely to use weather and climate information may have experienced weather and climate problems in the recent past—their heightened feelings of vulnerability are the result of negative experiences with weather or climate. The implication of this finding is that simply delivering weather and climate information to potential users may be insufficient in those cases in which the manager does not perceive climate to be a hazard, at least in

Understanding the social dimensions of vulnerability and related risks is therefore crucial to determining how climate variation and change will affect water resources.

---

[7] A slightly different view of adaptive capacity favored by the hazards and disaster research community is that it consists of two subcomponents: coping capacity and resilience. The former is the ability of people and systems to endure the harm; the latter is the ability to bounce back after exposure to harmful events. In both cases, water resource systems can take measures to increase their ability to cope and recover, again depending on the physical, social, economic, spiritual, and other resources they possess or have access to.

[8] Based on findings from two surveys of community water system managers (more than 400 surveyed in each study) in Pennsylvania's Susquehanna River Basin. The second survey compared Pennsylvania community water system managers to their counterparts in South Carolina (more than 250 surveyed) and found that managers who find climate forecasts and projections to be reliable are no more likely to use them than are those who find them to be unreliable. Thus, unless managers feel vulnerable (vulnerability being a function of whether they have had adverse experience with weather or climate), they are statistically less likely to use climate forecasts.

humid, water rich regions of the United States that we have studied[9]. Purveyors of weather and climate information may need to convince potential users that, despite the absence of recent adverse events, their water resources have suffered historically from, and therefore are vulnerable to, weather and climate.

The second barrier is that managers' perceptions about the usefulness of climate information varies not only with their exposure to adverse events, but also with the financial, regulatory, and management contexts of their decisions (Yarnal *et al.*, 2006; Dow *et al.*, 2007). The implication of this finding is that assessments of weather and climate vulnerability and of climate information needs must consider the institutional contexts of the resource systems and their managers. Achieving a better understanding of these contexts and of the informational needs of resource managers requires working with them directly.

Communication of the risks of climate change and variablility differs among scientific, political, and mass media elites—each selecting aspects of these issues that are most relevant to their conception of risk.

The third barrier is that managers expect more difficulties to come from associated financial and water quality impacts of climate challenges associated with floods and droughts than from their ability to find water and supply it to their customers (Yarnal *et al.*, 2006; Dow *et al.*, 2007). Combined with the second barrier, the implication is that managers view weather and climate forecasts as more salient when put into the context of system operations and management needs. Presenting managers with a climate forecast for the United States showing the regional probability of below-normal precipitation for the coming season may not generate much interest; presenting those managers with a Palmer Drought Severity Index tailored to their state that suggests a possible drought watch, warning, or emergency will grab their attention (Carbone and Dow, 2005). The Southwest drought case discussed at the end of this Section exemplifies how this salience worked

to prod decision makers to partner closely with water managers, and how the latter embraced climate knowledge in improving forecasts and demand estimates.

The fourth barrier is the way climate variability and change are framed as public policy issues, and how their risks are publically communicated. Regardless of the "actual" (if indeterminate) risks from climate change and variability, communication of the risks differs among scientific, political, and mass media elites—each systematically selecting aspects of these issues that are most relevant to their conception of risk, and thus, socially constructing and communicating its aspects most salient to a particular perspective. Thus, climate variability can be viewed as: a phenomenon characterized by probabilistic and consequential uncertainty (science); an issue that imposes fiduciary or legal responsibility on government (politics); or, a sequence of events that may lead to catastrophe unless immediate action is taken (Weingart *et al.*, 2000).

Related to this is considerable research that suggests that when risk information, such as that characteristic of climate change or variability modeling and forecasting, is generated by select groups of experts who work in isolation from the public (or from decision makers), the risks presented may sometimes be viewed as untrustworthy or as not credible and worthy of confidence. This research also suggests that building trust requires the use of public forums designed to facilitate open risk communication that is clear, succinct, and jargon-free, and that provide groups ample opportunity for ques-

---

[9] Additional research on water system manager perceptions is needed, in regions with varying hydrometeorological conditions, to discern if this finding holds true in other regions.

tions, discussion, feedback, and reaction (*e.g.*, Freudenburg and Rursch, 1994; Papadakis, 1996; Jasanoff, 1987; Covello *et al.*, 1990; NRC, 1989).

Research on these barriers also shows that personal experience has a powerful influence on perceptions of risk and vulnerability. They suggest that socioeconomic context is important in shaping perceptions, and, thus, the perceptions they produce are very specific. They also show that climate information providers must present their information in ways salient to potential users, necessitating customizing information for specific user groups. Finally, they suggest ways that perceptions can be changed.

Research on the influence of climate science on water management in western Australia (Power *et al.*, 2005) suggests that water resource decision makers can be persuaded to act on climate variability information if a strategic program of research in support of specific decisions (*e.g.*, responses to extended drought) can be wedded to a dedicated, timely risk communication program. In this instance, affected western Australian states formed a partnership between state agencies representing economic interests affected by drought, national research institutions engaged in meteorology and hydrology modeling, and water managers. This partnership succeeded in influencing decision making by: being sensitive to the needs of water managers for advice that was seen as "independent" ,in order to assure the public that water use restrictions were actually warranted; providing timely products and services to water users in an accessible way, and, directly involving water managers in the process of generating forecast information. The Georgia drought case (Box 3.1) also illustrates the need to be sensitive and responsive to decision-maker needs. As in Australia, ensuring scientific "independence" facilitated the efforts of managers to consider climate science in their decisions, and helped ensure that climate forecast information was "localized" through presentation at public meetings and other forums so that residents could apply it to local decisions (Power *et al.*, 2005). In sum, to overcome barriers to effective climate information communication, information must be specific to the sectoral context of managers and enhance their ability to realize

management objectives threatened by weather and climate.

We now examine three particularly vulnerable areas to climate variability: water quality, groundwater depletion, and energy production. Following this discussion, we feature a case study on *drought responses in the Southwest United States* which is instructive about the role that perceived vulnerability has played in adaptive responses.

Water Quality: Assessing the vulnerability of water quality to climate variability and change is a particularly challenging task, not only because quality is a function (partly) of water quantity, but because of the myriad physical, chemical and biological transformations that non-persistent pollutants undergo in watersheds and water bodies including fire hazards (*e.g.*, Georgia Forestry Commission, 2007). One of the most comprehensive literature reviews of the many ways in which water quality can be impacted by climate variability and change was undertaken by Murdoch *et al.* (2000). A synopsis of their major findings is depicted in Table 3.1.

One conclusion to be drawn from Table 3.1 is that climate variability and change can have both negative and positive impacts on water quality. In general, warmer surface-water temperatures and lower flows tend to have a negative impact through decreases in dissolved oxygen (DO). In contrast, decreased flows to receiving water bodies, especially estuaries and coastal waters, can improve water quality, while increased flows can degrade water quality of the receiving water bodies, particularly if they carry increased total loads of nutrients and sediments. In healthy watersheds that are relatively unimpacted by disturbances to the natural vegetation cover, increased stream flow may increase water quality in the given stream by increasing dilution and DO.

Increased runoff and flooding in urbanized areas can lead to increased loads of nonpoint source pollutants (Kirshen *et al.*, 2006) such as pesticides and fertilizer from landscaped areas, and point source pollutants, from the overflow of combined sewer systems (Furlow, 2006). In addition to increasing pesticide and

Climate variability and change can have both negative and positive impacts on water quality.

Table 3.1 Water Quality, Climate Variability, and Climate Change*

Impacts associated with increases in temperature alone
• Decreased oxygen-holding capacity due to higher surface-water temperatures. • In Arctic regions, the melting of ice and permafrost resulting in increased erosion, runoff, and cooler stream temperatures. • Changes in the seasonal timing and degree of stratification of temperate lakes. • Increased biomass productivity leading to increased rates of nutrient cycling, eutrophication and anoxia. • Increased rates of chemical transformation and bioaccumulation of toxins. • Changes in the rates of terrestrial nutrient cycling and the delivery of nutrients to surface waters.
Impacts associated with drought and decreases in streamflow
• Increased concentration of pollutants in streams, but decreased total export of those pollutants to the receiving water body. • Decreases in the concentration of pollutants that are derived from the flushing of shallow soils and by erosion. • Increases in the concentration of pollutants that are derived from deeper flow paths and from point sources. • Decreased stratification and increased mixing in estuaries and other coastal waters, leading to decreased anoxia of bottom waters and decreased nutrient availability (and eutrophication). • Movement of the freshwater-saltwater boundary up coastal river and intrusion of salt water into coastal aquifers—impacts which would be exacerbated by sea-level rise.
Impacts associated with flooding and increases in streamflow
• In general, mitigation of the impacts associated with drought and decreases in streamflow. • Increases in the spatial extent of source areas for storm flow, leading to the increased flushing of pollutants from both point and non-point sources of pollution. • Increased rates of erosion. • Increased rates of leaching of pollutants to groundwater. • Greater dilution of pollutants being countervailed by decreased rates of chemical and biological transformations owing to shorter residence times in soils, groundwater and surface waters.
* From Murdoch, et al., 2000

nutrient loads (Chang *et al.*, 2001), increase in runoff from agricultural lands can lead to greater sediment loads from erosion and pathogens from animal waste (Dorner *et al.*, 2006). Loads of non-point pollution may be especially large during flooding if the latter occurs after a prolonged dry period in which pollutants have accumulated in the watershed.

The natural vegetation cover that is integral to a healthy watershed can be disturbed not only by land-use but by the stresses of climate extremes directly (e.g., die off during drought and blow down of trees during tropical storms and hurricanes) and climate-sensitive disturbances indirectly (e.g., pest infestations and wildfire). Climate change and variability can also lead to both adaptive human changes in land use and land cover that can impact water quality (e.g. changes in cropping patterns and fertilizer use), as well as to mitigative ones (e.g., increased planting of low water use native plants). Hence there is a tight and complex coupling between

land use changes and the potential impacts of climate variability and change on water quality.

Water quality can also be indirectly impacted by climate variability and change through changes in water use. Withdrawals from streams and reservoirs may increase during a drought thereby degrading stream water quality through lower in-stream flows, polluted return flows, or both. Under the water rights system of the western United States, junior agricultural users may be cut off during drought, thereby actually reducing return flows from agricultural lands and further lowering in-stream flows.

Perhaps the most common water quality related, climate-sensitive decisions undertaken by water resource managers in the United States are in relation to the regulation of dams and reservoirs. Very often, reservoir releases are made to meet low flow requirements or maintain stream temperatures in downstream river reaches. Releases can also be made to improve water quality in downstream reservoirs, lakes and estuaries. Any operating decisions based on water quality usually occur in the context of the purpose(s) for which the dam and reservoir were constructed—typically some combination of hydropower, flood control, recreation, and storage for municipal supply and irrigation. Thus, decision-support systems for reservoir operation that include water quality usually do so in a multi-objective framework (e.g., Westphal *et al.*, 2003).

Municipal water providers would also be expected to respond to water quality degradation forecasts. Some decisions they might undertake include stockpiling treatment chemicals, enhanced treatment levels, *ad hoc* sediment control, preparing to issue water quality alerts, increasing water quality monitoring, and securing alternative supplies (see Denver and New York City case studies in Miller and Yates [2005] for specific examples of climate-sensitive water quality decision making by water utilities). Managers of coastal resources such as fisheries and beaches also respond to water-quality forecasts.

Decision making with regards to point sources will necessarily occur within the context of the permitting process under the National Pollution Discharge Elimination System and the in-stream water quality standards mandated by the Clean Water Act (Jacoby, 1990). Regulation of nonpoint sources falls entirely to the states and is therefore highly variable across the nation, but is in general done to a lesser degree than the regulation of point sources. Examples of actions, either voluntary or mandatory, that could be taken in response to a seasonal forecast of increased likelihood of flooding include: decreased fertilizer and pesticide application by farmers, measures for greater impoundment of runoff from feedlots, and protection of treatment ponds of all kinds from overflow.

Groundwater Depletion: The vulnerability of groundwater resources to climate variability and change is very much dependent on the hydrogeologic characteristics of a given aquifer. In general, the larger and deeper the aquifer, the less interannual climate variability will impact groundwater supplies. On the other hand, shallow aquifers that are hydraulically connected

to surface waters tend to have shorter residence times and therefore respond more rapidly to climate variability. The vulnerability of such aquifers should be evaluated within the context of their conjunctive use with surface waters.

Seasonal and interannual variability in water-table depths are a function of natural climate variability as well as variations in human exploitation of the resource. During periods of drought, water tables in unconfined aquifers may drop because of both reduced recharge and increased rates of pumping. Reduced hydraulic head at well intakes then decreases the potential yield of the given well or well field and increases the energy required for pumping. In extreme cases, the water table may drop below the well intake, resulting in complete drying of the well. Municipal supply and irrigation wells tend to be developed in larger aquifers and at depths greater than wells supplying individual domestic users. Therefore, they are in general less vulnerable to interannual climate variability. In addition to the reduction in the yield of water-supply wells, drops in water table depths during droughts may result in the drying of springs and worsening of low flow conditions in streams. Greater withdrawals may result because of the shifting of usage from depleted surface waters, as well as because of an overall increase in demand due to lower precipitation and greater evapotranspirative demand from the land surface and water bodies. Morehouse *et al.* (2002) find this to be the case in southern Arizona. To the extent that climate change reduces surface water availability in the U.S. Southwest, it can be anticipated that pressure on groundwater supplies will increase as a result.

When long-term average pumping rates exceed recharge rates the aquifer is said to be in *overdraft*. Zekster *et al.* (2005) identify four major impacts associated with groundwater extraction and overdraft: (1) reduction of stream flow and lake levels, (2) reduction or elimination of vegetation, (3) land subsidence, and (4) seawater intrusion. Additional impacts include changes in water quality due to pumping from different levels in aquifers and increased pumping costs. The Edwards Aquifer in south-central Texas, which supplies over two million people in the San Antonio metropolitan area, is identified by Loáiciga (2003) as particularly vulnerable to

The vulnerability of groundwater resources to climate variability and change is very much dependent on the hydrogeologic characteristics of a given aquifer. In general, the larger and deeper the aquifer, the less interannual climate variability will impact groundwater supplies.

climate change and variability because it is subject to highly variable rates of recharge and has undergone a steady increase in pumping rates over the last century. While groundwater overdraft is most common in the arid and semi-arid western United States (Roy *et al.*, 2005; Hurd *et al.*, 1999), it is not uncommon in the more humid East. Lyon *et al.* (2005) study the causes of the three drought emergencies that have been declared in Rockland County, New York since 1995. Seventy-eight percent of the county's public water supply is from small regional aquifers. Rather than increased frequency or intensity of meteorologic or hydrologic drought, the authors attribute drought emergencies to development and population growth overtaxing local supplies and to failure of aging water-supply infrastructure. The former is an example of *demand-driven* drought. The Ipswich River Basin in northeast Massachusetts is another example in the East where population growth is taxing groundwater resources. Because of reliance on ground water and in-stream flows for municipal and industrial supply, summer low flows in the Ipswich frequently reach critical levels (Zarriello and Ries, 2000).

A few researchers have studied the potential application of SI climate forecasting to forecasting of groundwater recharge and its implications for water management. For example, using U.S. Geological Survey recharge estimates for the Edwards Aquifer from 1970 to 1996, Chen *et al.* (2005) find that recharge rates during La Niña years average about twice those during El Niño years. Using a stochastic dynamic programming model, they show that optimal water use and allocation decision making based on El Niño-Southern Oscillation (ENSO)[10] forecasts could result in benefits of $1.1 to $3.5 million per year, mainly to agricultural users as a result of cropping decisions.

Hanson and Dettinger (2005) evaluate the SI predictability of groundwater levels in the Santa Clara-Calleguas Basin in coastal Southern California using a regional groundwater model (RGWM) as driven by a general circulation model (GCM). In agreement with other studies, they find a strong association between groundwater levels and the Pacific Decadal Oscillation (PDO) and ENSO. Their results led them to conclude that coupled GCM-RGWM modeling is useful for planning and management purposes, particularly with regard to conjunctive use of surface and ground water and the prevention of saltwater intrusion. They also suggest that GCM forecast skill may at times be strong enough to predict groundwater levels. Forecasts of greater surface water availability may allow utilities to reduce reliance on over-utilized and expensive groundwater resources. Bales *et al.* (2004) note that a forecast for heavy winter snowpack during the 1997/1998 El Niño led the Salt River Project in Arizona to reducing groundwater pumping in the fall and winter in favor of greater releases from reservoirs, thereby saving about $1 million.

Water Supply and Energy Production: Adequate water supplies are an essential part of energy production, from energy resource extraction (mining) to electric-power generation (DOE, 2006). Water withdrawals for cooling and scrubbing in thermoelectric generation now exceed those for agriculture in the United States (Hutson *et al.*, 2004), and this difference becomes much greater when hydropower uses are considered. Emerging energy sources, such as biofuels, synfuels, and hydrogen, will add to future water demands. Another new energy-related stress on water resource systems will be the integration of hydropower with other intermittent renewables, such as wind and solar, at the power system level. Hydropower is a very flexible, low-cost generating source that can be used to balance periods when other renewables are not available (*e.g.*, times of calm winds) and thus maintain electricity transmission reliability. As more non-hydro renewables are added to transmission grids, calls for fluctuating hydropower operation may become more frequent and economically valuable, and may compete with other water demands. If electricity demand increases by 50 percent in the next

Emerging energy sources, such as biofuels, synfuels, and hydrogen, will add to future water demands.

[10] The Southern Oscillation Index (SOI) is a calculation of monthly or seasonal fluctuations in the air pressure difference between Tahiti and Darwin, Australia. When the air pressure in Tahiti is below normal and the air pressure in Darwin is above normal, the SOI is in a negative phase. Prolonged periods of negative SOI values often occur with abnormally warm ocean waters across the eastern tropical Pacific resulting in a period called an El Niño. Conversely, prolonged periods of positive SOI values (air pressure in Tahiti is above normal and in Darwin it is below normal) coincides with abnormally cold ocean waters across the eastern tropical Pacific and is called a La Niña.

25 years, as predicted by the Energy Information Administration, then energy-related water uses can also be expected to expand greatly—an ominous trend, especially where available water resources are already over-allocated.

The Climate Change Science Program's Synthesis and Analysis Product 4.5 examined how climate change will affect the energy sector (CCSP, 2007). Some of the most direct effects of climate change on the energy sector will occur via water cycle processes (CCSP, 2007). For instance, changes in precipitation could affect prospects for hydropower, either positively or negatively, at different times and locations. Increases in storm intensity could threaten further disruptions of the type experienced in 2005 with Hurricane Katrina. Also, average warming can be expected to increase energy needs for cooling and reduce those for warming. Concerns about climate change impacts could change perceptions and valuations of energy technology alternatives. Any or all of these types of effects could have very real meaning for energy policies, decisions, and institutions in the United States, affecting discussions of courses of action and appropriate strategies for risk management and energy's water demands will change accordingly.

The energy-related decisions in water management are especially complex because they usually involve both water quality and quantity aspects, and they often occur in the context of multiple-use river basins. The Tennessee Valley is a good example of these complexities. The Tennessee Valley Authority (TVA) operates an integrated power system of nuclear, coal, and hydropower projects along the full length of the Tennessee River. TVA's river operations include upstream storage reservoirs and mainstem locks and dams, most of which include hydropower facilities. Cold water is a valuable resource that is actively stored in the headwater reservoirs and routed through the river system to maximize cooling efficien-

cies of the downstream thermoelectric plants. Reservoir releases are continuously optimized to produce least-cost power throughout the river basin, with decision variables of both water quantity and quality.

Case Study: Southwest drought—climate variability, vulnerability, and water management

Introduction

Climate variability affects water supply and management in the Southwest through drought, snowpack runoff, groundwater recharge rates, floods, and temperature-driven water demand. The region sits at a climatic crossroads, at the southern edge of reliable winter storm tracks and at the northern edge of summer North American monsoon penetration (Sheppard *et al.*, 2002). This accident of geography, in addition to its continental location, drives the region's characteristic aridity. Regional geography also sets the region up for extreme vulnerability to subtle changes in atmospheric circulation and the impacts of temperature trends on snowmelt, evaporation, moisture stress on ecosystems, and urban water demands. The instrumental climate record provides ample evidence of persistent regional drought during the 1950s (Sheppard *et al.*, 2002; Goodrich and Ellis, 2006), and its influence on Colorado River runoff (USGS, 2004); in addition the impact of the 1950s drought on regional ecosystems is well documented (Allen and Breshears, 1998; Swetnam and Betancourt, 1998). Moreover, it has been well known for close to a decade that

The energy-related decisions in water management are especially complex because they usually involve both water quality and quantity aspects, and they often occur in the context of multiple-use river basins.

June 29, 2002

December 23, 2003

Interest in the effects of climate variability on water supplies in the Southwest has been limited by dependence on seemingly unlimited groundwater resources, which are largely buffered from interannual climate fluctuations.

interannual and multi-decadal climate variations, forced by persistent patterns of ocean-atmosphere interaction, lead to sustained wet periods and severe sustained drought (Andrade and Sellers, 1988; D'Arrigo and Jacoby, 1991; Cayan and Webb, 1992; Meko *et al.*, 1995; Mantua *et al.*, 1997; Dettinger *et al.*, 1998).

Sources of vulnerability

Despite this wealth of information, interest in the effects of climate variability on water supplies in the Southwest has been limited by dependence on seemingly unlimited groundwater resources, which are largely buffered from interannual climate fluctuations. Evidence of extensive groundwater depletion in Arizona and New Mexico, from a combination of rapid urban expansion and sustained pumping for irrigated agriculture, has forced changes in water policy, resulting in a greater reliance on renewable surface water supplies (Holway, 2007; Anderson and Woosley, Jr., 2005; Jacobs and Holway, 2004). The distance between the Southwest's urban water users and the sparsely-populated mountain sources of their surface water in Wyoming, Utah, and Colorado, reinforces a lack of interest in the impacts of climate variations on water supplies (Rango, 2006; Redmond, 2003). Until Southwest surface water supplies were substantially affected by sustained drought, beginning in the late 1990s, water management interest in climate variability seemed to be focused on the increased potential for flood damage during El Niño episodes (Rhodes *et al.*, 1984; Pagano *et al.*, 2001).

Observed vulnerability of Colorado River and Rio Grande water supplies to recent sustained drought, has generated profound interest in the effects of climate variability on water supplies and management (e.g., Sonnett *et al.*, 2006). In addition, extensive drought-driven stand-replacing fires in Arizona and New Mexico watersheds have brought to light indirect impacts of climate variability on water quality and erosion (Neary *et al.*, 2005; Garcia *et al.*, 2005; Moody and Martin, 2001). Prompted by these recent dry spells and their impacts, New Mexico and Arizona developed their first drought plans (NMDTF, 2006; GDTF, 2004); in fact, repeated drought episodes, combined with lack of effective response, compelled New Mexico to twice revise its drought plan (NMDTF, 2006; these workshops are discussed in Chapter 4 in Case Study H). Colorado River Basin water managers have commissioned tree ring reconstructions of streamflow, in order to revise estimates of record droughts, and to improve streamflow forecast performance (Woodhouse and Lukas, 2006; Hirschboeck and Meko, 2005). These reconstructions and others (Woodhouse *et al.*, 2006; Meko *et al.*, 2007) reinforce concerns over surface water supply vulnerability, and the effects of climate variability and trends (e.g., Cayan *et al.*, 2001; Stewart *et al.*, 2005) on streamflow.

Decision-support tools

Diagnostic studies of the associations between ENSO teleconnections, multi-decadal variations in the Pacific Ocean-atmosphere system, and Southwest climate demonstrate the potential predictability of seasonal climate and hydrology in the Southwest (Cayan *et al.*, 1999; Gutzler, *et al.*, 2002; Hartmann *et al.*, 2002; Hawkins *et al.*, 2002; Clark *et al.*, 2003; Brown and Comrie, 2004; Pool, 2005). ENSO teleconnections currently provide an additional source of information for ensemble streamflow predictions by the National Weather Service (NWS) Colorado Basin River Forecast Center (Brandon *et al.*, 2005). The operational use of ENSO teleconnections as a primary driver in Rio Grande and Colorado River streamflow forecasting, however, is hampered by high variability (Dewalle *et al.*, 2003), and poor skill in the headwaters of these rivers (Udall and Hoerling, 2005; FET, 2008).

Future prospects

Current prospects for forecasting beyond ENSO time-scales, using multi-decadal "regime shifts" (Mantua, 2004) and other information (McCabe *et al.*, 2004) are limited by lack of spatial resolution, the need for better understanding of land-atmosphere feedbacks, and global atmosphere-ocean interactions (Dole, 2003; Garfin *et al.*, 2007). Nevertheless, Colorado River and Rio Grande water managers, as well as managers of state departments of water resources have embraced the use of climate knowledge in improving forecasts, preparing for infrastructure enhancements, and estimating demand (Fulp, 2003; Shamir *et al.*, 2007). Partnerships among water managers, forecasters, and researchers hold the most promise for reducing water supply vulnerabilities and other water management risks through the incorporation of climate knowledge (Wallentine and Matthews, 2003).

3.2.4 Institutional Factors That Inhibit Information Use in Decision-Support Systems

In Section 3.1, decision support was defined as a process that generates climate science products *and* translates them into forms useful for decision makers through dissemination and communication. This process, when successful, leads to institutional transformation (NRC, 2008). Five factors are cited as impediments to optimal use of decision-support systems' information: (1) lack of integration of systems with expert networks; (2) lack of institutional coordination; (3) insufficient stakeholder engagement in product development; (4) insufficient cross-disciplinary interaction; and, (5) expectations that the expected "payoff" from forecast use may be low. The *Red River flooding and flood management case* following this discussion exemplifies some of these problems, and describes some promising efforts being expended in overcoming them.

Some researchers (Georgakakos *et al.*, 2005) note that because water management decisions are subject to gradual as well as rapid changes

in data, information, technology, natural systems, uses, societal preferences, and stakeholder needs, effective decision-support processes regarding climate variability information must be adaptive and include self-assessment and improvement mechanisms in order to be kept current (Figure 3.2).

These assessment and improvement mechanisms, which produce transformation, are denoted by the upward-pointing feedback links shown in Figure 3.2, and begin with monitoring and evaluating the impacts of previous decisions. These evaluations ideally identify the need for improvements in the effectiveness of policy outcomes and/or legal and institutional frameworks. They also embrace assessments of the quality and completeness of the data and information generated by decision-support systems and the validity and sufficiency of current knowledge. Using this framework as a point of departure makes discussing our five barriers to information use easier to comprehend.

First, the lack of integrated decision-support systems and expert networks to support planning and management decisions means that decision-support experts and relevant climate information are often not available to decision makers who would otherwise use this information. This lack of integration is due to several factors, including resources (e.g., large agencies can better afford to support modeling efforts, consultants, and large-scale data management efforts than can smaller, less-well funded ones), organizational design (expert networks and support systems may not be well-integrated administratively from the vantage point of connecting information with users' "decision routines"), and opportunities for interaction between expert system designers and managers (the strength of communication networks to permit decisions and the information used for them to be challenged, adapted, or modified— and even to frame scientific questions). This challenge embraces users and producers of climate information, as well as the boundary organizations that can serve to translate information (Hartmann, 2001; NRC, 1996; Sarewitz and Pielke, 2007; NRC, 2008).

Second, the lack of coordination of institutions responsible for water resources management

Partnerships among water managers, forecasters, and researchers hold the most promise for reducing water supply vulnerabilities and other water management risks through the incorporation of climate knowledge.

Limited stakeholder participation and political influence in decision-making processes means that decision-support products may not equitably penetrate to all relevant audiences.

means that information generated by decision-support networks must be communicated to various audiences in ways relevant to their roles and responsibilities (Section 3.2.1). Figure 3.2 and discussion of the factors that led to development of better decision support for flood hazard alleviation on the *Red River of the North* reveal how extreme environmental conditions compound the challenge in conveying information to different audiences given the dislocation and conflict that may arise.

Third, limited stakeholder participation and political influence in decision-making processes means that decision-support products may not equitably penetrate to all relevant audiences. It also means that because water issues typically have low visibility for most of the public, the economic and environmental dislocations caused by climate variability events (*e.g.*, drought, floods), or even climate change, may exacerbate these inequities and draw sudden, sharp attention to the problems resulting from failure to properly integrate decision-support models and forecast tools, since disasters often strike disadvantaged populations dispropor-

tionately (*e.g.*, Hurricane Katrina in 2005) (Hartmann *et al.*, 2002; Carbone and Dow, 2005; Subcommittee on Disaster Reduction, 2005; Leatherman and White, 2005).

Fourth, the lack of adequate cross-disciplinary interaction between science, engineering, public policy-making, and other knowledge and expertise sectors, as well as across agencies, academic institutions, and private sector organizations, exacerbates these problems by making it difficult for decision-support information providers to communicate with one another. It also exacerbates the problem of information overload by inhibiting use of incremental additional tools, the sources and benefits of which are unclear to the user. In short, certain current decision-support services are often narrowly focused, developed by over-specialized professionals working in a "stovepipe" system of communication within their organizations. While lack of integration can undermine the effectiveness of decision-support tools and impede optimal decisions, it may create opportunities for design, development and use of effective decision-support services.

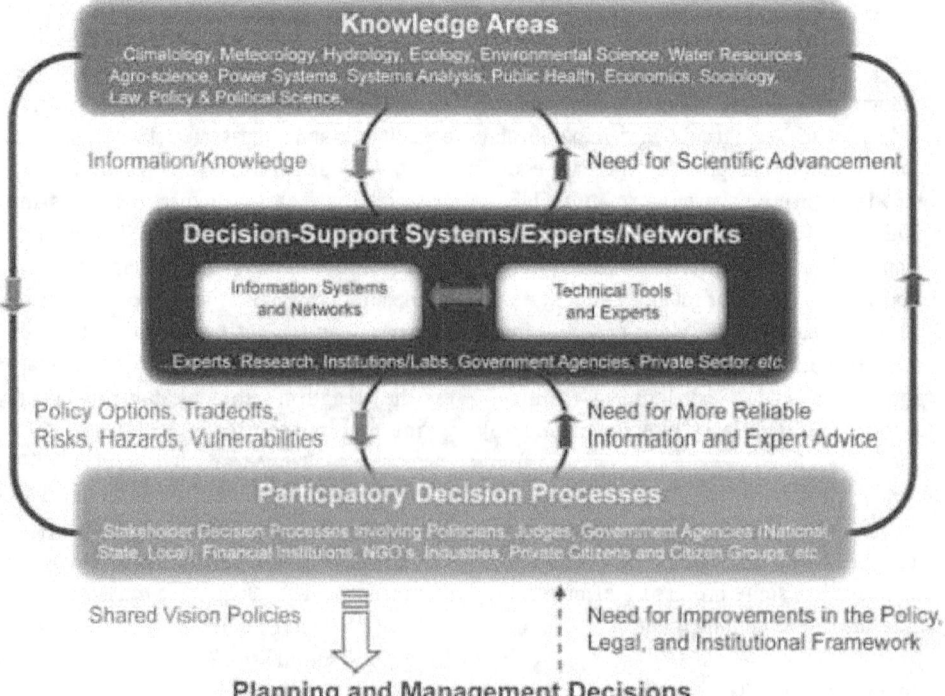

Figure 3.2 Water resources decision processes.

3.2.5 Reliability and Trustworthiness as Problems in Collaboration

The collaborative process for decision support must be believable and trustworthy, with benefits to all engaged in it. One of the challenges in ensuring that information is perceived by decision makers as trustworthy is that trust is the result of an interactive process of long-term, sustained effort by scientists to respond to, work with, and be sensitive to the needs of decision makers and users, and of decision makers becoming sensitive to, and informed about, the process of research. In part, trust is also a matter of the perceived credibility of the outcomes generated by decision-support systems.

The *Red River Flood warning case* (Section 3.2.4) provides an excellent example of this problem—users had become comfortable with single-valued forecasts and thus had applied their own experience in determining how much confidence to place in the forecasts they received. Coupled with the dependence on media as the tool for conveying weather information, the inclusion of uncertainty information in a forecast was viewed by some as a weakness, or disadvantage, in providing adequate warning of impending flood conditions, instead of an advantage in ensuring a more sound and useful forecast product.

Two other case vignettes featured below, *the Yakima and Upper Colorado River basins*, reveal the inverse dimensions of this problem. In effect, what happens if forecast information proves to be incorrect in its predictions, because predictions turned out to be technically flawed, overly (or not sufficiently) conservative in their estimate of hazards, contradictory in the face of other information, or simply insufficiently sensitive to the audiences to whom forecasts were addressed?

As these cases suggest, given the different expectations and roles of scientists and decision makers, what constitutes credible information to a scientist involved in climate prediction or evaluation may differ from what is considered credible information by a decision maker. To a decision maker, forecast credibility is often perceived as hinging upon its certainty. The more certain and exact a forecast, the more trusted it will be by decision makers, and the more trust-worthy the developers of that information will be perceived. As shown below, improvements in forecast interpretation and translation, communication and institutional capacity to adjust to changing information and its consequences, are essential to addressing this problem. A basic characteristic of much forecast information is that even the best forecasts rarely approach close to absolute certainty of prediction—this issue is discussed in Section 3.3.2.

3.2.5.1 OTHER RELIABILITY AND TRUSTWORTHINESS ISSUES: THE NEED FOR HIGH RESOLUTION DATA

Research on the information needs of water decision makers has increasingly brought attention to the fact that use of climate-related decision-support tools is partly a function of the extent to which they can be made relevant to site-specific conditions and specific managerial resource needs, such as flow needs of aquatic species; the ability to forecast the impact of climate variability on orographic precipitation; and, the ability to fill in gaps in hydrologic monitoring (CDWR, 2007). In effect, proper integration of climate information into a water resource management context means developing high-resolution outputs able to be conveyed at the watershed level. It also means predicting changes in climate forecasts through the season and year, and regularly updating predictions. Specificity of forecast information can be as important as reliability for decision making at the basin and watershed level (CDWR, 2007). The Southwest drought case discussed in Section 3.2.3 illustrates the importance of information specificity in the context of water managers' responses, particularly within the Colorado River basin.

3.2.5.2 UNCERTAINTY IN THE REGULATORY PROCESS

While uncertainty is an inevitable part of the water resource decision makers' working environment, one source of lack of trust revolves around multi-level, multi-actor governance (Section 3.2.1). Shared governance for water management, coupled with the risk-averse character of traditional public works-type water agencies in particular, leads to situations where, while parties may act together for purposes of shared governance, "they may not have common goals or respond to common incentives"

What constitutes credible information to a scientist involved in climate prediction or evaluation may differ from what is considered credible information by a decision maker.

(NRC, 2008). Moreover, governance processes that cross various agencies, jurisdictions, and stakeholder interests are rarely straightforward, linear, or predictable because different actors are asked to provide information or resources peripheral to their central functions. In the absence of clear lines of authority, trust among actors and open lines of communication are essential (NRC, 2008).

As shown in Chapter 4 in the discussion of the *South Florida water management* case, a regulatory change introduced to guide water release decisions helped increase certainty and trust in the water allocation and management process. The South Florida Water Management District uses a Water Supply and Environment (WSE) schedule for Lake Okeechobee that employs seasonal and multi-seasonal climate outlooks as guidance for regulatory releases (Obeysekera *et al.*, 2007). The WSE schedule, in turn, uses ENSO and Atlantic Multi-decadal Oscillation (AMO; Enfield *et al.*, 2001) to estimate net inflow. The discussion of this case shows how regulatory changes initially intended to simply guide water release decisions can also help build greater certainty and trust in the water allocation and management process by making decisions predictable and transparent.

3.2.5.3 DATA PROBLEMS
Lack of information about geographical and temporal variability in climate processes is one of the primary barriers to adoption and use of specific products. An important dimension of this lack of information problem, relevant to discussions of reliability and trust, revolves around how decision makers make decisions when they have poor, no, or little data. Decision research from the social and behavioral sciences suggests

that when faced with such problems, individual decision makers typically omit or ignore key elements of good decision processes. This leads to decisions that are often ineffective in bringing about the results they intended (Slovic *et al.*, 1977). Furthermore, decision makers, such as water managers responsible for making flow or allocation decisions based on incomplete forecast data, may respond to complex tasks by employing professional judgment to simplify them in ways that seem adequate to the problem at hand, sometimes adopting "heuristic rules" that presume different levels of risk are acceptable based on their prior familiarity with a similar set of problems (Tversky and Kahneman, 1974; Payne *et al.*, 1993).

Decision makers and the public also may respond to probabilistic information or questions involving uncertainty with predictable biases that ignore or distort important information (Kahneman *et al.*, 1982) or exclude alternative scenarios and possible decisions (*e.g.*, Keeney, 1992; NRC, 2005). ENSO forecasts illustrate some of these problems[11]. Operational ENSO-based forecasts have only been made since the late 1980s while ENSO-related products that provide information about which forecasts are likely to be most reliable for what time periods and in which areas, have an even shorter history. Thus, decision-maker experience in their use has been limited. Essential knowledge for informed use of ENSO forecasts includes understanding of the temporal and geographical domain of ENSO impacts. Yet, making a decision based only on this information may expose a manager unnecessarily to consequences from that decision such as having to having to make costly decisions regarding supplying water to residents when expected rains from an ENSO event do not materialize.

3.2.5.4 CHANGING ENVIRONMENTAL, SOCIAL AND ECONOMIC CONDITIONS
Over the past three decades, a combination of economic changes (*e.g.*, reductions in federal spending for large water projects), environ-

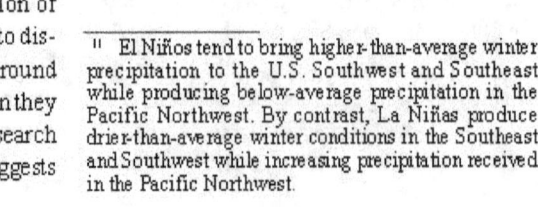

---

[11] El Niños tend to bring higher-than-average winter precipitation to the U.S. Southwest and Southeast while producing below-average precipitation in the Pacific Northwest. By contrast, La Niñas produce drier-than-average winter conditions in the Southeast and Southwest while increasing precipitation received in the Pacific Northwest.

mental conditions (e.g., demands for more non-structural measures to address water problems, population growth, and heightened emphasis on environmental restoration practices), and public demands for greater participation in water resource management have led to new approaches to water management. In Chapter 4 we address two of these approaches: adaptive management and integrated resource management. These approaches emphasize explicit commitment to environmentally-sound, socially-just outcomes; greater reliance upon drainage basins as planning units; program management via spatial and managerial flexibility, collaboration, participation, and peer-reviewed science (Hartig *et al.*, 1992; Landre and Knuth, 1993; Cortner and Moote, 1994; Water in the West, 1998; May *et al.*, 1996; McGinnis, 1995; Miller *et al.*, 1996; Cody, 1999; Bormann *et al.*, 1993; Lee, 1993). As shall be seen, these approaches place added demands on water managers regarding use of climate variability information, including adding new criteria to decision processes such as managing in-stream flows/low flows, climate variability impacts on runoff, water quality, fisheries, and water uses.

3.2.5.5 PUBLIC PERCEPTION AND POLITICS MAY OUTWEIGH FACTS AND PROFESSIONAL JUDGMENT

Climate variability and its risks are viewed through perceptual frames that affect not only decision makers and other policy elites, but members of the general public. Socialization and varying levels of education contribute to a social construction of risk information that may lead the public to view extreme climate variability as a sequence of events that may lead to catastrophe unless immediate action is taken (Weingart *et al.*, 2000). Extreme events may heighten the influence of sensational reporting, impede reliance upon professional judgment, lead to sensationalized reporting, and affect a sudden rise in public attention that may even shut off political discussion of the issue (Weingert *et al.*, 2000).

3.2.5.6 DECISION MAKERS MAY BE VULNERABLE WHEN THEY USE INFORMATION

Decision makers can lose their jobs, livelihoods, stature, or reputation by relying on forecasts that are wrong. Likewise, similar consequences can come about from untoward outcomes of decisions based on *correct* forecasts. This fact tends to make decision makers risk averse, and sometimes politically over-sensitive when using information, as noted in Chapter 4. As Jacobs (2002) notes in her review, much has been written on the reasons why decision makers and scientists rarely develop the types of relationships and information flows necessary for full integration of scientific knowledge into the decision-making process (Kirby, 2000; Pagano *et al.*, 2001; Pulwarty and Melis, 2001 Rayner *et al.*, 2005). The primary reasons are problems with relevance (are the scientists asking and answering the right questions?), accessibility of findings (are the data and the associated value-added analysis available to and understandable by the decision makers?), acceptability (are the findings seen as accurate and trustworthy?) conclusions being drawn from the data (is the analysis adequate?) and context (are the findings useful given the constraints in the decision process?).

Scientists have some authority to overcome some of these sources of uncertainty that result in distrust (e.g., diagnosing problems properly, providing adequate data, updating forecasts regularly, and drawing correct forecast conclusions). Other constraints on uncertainty, however, may be largely out of their control. Sensitivity to these sources of uncertainty, and their influence upon decision makers, is important.

The *Yakima case*, discussed earlier in the context of forecast credibility, further illustrates how decision makers can become vulnerable by relying on information that turns out to be inaccurate or a poor predictor of future climate variability events. It underscores the need for trust-building mechanisms to be built into forecast translation projects, such as issuing forecast confidence limits, communicating better with the public and agencies, and considering the consequences of potential actions taken by users in the event of an erroneous forecast. The next section discusses particular challenges related to translation.

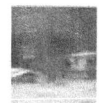

Decision makers can lose their jobs, livelihoods, stature, or reputation by relying on forecasts that are wrong.

3.3 WHAT ARE THE CHALLENGES IN FOSTERING COLLABORATION BETWEEN SCIENTISTS AND DECISION-MAKERS?

This Section examines problems in translating climate forecasts and hydrology information into integrated water management decisions, forecast communication, and operationalizing decision-support systems. This discussion focuses on translation of scientific information into forms useful and useable by decision makers.

3.3.1 General Problems in Fostering Collaboration

The social and decision sciences have learned a great deal about the obstacles, impediments, and challenges in translating scientific information, especially forecasts, for decision makers generally, and resource managers in particular. Simply "doing research" on a problem does not assure in any way that the research results can or will contribute to solving a societal problem; likewise "more research does not necessarily lead to better decisions" (e.g., Cash et al., 2003; Jacobs et al., 2005; Sarewitz and Pielke, 2007; Rayner et al., 2005). Among the principal reasons information may not be used by decision makers are that they do fit the setting or timing in which the decision occurs and that there are external constraints that preclude its use. A further explanation follows.

The information may be viewed as irrelevant to the user or inappropriate to the decision context: While scientists' worldviews are strongly influenced and affected by the boundaries of their own research and disciplines, decision makers' worldviews are conditioned by the "decision space" (Jacobs et al., 2005). Decision space refers to the range of realistic options available to a given decision maker to resolve a particular problem. While a new scientifically-derived tool or source of information may have obvious applications when viewed from a theoretical perspective, a decision maker may be constrained from using a tool or information by external factors.

External constraints such as laws and regulations may limit the range of options available

to the decision maker: Policies, procedures, and precedents relevant to a given decision—including decisional rules and protocols, expectations imposed by decision makers through training and by peer and supervisory expectations, sufficiency of resources (e.g., time and money) within organizations to properly integrate information and tools into decision making, and the practicality of implementing various options prescribed by tools and/or information given the key questions the decision maker must manage on a daily basis—are all factors that limit decision makers' use of information. These factors can also limit the range of options available to decision makers.

Political scientists who study administrative organizations cite three principal ways the rule-making culture of administrative organizations hinders information use, ranging from the nature of policy "attentiveness" in administrative organizations in which awareness of alternatives is often driven by demands of elected officials instead of newly available information (e.g., Kingdon, 1995), to organizational goals and objectives which often frame or restrict the flow of information and "feedback". Another set of reasons revolves around the nature of indirect commands within organizations that evolve through trial and error. Over time, these commands take the form of rules and protocols which guide and prescribe appropriate and inappropriate ways of using information in bureaucracies (Stone, 1997; Torgerson, 2005).

The following case, relating to the translation of drought information in the southeastern United States, describes the influence of institutional constraints on information use. In this instance, the problem of drought is nested within a larger regional water dispute among three states. By describing the challenges in incorporating drought and water shortage information into basin-wide water planning, this case also helps clarify a number of salient problems faced by water managers working with complex information in a contentious political or legal context. In short, information usefulness is determined in part by social and political context or "robustness". To be "socially robust", information must first be valid outside, as well as inside, the laboratory where it is developed; and secondly,

Simply "doing research" on a problem does not assure in any way that the research results can or will contribute to solving a societal problem; likewise "more research does not necessarily lead to better decisions".

it must involve an extended group of experts, including lay "experts" (Gibbons, 1999).

Case Study: The Southeast Drought: Another Perspective on Water Problems in the Southeastern United States

Introduction and context
As mentioned earlier, drought risk consists of a hazard component (e.g., lack of precipitation, along with direct and indirect effects on runoff, lake levels and other relevant parameters) and a vulnerability component. Some aspects of vulnerability include the condition of physical infrastructure; economics, awareness and preparedness; institutional capability and flexibility; policy, demography, and access to technology (Wilhite *et al.*, 2000). Thus, there are clearly non-climatic factors that can enhance or decrease the likelihood of drought impacts. Laws, institutions, policies, procedures, precedents and regulations, for instance, may limit the range of options available to the decision maker, even if he or she is armed with a perfect forecast.

In the case of the ongoing drought in the southeastern United States, the most recent episode, beginning in 2006 and intensifying in 2007 (see Box Figure 3.1), impacts to agriculture, fisheries, and municipal water supplies were likely exacerbated by a lack of action on water resources compacts between Georgia, Alabama, and Florida (Feldman, 2007). The hazard component was continuously monitored at the state, regional, and national level by a variety of institutions, including state climatologists, the Southeast Regional Climate Center, the Southeast Climate Consortium, the USGS, the NWS, the U.S. Drought Monitor and others. In some cases, clear decision points were specified by state drought plans (Steinemann and Cavalcanti, 2006; Georgia DNR, 2003). (Florida lacks a state drought plan.) During the spring of 2007 the situation worsened as record precipitation deficits mounted, water supplies declined, and drought impacts, including record-setting wildland fires, accumulated (Georgia Forestry Commission, 2007). Georgia decision makers faced the option of relying on a forecast for above-average Atlantic hurricane frequency, or taking more cautious, but decisive, action to stanch potentially critical water shortages.

Public officials allowed water compacts to expire, because they could not agree on water allocation formulae. As a result, unresolved conflicts regarding the relative priorities of upstream and downstream water users (e.g., streamflows intended to preserve endangered species and enrich coastal estuaries vied for the same water as reservoir holdings intended to drought-proof urban water uses) impeded the effective application of climate information to mitigate potential impacts.

The Apalachicola–Chattahoochee–Flint River basin compact negotiations
The Apalachicola–Chattahoochee–Flint River Basin Compact was formed to address the growing demands for water in the region's largest city, Atlanta, while at the same time balancing off-stream demands of other users against in-stream needs to support fisheries and minimum flows for water quality (Hull, 2000). While the basin is rapidly urbanizing, farming, and the rural communities that depend upon it, remain important parts of the region's economy. Conflicts between Georgia, Florida, and Alabama over water rights in the basin began in the late 1800s. Today, metro-Atlanta daily draws more than 400 million gallons of water from the river and discharges into it more than 300 million gallons of wastewater.

Following protracted drought in the region in the 1990s, decision makers in Alabama, Florida, and Georgia dedicated themselves to avoiding lengthy and expensive litigation that likely would have led to a decision that would have pleased no one. In 1990, the three states began an 18-month negotiation process that resulted, first, in a *Letter of Agreement* (April, 1991) to address short term issues in the basin and then, in January 1992, a *Memorandum of Agreement* that, among other things, stated that the three states were in accord on the need for a study of the water needs of the three states. The three states' governors also agreed to initiate a comprehensive study by the Army Corps of Engineers (Kundell and Tetens, 1998).

At the conclusion of the 1998 compact summit, chaired by former Representative Gingrich, the three states agreed to: protect federal regulatory discretion and water rights; assure public participation in allocation decisions; consider

> Drought risk consists of a hazard component and a vulnerability component. Some aspects of vulnerability include the condition of physical infrastructure; economics, awareness and preparedness; institutional capability and flexibility; policy, demography, and access to technology.

environmental impacts in allocation; and develop specific allocation numbers—in effect, guaranteeing volumes "at the state lines". Water allocation formulas were to be developed and agreed upon by December 31, 1998. However, negotiators for the three states requested at least a one-year extension of this deadline in November of 1998, and several extensions and requests for extensions have subsequently been granted over the past dozen years, often at the 11th hour of stalemated negotiations.

Opportunities for a breakthrough came in 2003. Georgia's chief negotiator claimed that the formulas posted by Georgia and Florida, while different, were similar enough to allow the former to accept Florida's numbers and to work to resolve language differences in the terms and conditions of the formula. Alabama representatives concurred that the numbers were workable and that differences could be resolved. Nonetheless, within days of this tentative settlement, negotiations broke off once again (Georgia Environmental Protection Division, 2002). In August 2003, Governors Riley, Bush, and Perdue from Alabama, Florida, and Georgia, respectively, signed a memorandum of understanding detailing the principles for allocating water for the ACF over the next 40 years; however, as of this writing, Georgia has lost an appeal in the Appellate Court of the District of Columbia to withdraw as much water as it had planned to do, lending further uncertainty to this dispute (Goodman, 2008).

Policy impasse

Three issues appear to be paramount in the failure to reach accord. First, various demands imposed on the river system may be incompatible, such as protecting in-stream flow while permitting varied off-stream uses. Second, many of the prominent user conflicts facing the three states are up- *versus* down-stream disputes. For example, Atlanta is a major user of the Chattahoochee. However, it is also a "headwaters" metropolis. The same water used by Atlanta for water supply and wastewater discharge is used by "up-streamers" for recreation and to provide shoreline amenities such as high lake levels for homes (true especially along the shoreline of Lake Lanier), and provides downstream water supply to other communities. Without adequate drawdown from Lanier,

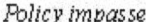

for example, water supplies may be inadequate to provide for all of Atlanta's needs. Likewise, water quality may be severely degraded because of the inability to adequately dilute pollution discharges from point and non-point sources around Atlanta. This is especially true if in-stream water volumes decline due to growing off-stream demands.

Finally, the compact negotiating process itself lacks robustness; technically, the compact does not actually take effect until an allocation formula can be agreed upon. Thus, instead of agreeing on an institutional framework that can collect, analyze, translate, and use information to reach accord over allocation limits and water uses, the negotiations have been targeted on first determining a formula for allocation based on need (Feldman, 2007). As we have seen in the previous case on drought management in Georgia, climate forecast information is being used to enhance drought preparedness and impact mitigation. Nevertheless, as noted in that case, conservation measures in one state alone cannot mitigate region-wide problems affecting large, multi-state watersheds. The same holds true for regional water supply dispute-resolution. Until a cooperative decision-making platform emerges whereby regional climate forecast data can be used for conjoint drought planning, water allocation prescriptions, and incorporation of

Conservation measures in one state alone cannot mitigate region-wide problems affecting large, multi-state watersheds.

regional population and economic growth (not currently done on an individual state-level), effective use of decision-support information (*i.e.*, transformation) will remain an elusive goal.

3.3.1.1 RESEARCHERS OFTEN DEVELOP PRODUCTS AND TOOLS THAT THEY BELIEVE WILL BE USEFUL, AND MAKE THEM AVAILABLE FOR USE WITHOUT VERIFYING WHETHER THEY ARE NEEDED

This is sometimes referred to as the "loading dock" phenomenon (Cash *et al.*, 2006). It generally results from one-way communication, without sufficient evaluation of the needs of stakeholders. The challenge of integrating information and tools into decision making is a problem endemic to all societies, particularly, as this Product presents, in the case of climate variability and water management. Developing nations are faced with the additional impediment of facing these problems without adequate resources. The following case study of Northeast Brazil is one example of this struggle.

Case Study: Policy learning and seasonal climate forecasting application in Northeast Brazil—integrating information into decisions

Introduction
The story of climate variability forecast application in the state of Ceará (Northeast Brazil) chronicles a policy process in which managers have deployed seasonal climate forecasting experimentally for over ten years for water and agriculture, and have slowly learned different ways in which seasonal forecasting works, does not work, and could be improved for decision making (Lemos *et al.*, 2002; Lemos, 2003; Lemos and Oliveira, 2004; Taddei 2005; Pfaff *et al.*, 1999).

The *Hora de Plantar* ("Time to Plant") Program, begun in 1988, aimed at distributing high-quality, selected seed to poor subsistence farmers in Ceará and at maintaining a strict planting calendar to decrease rain-fed farmers sensitivity to climate variability (Lemos, 2003). In exchange for selected seeds, farmers "paid" back the government with grain harvested during the previous season or received credit to be paid the following year. The rationale for the program was to provide farmers with high quality seeds (corn, beans, rice, and cotton), but to distribute them only when planting conditions were appropriate. Because farmers tend to plant with the first rains (sometimes called the "pre-season") and often have to replant, the goal of this program was to use a simplified soil/climate model, developed by the state meteorology agency (FUNCEME) to orient farmers with regard to the actual onset of the rainy season (Andrade, 1995).

While the program was deemed a success (Golnaraghi and Kaul, 1995), a closer look revealed many drawbacks. First, it was plagued by a series of logistical and enforcement problems (transportation and storage of seed, lack of enough distribution centers, poor access to information and seeds by those most in need, fraud, outdated client lists) (Lemos *et al.*, 1999). Second, local and lay knowledge accumulated for years to inform its design was initially ignored. Instead, the program relied on a model of knowledge use that privileged the use of technical information imposed on the farmers in an exclusionary and insulated form that alienated stakeholders and hampered buy-in from clients (Lemos, 2003). Third, farmers strongly resented *Hora de Plantar*'s planting calendar and its imposition over their own best judgment. Finally, there was the widespread perception among farmers (and confirmed by a few bank managers) that a "bad" forecast negatively affected the availability of rural credit (Lemos *et al.*, 1999). While many of the reasons farmers disliked the program had little to do with climate forecasting, the overall perception was that FUNCEME was to blame for its negative impact on their livelihoods (Lemos *et al.*, 2002; Lemos, 2003; Meinke *et al.*, 2006). As a result, there was both a backlash against the program and a relative discredit of FUNCEME as a technical agency and of the forecast by association. The program is still active, although by 2002, the strict coupling of seed distribution and the planting calendar had been phased out (Lemos, 2003).

In 1992, as part of Ceará's modernizing government administration, and in response to a long period of drought, the State enacted Law 11.996 that defined its policy for water resources management. This new law created several levels of water management, including watershed

The challenge of integrating information and tools into decision making is a problem endemic to all societies, particularly, in the case of climate variability and water management.

Users' Commissions, Watershed Committees and a state level Water Resources Council. The law also defined the watershed as the planning unit of action; spelled out the instruments of allocation of water permits and fees for the use of water resources; and regulated further construction in the context of the watershed (Lemos and Oliveira, 2004; Formiga-Johnsson and Kemper, 2005; Pfaff *et al.*, 1999).

Innovation—Using Information More Effectively

One of the most innovative aspects of water reform in Ceará was creation of an interdisciplinary group within the state water management agency (COGERH) to develop and implement reforms. The inclusion of social and physical scientists within the agency allowed for the combination of ideas and technologies that critically affected the way the network of *técnicos* and their supporters went about implementing water reform in the State. From the start, COGERH sought to engage stakeholders, taking advantage of previous political and social organization within the different basins to create new water organizations (Lemos and Oliveira, 2005). In the Lower Jaguaribe-Banabuiú River basin, for example, the implementation of participatory councils went further than the suggested framework of River Basin Committees to include the Users Commission to negotiate water allocation among different users directly (Garjulli, 2001; Lemos and Oliveira, 2004; Taddei, 2005; Pfaff *et al.*, 1999). COGERH *técnicos* specifically created the Commission independently of the "official" state structure to emphasize their autonomy *vis-à-vis* the State (Lemos and Oliveira, 2005). This agenda openly challenged a pattern of exclusionary water policymaking prevalent in Ceará and was a substantial departure from the top-down, insulated manner of water allocation in the past (Lemos and Oliveira, 2004). The ability of these *técnicos* to implement the most innovative aspects of the Ceará reform can be explained partly by their insertion into policy networks that were instrumental in overcoming the opposition of more conservative sectors of the state apparatus and their supporters in the water user community (Lemos and Oliveira, 2004).

The role of knowledge in building adaptive capacity in the system was also important because it helped democratize decision making. In Ceará, the organization of stakeholder councils and the effort to use technical knowledge, especially reservoir scenarios to inform water release, may have enhanced the system's adaptive capacity to climate variability as well as improved water resources sustainability (Formiga-Johnson and Kemper, 2005; Engle, 2007). In a recent evaluation of the role of governance institutions in influencing adaptive capacity building in two basins in northeastern Brazil (Lower Jaguaribe in Ceará and Pirapama in Pernambuco), Engle (2007) found that water reform played a critical role in increasing adaptive capacity across the two basins. And while the use of seasonal climate knowledge has been limited so far (the scenarios assume zero inflows from future rainfall), there is great potential that use of seasonal forecasts could affect several aspects of water management and use in the region and increase forecast value.

In the context of Ceará's Users Commissions, the advantages are twofold. First, by making simplified reservoir models available to users, COGERH is not only enhancing public knowledge about the river basin but also is crystallizing the idea of collective risk. While individual users may be willing to go along with the status quo, collective decision-making processes may be much more effective in curbing overuse. Second, information can play a critical role in democratization of decision making at the river basin level by training users to make decisions, and dispelling the widespread distrust that has developed as a result of previous applications of climate information. Finally, the case suggests that incorporating social science into processes that are being designed to optimize the use of climate forecast tools in specific water management contexts can enhance outcomes by helping poorer communities better adapt to, and build capacity for, managing climate variability impacts on water resources. Building social capital can be advantageous for other environmental issues as well, including an increasing likelihood of public attentiveness, participation, awareness, and engagement in monitoring of impacts.

Incorporating social science into processes that are being designed to optimize the use of climate forecast tools in specific water management contexts can enhance outcomes by helping poorer communities better adapt to, and build capacity for, managing climate variability impacts on water resources.

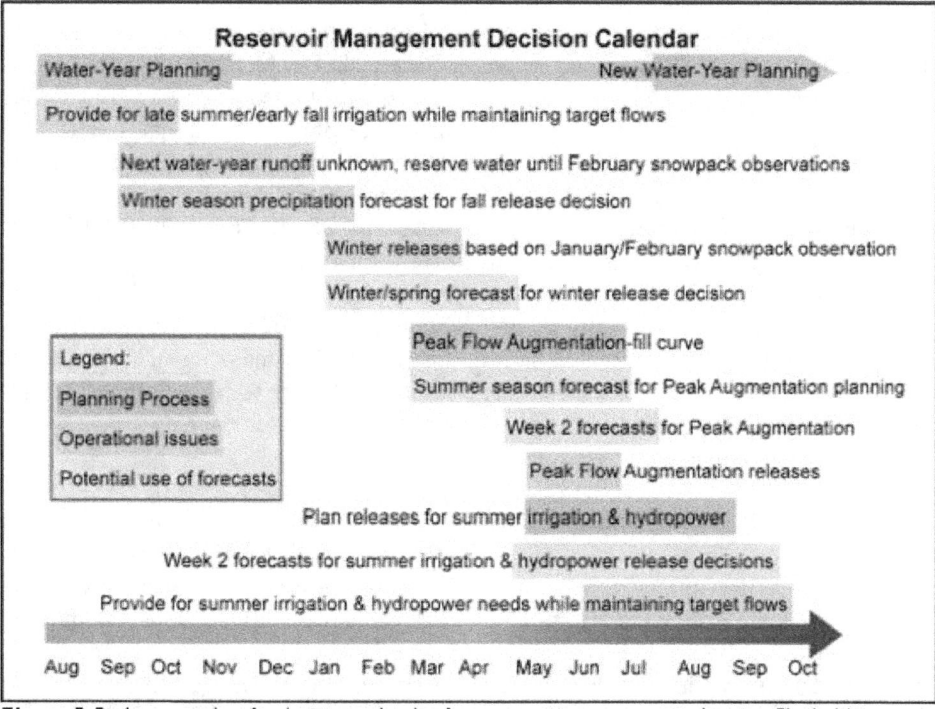

Figure 3.3 An example of a decision calendar for reservoir management planning. Shaded bars indicate the timing of information needs for planning and operational issues over the year (Source: Ray and Webb, 2000).

3.3.1.2 INFORMATION MAY NOT BE AVAILABLE AT THE TIME IT COULD BE USEFUL

It is well established in the climate science community that information must be timely in order to be useful to decision makers. This requires that researchers understand and be responsive to the time frames during the year for which specific types of decisions are made. Pulwarty and Melis (2001), Ray and Webb (2000), and Wiener *et al.* (2000) have developed and introduced the concept of "decision calendars" in the context of the Western Water Assessment in Boulder, Colorado (Figure 3.3). Failure to provide information at a time when it can be inserted into the annual series of decisions made in managing water levels in reservoirs, for example, may result in the information losing virtually all of its value to the decision maker. Likewise, decision makers need to understand the types of predictions that can be made and trade-offs between longer-term predictions of information at the local or regional scale and potential decreases in accuracy. They also need to help scientists in formulating research questions.

The importance of leadership in initiating change cannot be overstated (Chapter 4), and its importance in facilitating information ex-

change is also essential; making connections with on-the-ground operational personnel and data managers in order to facilitate information exchange is of particular importance. The presence of a "champion" within stakeholder groups or agencies may make the difference in successful integration of new information. Identifying people with leadership qualities and working through them will facilitate adoption of new applications and techniques. Recently-hired water managers have been found to be more likely to take risks and deviate from precedent and "craft skills" that are unique to a particular water organization (Rayner *et al.*, 2005).

The following vignette on the Advanced Hydrologic Prediction System (AHPS), established in 1997, exemplifies a conscious effort by the National Weather Service to respond to many of these chronic relational problems in a decisional context. AHPS is an effort to go beyond traditional river stage forecasts which are short-term (one to three days), and are the product of applied historical weather data, stream gage data, channel cross-section data, water supply operations information, and hydrologic model characteristics representing large regions. It is an effort that has worked, in part, because it has

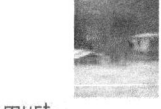

Information must be timely in order to be useful to decision makers. This requires that researchers understand and be responsive to the time frames during the year for which specific types of decisions are made.

many "champions"; however, questions remain about whether resources for the initiative have been adequate.

AHPS responds directly to the problem of timely information availability by trying to provide forecasting information sooner, particularly on potential flooding; linking it directly to local decision makers, providing the information in a visual format; and, perhaps most of all, providing a dedicated program within NOAA (and the NWS) that has the capacity to work directly with the user community and monitor ongoing, evolving decision-support needs.

Vignette: AHPS—Advantages over conventional forecasting

Applying the same hydrologic data used in current methods, AHPS also employs advanced hydrologic models with characteristics specific to local watersheds and tributaries. These advanced, localized hydrologic models increase forecast accuracy by 20 percent over existing models. Its outputs are more accurate, detailed, and visually oriented, and are able to provide decision makers and the public with information on, among other variables: how high a river will rise, when it will reach its peak, where properties will be subject to flooding, and how long a flood event will continue. It is estimated that national implementation of AHPS will save at least $200 million per year in reduced flood losses and contribute an additional $400 million a year in economic benefits to water resource users (Advanced Hydrologic Prediction Service/ <http://www.state.nj.us/drbc/Flood_Website/AHPS.htm>).

Benefits and application
AHPS provides detailed products in an improved format. Because it is visually oriented, it provides information in a format that is easier to understand and use by the general public as well as planners and scientists. AHPS depicts the magnitude and probability of hydrologic events, and gives users an idea of worst case scenario situations. Finally, AHPS provides forecasts farther in advance of current methods, allowing people additional time to protect themselves, their families, and their property from floods.

Following the Great Flood of 1993 in the Midwest, the Des Moines River Basin in Iowa was selected to be a location to test for the first phase toward national implementation of AHPS. Residents, via the Internet, can now access interactive maps displaying flood forecast points. Selecting any of the flood forecast points on the map allows Internet users to obtain river stage forecast information for the point of interest. Available information includes: river flood stages, flow and volume probabilities, site maps, and damage tables projecting areas are likely to be subject to flooding.

Status and assessment
A 2006 NRC report found AHPS to be an ambitious climate forecast program that promises to provide services and products that are timely and necessary. However, it expressed concerns about "human and fiscal resources", recommending that there is a need for trained hydrologic scientists to conduct hydrologic work in the NWS. Regarding fiscal resources, "the budgetary history and current allocation seem misaligned with the ambitious goals of the program". Thus, the program's goals and budget should be brought into closer alignment (NRC, 2006).

3.3.2 Scientists Need to Communicate Better and Decision-Makers Need a Better Understanding of Uncertainty—it is Embedded in Science

Discussions of uncertainty are at the center of many debates about forecast information and its usefulness. Uncertainties result from: the relevance and reliability of data, the appropriateness of theories used to structure analyses, the completeness of the specification of the problem, and in the "fit" between a forecast and the social and political matters of fact on the ground (NRC, 2005). While few would disagree that uncertainties are inevitable, there is less agreement as to how to improve ways of describing uncertainties in forecasts to provide widespread benefits (NRC, 2005). It is important to recognize that expectations of certainty are unrealistic in regards to climate variability. Weather forecasts are only estimates; the risk tolerance (Section 3.2.3) of the public is often unrealistically low. As we have seen in multiple cases, one mistaken forecast (e.g., the Yakima basin case) can have an impact out of proportion

While few would disagree that uncertainties are inevitable, there is less agreement as to how to improve ways of describing uncertainties in forecasts to provide widespread benefits.

to the gravity of its consequences. Some starting points from the literature include helping decision makers understand that uncertainty does not make a forecast scientifically flawed, only imperfect. Along these lines, decision makers must understand the types of predictions that can be made and trade-offs between predictions of information at the local or regional scale that are less accurate than larger scale predictions (Jacobs *et al.*, 2005). They also need to help scientists formulate research questions that result in relevant decision-support tools.

Second, uncertainty is not only inevitable, but necessary and desirable. It helps to advance and motivate scientific efforts to refine data, analysis, and forecaster skills; replicate research results; and revise previous studies, especially through peer review (discussed below) and improved observation. As one observer has noted, "(un)certainty is not the hallmark of bad science, it is the hallmark of honest science (when) we know enough to act is inherently a policy question, not a scientific one" (Brown, 1997).

Finally, the characterization of uncertainty should consider the decision relevance of different aspects of the uncertainties. Failure to appreciate such uncertainties results in poor decisions, misinterpretation of forecasts, and diminished trust of analysts. Considerable work on uncertainty in environmental assessments and models make this topic ripe for progress (*e.g.*, NRC, 1999).

Vignette: Interpreting Climate Forecasts—uncertainties and temporal variability

Introduction
Lack of information about geographical and temporal variability in climate processes is one of the primary barriers to adoption and use of specific products. ENSO forecasts are an excellent example of this issue. While today El Niño (EN) and La Niña (LN) are part of the public vocabulary, operational ENSO-based forecasts have only been made since the late 1980s. Yet, making a decision based only on the forecasts themselves may expose a manager to unanticipated consequences. Additional information can mitigate such risk. ENSO-related ancillary products, such as those illustrated in Figures 3.4

and 3.5, can provide information about which forecasts are likely to be most reliable for what time periods and in which areas. As Figure 3.4 shows, informed use of ENSO forecasts requires understanding of the temporal and geographical domain of ENSO impacts. EN events tend to bring higher than average winter precipitation to the U.S. Southwest and Southeast while producing below-average precipitation in the Pacific Northwest. LN events are the converse, producing above-average precipitation in the Pacific Northwest and drier patterns across the southern parts of the country. Further, not all ENs or LNs are the same with regard to the amount of precipitation they produce. As illustrated in Figure 3.6, which provides this kind of information for Arizona, the EN phase of ENSO tends to produce above-average winter precipitation less dependably than the LN phase produces below-average winter precipitation.

An example of the value of combining ENSO forecasts with information about how ENSO tended to affect local systems arose during the 1997/1998 ENSO event. In this case, the Arizona-based Salt River Project (SRP) made a series of decisions based on the 1997/1998 EN forecast plus analysis of how ENs tended to affect their system of rivers and reservoirs. Knowing that ENs tended to produce larger streamflows late in the winter season, SRP managers reduced groundwater pumping in August 1997 in anticipation of a wet winter. Their contingency plan called for resuming groundwater pumping if increased streamflows did not materialize by March 1, 1998. As the winter progressed, it became apparent that the EN had produced a wet winter and plentiful water supplies in SRP's reservoirs. The long-lead decision to defer groundwater pumping in this instance saved SRP $1 million (Pagano *et al.*, 2001). SRP was uniquely well positioned to take this kind of risk because the managers making the decisions had the support of upper-level administrators and because the organization had unusually straightforward access to information. First, a NWS office is co-located in the SRP administrative headquarters, and second, key decision makers had been interacting regularly with climate and hydrology experts associated with the NOAA-funded Climate Assessment for the Southwest (CLIMAS) project, located at the University of Arizona. Relatively few decision

Uncertainty helps to advance and motivate scientific efforts to refine data, analysis, and forecaster skills; replicate research results; and revise previous studies, especially through peer review and improved observation.

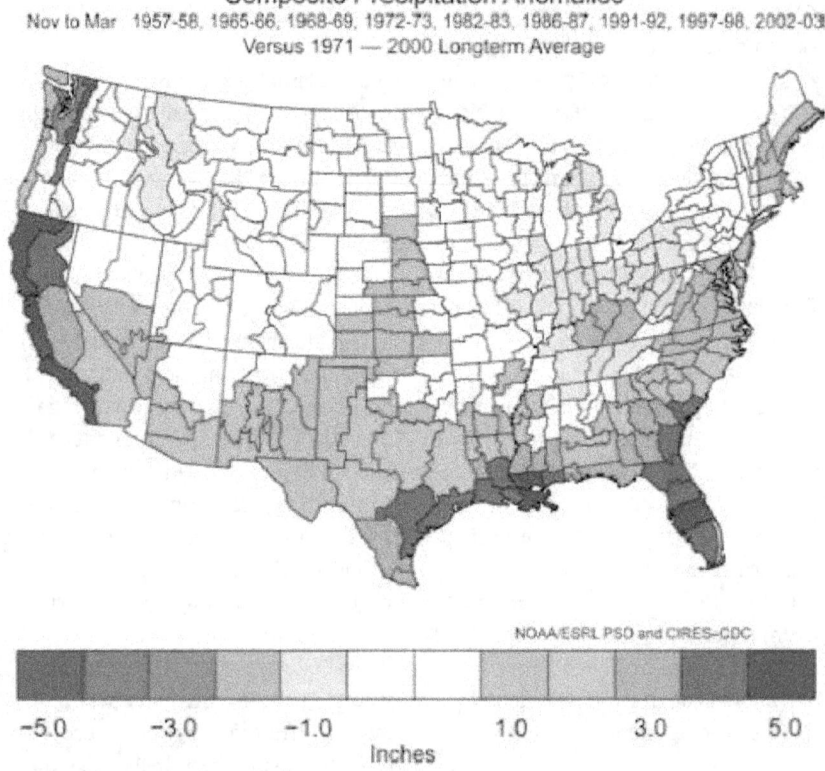

Figure 3.4 El Niño precipitation anomalies in inches (Source: NOAA Earth System Research Laboratory)

makers have this level of support for using climate forecasts and associated information. The absence of such support systems may increase managers' exposure to risk, in turn generating a strong disincentive to use climate forecasts.

3.4 SUMMARY

Decision-support systems are not often well integrated into policy networks to support planning and management, making it difficult to convey information. Among the reasons for this are a tendency toward institutional conservatism by water agencies, a decision-making climate that discourages innovation, lack of national-scale coordination of decisions, difficulties in providing support for decisions at varying spatial and temporal scales due to vast variability in "target audiences" for products, and growing recognition that rational choice models of information transfer are overly simplistic. The case of information use in response to Georgia's recent drought brings to light problems that students of water decision making have long described about resistance to innovation.

Ensuring information relevance requires overcoming the barriers of over-specialization by encouraging inter-disciplinary collaboration in product and tool development. Decision makers need to learn to appreciate the inevitability and desirability of forecast uncertainties at a regional scale on the one hand, and potential decreases in accuracy on the other. Scientists must understand both internal institutional impediments (agency rules and regulations) as well as external ones (e.g., political-level conflicts over water allocation as exemplified in the Southeast United States, asymmetries in information access in the case of Northeast Brazil) as factors constraining decision-support translation and decision transformation. While the nine cases discussed here have been useful and instructive, more generalizable findings are needed in order to develop a strong, theoretically-grounded understanding of processes that facilitate information dissemination, communication, use, and evaluation—and to predict effective methods of boundary spanning between decision makers and information generators. We discuss this set of problems in Chapter 4.

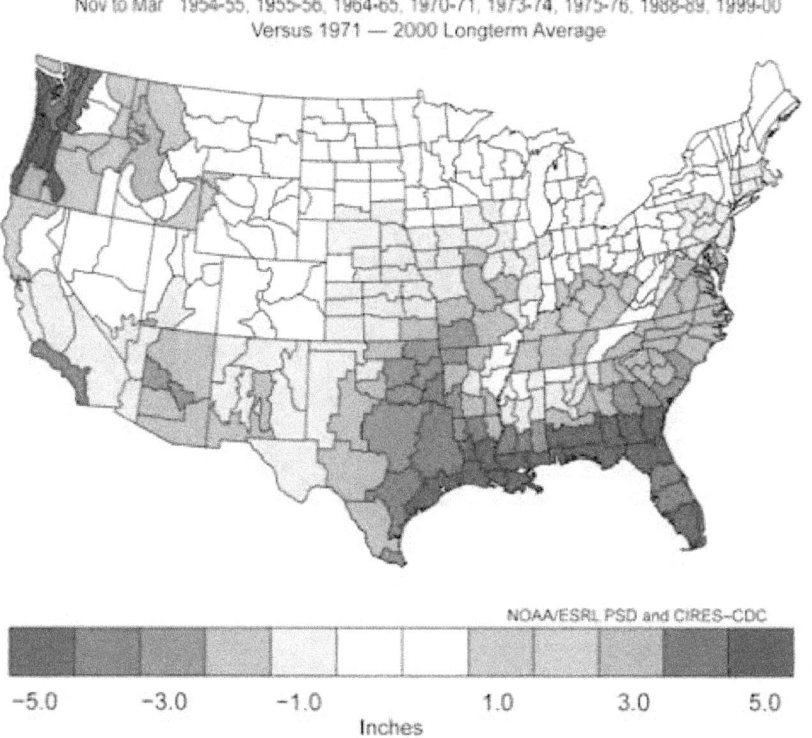

Figure 3.5 La Niña precipitation anomalies in inches (Source: NOAA Earth System Research Laboratory)

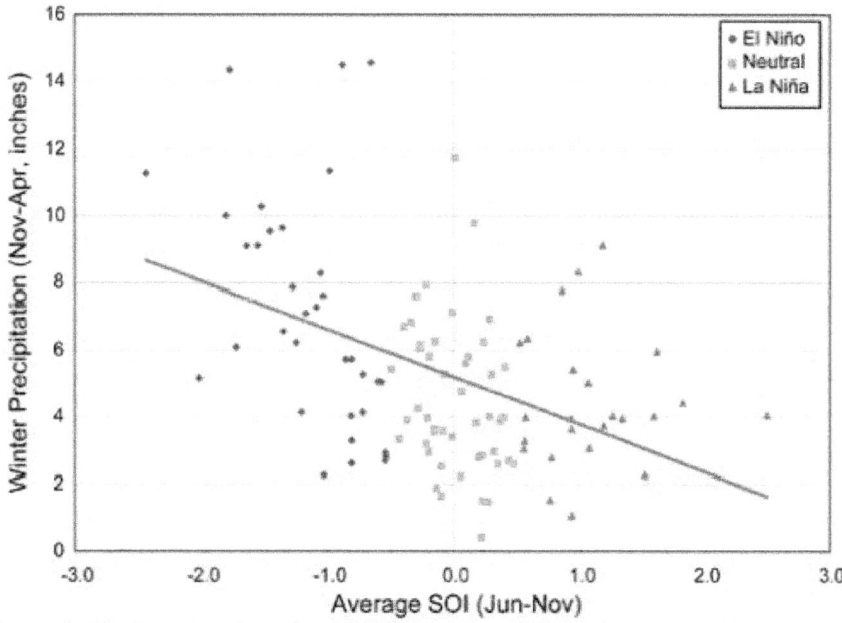

Figure 3.6 Southern Oscillation Index (SOI) June through November, *versus* Winter precipitation November through April for 1896 to 2001 for three phases of ENSO; El Niño, La Niña, and Neutral, for Arizona climate division 6. Note the greater variation in El Niño precipitation (blue) than in La Niña precipitation (red).

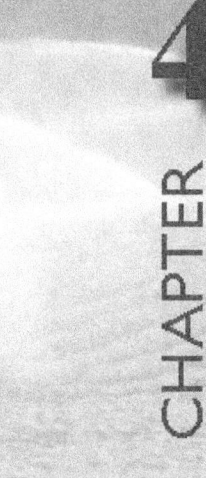

CHAPTER 4

Making Decision-Support Information Useful, Useable, and Responsive to Decision-Maker Needs

Convening Lead Authors: David L. Feldman, Univ. of California, Irvine; Katharine L. Jacobs, Arizona Water Institute

Lead Authors: Gregg Garfin, Univ. of Arizona; Aris Georgakakos, Georgia Inst. of Tech.; Barbara Morehouse, Univ. of Arizona; Pedro Restrepo, NOAA; Robin Webb, NOAA; Brent Yarnal, Penn. State Univ.

Contributing Authors: Dan Basketfield, Silverado Gold Mines Inc.; Holly Hartmann, Univ. of Arizona; John Kochendorfer, Riverside Technology, Inc.; Cynthia Rosenzweig, NASA; Michael Sale, ORNL; Brad Udall, NOAA; Connie Woodhouse, Univ. of Arizona

KEY FINDINGS

Decision-support experiments that apply seasonal and interannual climate variability information to basin and regional water resource problems serve as test beds that address diverse issues faced by decision makers and scientists. They illustrate how to identify user needs, overcome communication barriers, and operationalize forecast tools. They also demonstrate how user participation can be incorporated into tool development.

Five major lessons emerge from these experiments and supporting analytical studies:

- The effective integration of seasonal-to-interannual climate information in decisions requires long-term collaborative research and application of decision support through identifying problems of mutual interest. This collaboration will require a critical mass of scientists and decision makers to succeed and there is currently an insufficient number of "integrators" of climate information for specific applications.

- Investments in long-term research-based relationships between scientists and decision makers must be adequately funded and supported. In general, progress on developing effective decision-support systems is dependent on additional public and private resources to facilitate better networking among decision makers and scientists at all levels as well as public engagement in the fabric of decision making.

- Effective decision-support tools must integrate national production of data and technologies to ensure efficient, cross-sector usefulness with customized products for local users. This requires that tool developers engage a wide range of participants, including those who generate tools and those who translate them, to ensure that specially-tailored products are widely accessible and are immediately adopted by users insuring relevancy and utility.

- The process of tool development must be inclusive, interdisciplinary, and provide ample dialogue among researchers and users. To achieve this inclusive process, professional reward systems that recognize people who develop, use and translate such systems for use by others are needed within water management and related agencies, universities and organizations. Critical to this effort, further progress is needed in boundary spanning—the effort to translate tools to a variety of audiences across institutional boundaries.

- Information generated by decision-support tools must be implementable in the short term for users to foresee progress and support further tool development. Thus, efforts must be made to effectively integrate public concerns and elicit public information through dedicated outreach programs.

4.1 INTRODUCTION

This Chapter examines a series of decision-support experiments that explore how information on seasonal-to-interannual (SI) climate variability is being used, and how various water management contexts serve as test beds for implementing decision-support outputs. We describe how these experiments are implemented and how SI climate information is used to assess potential impacts of and responses to climate variability and change. We also examine characteristics of effective decision-support systems, involving users in forecast and other tool development, and incorporating improvements.

Section 4.2 discusses a series of experiments from across the nation, and in a variety of contexts. Special attention is paid to the role of key leadership in organizations to empower employees, take risks, and promote inclusiveness. This Section highlights the role of organizational culture in building pathways for innovation related to boundary-spanning approaches.

Section 4.3 examines approaches to increasing user knowledge and enhancing capacity building. We discuss the role of two-way communication among multiple forecast and water resource sectors, and the importance of translation and integration skills, as well as operations staff incentives for facilitating such integration.

Section 4.4 discusses the development of measurable indicators of progress in promoting climate information access and effective use, including process measures such as consultations between agencies and potential forecast user communities. The role of efforts to enhance dialogue and exchange among researchers and users is emphasized.

Finally, Section 4.5 summarizes major findings, directions for further research, and recommendations, including: needs for better understanding of the role of decision-maker context for tool use, how to assess vulnerability to climate, communicating results to users, bottom-up as well as top-down approaches to boundary-spanning innovation,

and applicability of lessons from other resource management sectors (e.g., forestry, coastal zone management, hydropower) on decision-support use and decision maker/scientist collaboration.

We conclude that, at present, the weak conceptual grounding afforded by cases from the literature necessitates that we base measures to improve decision support for the water resources management sector, as it pertains to inclusion of climate forecasts and information, on best judgment extrapolated from case experience. Additional research is needed on effective models of boundary spanning in order to develop a strong, theoretically-grounded understanding of the processes that facilitate information dissemination, communication, use, and evaluation so that it is possible to generalize beyond single cases, and to have predictive value.

4.2 DECISION-SUPPORT TOOLS FOR CLIMATE FORECASTS: SERVING END-USER NEEDS, PROMOTING USER-ENGAGEMENT AND ACCESSIBILITY

This Section examines a series of decision-support experiments from across the United States. Our objective is to learn how the barriers to optimal decision making, including impediments to trust, user confidence, communication of information, product translation, operationalization of decision-support tools, and policy transformation discussed in Chapter 3, can be overcome. As shall be seen, all of these experiments share one characteristic: users have been involved, to some degree, in tool development—through active elicitation of their needs, involvement in tool design, evaluation of tool effectiveness (and feedback into product refinement as a result of tool use), or some combination of factors.

4.2.1 Decision-Support Experiments on Seasonal-to-Interannual Climate Variability

The following seven cases are important testbeds that examine how, and how effectively, decision-support systems have been used to manage diverse water management needs,

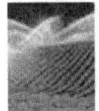

including ecological restoration, riparian flow management, urban water supply, agricultural water availability, coastal zone issues, and fire management at diverse spatial scales: from cities and their surrounding urban concentrations (New York, Seattle), to regions (Northern California, South Florida, Intermountain West); a comprehensively-managed river basin (CALFED); and a resource (forest lands) scattered over parts of the U.S. West and Southwest. These cases also illustrate efforts to rely on temporally diverse information (*i.e.*, predictions of future variability in precipitation, sea-level rise, and drought as well as past variation) in order to validate trends.

Most importantly, these experiments represent the use of different ways of integrating information into water management to enable better decisions to be made, including neural networks[1] in combination with El Niño-Southern Oscillation (ENSO) forecasting; temperature, precipitation and sea-level rise prediction; probabilistic risk assessment; integrated weather, climate and hydrological models producing short- and longer-term forecasts; weather and streamflow station outputs; paleoclimate records of streamflow and hydroclimatic variability; and the use of climate change information on precipitation and sea-level rise to address shorter-term weather variability.

Experiment 1:
How the South Florida Water Management District Uses Climate Information

The Experiment
In an attempt to restore the Everglades ecosystem of South Florida, a team of state and federal agencies is engaged in the world's largest restoration program (Florida Department of Environmental Protection and South Florida Water Management District, 2007). A cornerstone of this effort is the understanding that SI climate variability (as well as climate change) could have significant impacts on the region's hydrology over the program's 50-year lifetime.

---

[1] A neural network or "artificial neural network" is an approach to information processing paradigm that functions like a brain in processing information. The network is composed of a large number of interconnected processing elements (neurons) that work together to solve specific problems and, like the brain, the entire network learns by example.

The South Florida Water Management District (SFWMD) is actively involved in conducting and supporting climate research to improve the prediction and management of South Florida's complex water system (Obeysekera *et al.*, 2007). The SFWMD is significant because it is one of the few cases in which decade-scale climate variability information is being used in water resource modeling, planning, and operation programs.

Background/Context
Research relating climatic indices to South Florida climate started at SFWMD more than a decade ago (South Florida Water Management District, 1996). Zhang and Trimble (1996), Trimble *et al.* (1997), and Trimble and Trimble (1998) used neural network models to develop a better understanding of how ENSO and other climate factors influence net inflow to Lake Okeechobee. From that knowledge, Trimble (1998) demonstrated the potential for using ENSO and other indices to predict net inflow to Lake Okeechobee for operational planning. Subsequently, SFWMD was able to apply climate forecasts to its understanding of climate-water resources relationships in order to assess risks associated with seasonal and multi-seasonal operations of the water management system and to communicate the projected outlook to agency partners, decision makers, and other stakeholders (Cadavid *et al.*, 1999).

Implementation/Application
The SFWMD later established the Water Supply and Environment (WSE), a regulation schedule for Lake Okeechobee that formally uses seasonal and multi-seasonal climate outlooks as guidance for regulatory release decisions

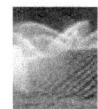

There are many different ways of integrating information into water management to enable better decisions.

Water management
in coastal urban
areas faces unique
challenges due
to vulnerabilities
of much of the
existing water supply
and treatment
infrastructure
to storm surges,
coastal erosion,
coastal subsidence,
and tsunamis.

(Obeysekera *et al.*, 2007). The WSE schedule uses states of ENSO and the Atlantic Multidecadal Oscillation (AMO) (Enfield *et al.*, 2001) to estimate the Lake Okeechobee net inflow outlook for the next six to 12 months. A decision tree with a climate outlook is a unique component of the WSE schedule and is considered a major advance over traditional hydrologic rule curves typically used to operate large reservoirs (Obeysekera *et al.*, 2007). Evaluation of the application of the WSE schedule revealed that considerable uncertainty in regional hydrology remains and is attributable to some combination of natural climatic variation, long-term global climate change, changes in South Florida precipitation patterns associated with drainage and development, and rainfall-runoff relationships altered by infrastructure changes (Obeysekera *et al.*, 2007).

Lessons Learned
From its experience with climate information and research, SFWMD has learned that to improve its modeling capabilities and contributions to basin management, it must improve its ability to: differentiate trends and discontinuities in basin flows associated with climate variation from those caused by water management; gauge the skill gained in using climate information to predict basin hydroclimatology; improve management; account for management uncertainties caused by climate variation and change; and evaluate how climate change projections may affect facility planning and operation of the SFWMD (Bras, 2006; Obeysekera *et al.*, 2007).

The district has also learned that, given the decades needed to restore the South Florida ecosystem, adaptive management is an effective way to incorporate SI climate variation into its modeling and operations decision-making processes, especially since longer term climate change is likely to exacerbate operational challenges. As previously stated, this experiment is also unique in being the only one that has been identified in which decadal climate

status (*e.g.*, state of the AMO) is being used in a decision-support context.

Experiment 2:
Long-Term Municipal Water Management Planning—New York City

The Experiment
Projections of long-term climate change, while characterized by uncertainty, generally agree that coastal urban areas will, over time, be increasingly threatened by a unique set of hazards. These include sea-level rise, increased storm surges, and erosion. Two important questions facing decision makers are: (1) How will long-term climate change increase these threats, which are already of concern to urban planners? and (2) Can information on the likely changes in recurrence intervals of extreme events (*e.g.*, tropical storms) be used in long term municipal water management planning and decision making?

Background and Context
Water management in coastal urban areas faces unique challenges due to vulnerabilities of much of the existing water supply and treatment infrastructure to storm surges, coastal erosion, coastal subsidence, and tsunamis (Jacobs *et al.*, 2007; OFCM, 2004). Not only are there risks due to extreme events under current and evolving climate conditions, but many urban areas rely on aging infrastructure that was built in the late nineteenth and early twentieth centuries. These vulnerabilities will only be amplified by the addition of global warming-induced sea-level rise due to thermal expansion of ocean water and the melting of glaciers, mountain ice caps and ice sheets (IPCC, 2007).

For example, observed global sea-level rise was ~1.8 millimeters (~0.07 inch) per year from 1961 to 2003, whereas from 1993 to 2003 the rate of sea-level rise was ~3.1 millimeters (~0.12 inch) per year (IPCC, 2007). The Intergovernmental Panel on climate Change (IPCC) projections for the twenty-first century (IPCC, 2007) are for an "increased incidence of extreme high sea level" which they define as the highest one percent of hourly values of observed sea level at a station for a given reference period. The New York City Department of Environmental Protection (NYCDEP) is one example of an urban agency that is adapting strategic and capital planning to take into account the potential effects of climate change—sea-level rise, higher temperature, increases in extreme events, and changing precipitation patterns—on the city's water systems. NYCDEP, in partnership with local universities and private sector consultants, is evaluating climate change projections, impacts, indicators, and adaptation and mitigation strategies to support agency decision making (Rosenzweig *et al.*, 2007).

Implementation/Application
In New York City (NYC), as in many coastal urban areas, many of the wastewater treatment plants are at elevations of 2 to 6 meters above present sea level and thus within the range of current surges for tropical storms and hurricanes and extra-tropical cyclones (or "Nor'easters") (Rosenzweig and Solecki, 2001; Jacobs, 2001). Like many U.S. cities along the northern Atlantic Coast, NYC's vulnerability to storm surges is predominantly from Nor'easters that occur largely between late November and March, and tropical storms and hurricanes that typically strike between July and October. Based on global warming-induced sea-level rise inferred from IPCC studies, the recurrence interval for the 100-year storm flood (probability of occurring in any given year = 1/100) may decrease to 60 years or, under extreme changes, a recurrence interval as little as four years (Rosenzweig and Solecki, 2001; Jacobs *et al.*, 2007).

Increased incidence of high sea levels and heavy rains can cause sewer back-ups and water treatment plant overflows. Planners have identified activities to address current and future concerns such as using sea-level rise forecasts as inputs to storm surge and elevation models to anticipate the impact of flooding on NYC coastal water resource-related facilities. Other concerns include potential water quality impairment from heavy rains that can increase pathogen levels and turbidity with the possible effects magnified by "first-flush" storms: heavy rains after weeks of dry weather. NYC water supply reservoirs have not been designed for rapid releases and any changes to operations to limit downstream damage through flood control measures will reduce water supply. In addition, adding filtration capacity to the water supply system would be a significant challenge.

Planners in NYC have begun to consider these issues by defining risks through probabilistic climate scenarios, and categorizing potential adaptations as related to (1) operations/management; (2) infrastructure; and (3) policy (Rosenzweig *et al.*, 2007). The NYCDEP is examining the feasibility of relocating critical control systems to higher floors/ground in low-lying buildings, building protective flood walls, modifying design criteria to reflect changing hydrologic processes, and reconfiguring outfalls to prevent sediment build-up and surging. Significant strategic decisions and capital investments for NYC water management will continue to be challenged by questions such as: How does the city utilize projections in ways that are robust to uncertainties? And, when designing infrastructure in the face of future uncertainty, how can these planners make infrastructure more robust and adaptable to changing climate, regulatory mandates, zoning, and population distribution?

Lessons Learned
When trends and observations clearly point to increasing risks, decision makers need to build support for adaptive action despite inherent uncertainties. The extent and effectiveness of adaptive measures will depend on building awareness of these issues among decision makers, fostering processes of interagency interaction and collaboration, and developing common standards (Zimmerman and Cusker, 2001).

New plans for regional capital improvements can be designed to include measures that will reduce vulnerability to the adverse effects of sea-level rise. Wherever plans are underway for

> When trends and observations clearly point to increasing risks, decision makers need to build support for adaptive action despite inherent uncertainties.

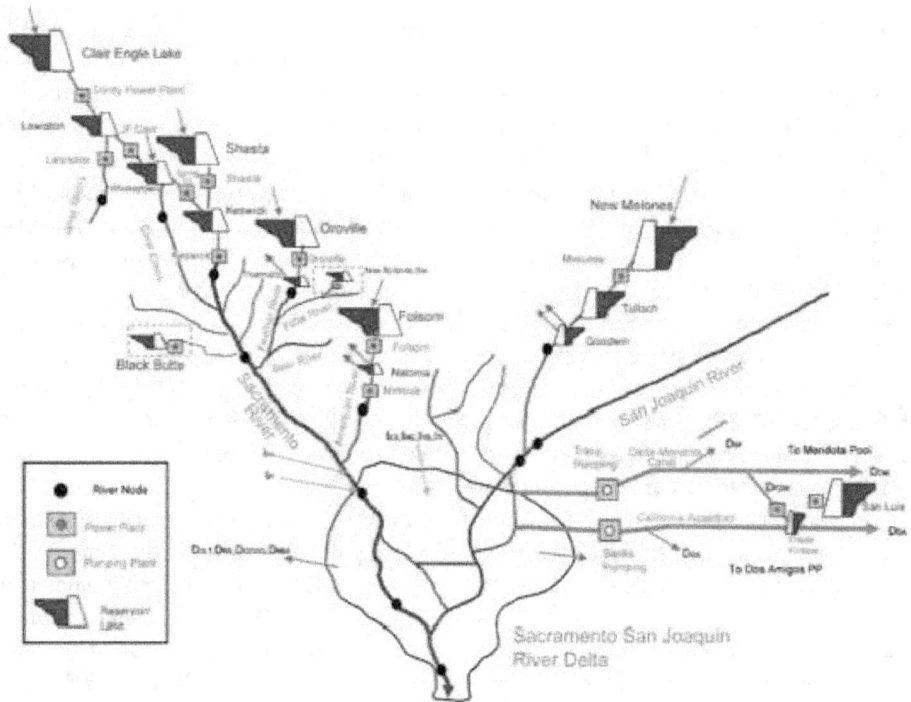

Figure 4.1 Map of Sacramento and San Joaquin River Delta.

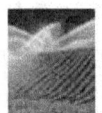

upgrading or constructing new roadways, airport runways, or wastewater treatment plants, which may already include flood protection; project managers now recognize the need to consider sea-level rise in planning activities (*i.e.*, OFCM, 2002).

In order to incorporate new sources of risk into engineering analysis, the meteorological and hydrology communities need to define and communicate current and increasing risks clearly, and convey them coherently, with explicit consideration of the inherent uncertainties. Research needed to support regional stakeholders include: further reducing uncertainties associated with sea-level rise, providing more reliable predictions of changes in frequency and intensity of tropical and extra-tropical storms, and determining how saltwater intrusion will impact freshwater. Finally, regional climate model simulations and statistical techniques being used to predict long-term climate change impacts could be down-scaled to help manage projected SI climate variability. This could be especially useful for adaptation planning (OFCM, 2007a).

Experiment 3:
Integrated Forecast and Reservoir Management (INFORM)—Northern California

The Experiment
The Integrated Forecast and Reservoir Management (INFORM) project aims to demonstrate the value of climate, weather, and hydrology forecasts in reservoir operations. Specific objectives are to: (1) implement a prototype integrated forecast-management system for the Northern California river and reservoir system in close collaboration with operational forecasting and management agencies, and (2) demonstrate the utility of meteorological/climate and hydrologic forecasts through near-real-time tests of the integrated system with actual data and management input.

Background and Context
The Northern California river system (Figure 4.1) encompasses the Trinity, Sacramento, Feather, American, and San Joaquin river systems, and the Sacramento-San Joaquin Delta (see: Experiment 7, CALFED)[2]. The Sacramento and San Joaquin Rivers join to form an extensive delta region and eventually flow out

---

[2]   CA. Gov. Welcome to Calfed Bay-Deltas Program. http://calwater.ca.gov/index.aspx

into the Pacific Ocean. The Northern California river and reservoir system serves many vital water uses, including providing two-thirds of the state's drinking water, irrigating seven million acres of the world's most productive farmland, and providing habitat to hundreds of species of fish, birds, and plants. In addition, the system protects Sacramento and other major cities from flood disasters and contributes significantly to the production of hydroelectric energy. The Sacramento-San Joaquin Delta provides a unique environment and is California's most important fishery habitat. Water from the delta is pumped and transported through canals and aqueducts south and west serving the water needs of many more urban, agricultural, and industrial users.

An agreement between the U.S. Department of the Interior, U.S. Bureau of Reclamation, and California Department of Water Resources provides for the coordinated operation of the federal and state facilities (Agreement of Coordinated Operation-COA). The agreement aims to ensure that each project obtains its share of water from the San Joaquin Delta and protects other beneficial uses in the Delta and the Sacramento Valley. Coordination is structured around the necessity to meet in-basin use requirements in the Sacramento Valley and the San Joaquin Delta, including delta outflow and water quality requirements.

Implementation/Application
The INFORM Forecast-Decision system consists of a number of diverse elements for data handling, model runs, and output archiving and presentation. It is a distributed system with on-line and off-line components. The system routinely captures real-time National Center for Environmental Predictions (NCEP) ensemble forecasts and uses both ensemble synoptic forecasts from NCEP's Global Forecast System (GFS) and ensemble climate forecasts from NCEP's Climate Forecast System (CFS). The former produces real-time short-term forecasts, and the latter produce longer-term forecasts as needed (HRC-GWRI, 2006).

The INFORM DSS is designed to support the decision-making process, which includes multiple decision makers, objectives, and temporal scales. Toward this goal, INFORM

DSS includes a suite of interlinked models that address reservoir planning and management at multi-decadal, interannual, seasonal, daily, and hourly time scales. The DSS includes models for each major reservoir in the INFORM region, simulation components for watersheds, river reaches, and the Bay Delta, and optimization components suitable for use with ensemble forecasts. The decision software runs off-line, as forecasts become available, to derive and assess planning and management strategies for all key system reservoirs. DSS is embedded in a user-friendly, graphical interface that links models with data and helps visualize and manage results.

Development and implementation of the INFORM Forecast-Decision system was carried out by the Hydrologic Research Center (in San Diego) and the Georgia Water Resources Institute (in Atlanta), with funding from NOAA, CALFED, and the California Energy Commission. Other key participating agencies included U.S. National Weather Service California–Nevada River Forecast Center, the California Department of Water Resources, the U.S. Bureau of Reclamation Central Valley Operations, and the Sacramento District of the U.S. Army Corps of Engineers. Other agencies and regional stakeholders (e.g., the Sacramento Flood Control Authority, SAFCA, and the California Department of Fish and Game) participated in project workshops and, indirectly, through comments conveyed to the INFORM Oversight and Implementation Committee.

Lessons Learned
The INFORM approach demonstrates the value of advanced forecast-decision methods for water resource decision making, attested to by participating agencies who took part in designing the experiments and who are now proceeding to incorporate the INFORM tools and products in their decision-making processes.

From a technical standpoint, INFORM served to demonstrate important aspects of integrated forecast-decision systems, namely that (1) seasonal climate and hydrologic forecasts benefit reservoir management, provided that they are used in connection with adaptive dynamic decision methods that can explicitly account for and manage forecast uncertainty;

Seasonal climate and hydrologic forecasts benefit reservoir management, provided that they are used in connection with adaptive dynamic decision methods that can explicitly account for and manage forecast uncertainty.

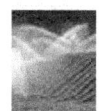

Ignoring forecast
uncertainty
in reservoir
regulation and
water management
decisions leads to
costly failures.

(2) ignoring forecast uncertainty in reservoir regulation and water management decisions leads to costly failures; and (3) static decision rules cannot take full advantage of and handle forecast uncertainty information. The extent to which forecasts benefit the management process depends on their reliability, range, and lead time, in relation to the management systems' ability to regulate flow, water allocation, and other factors.

Experiment 4:
How Seattle Public Utility District Uses
Climate Information to Manage Reservoirs

The Experiment
Seattle Public Utilities (SPU) provides drinking water to 1.4 million people living in the central Puget Sound region of Washington. SPU also has instream (*i.e.*, river flow), resource management, flood control management and habitat responsibilities on the Cedar and South Fork Tolt Rivers, located on the western slopes of the Cascade Mountains. Over the past several years SPU has taken numerous steps to improve the incorporation of climate, weather, and hydrologic information into the real-time and SI management of its mountain water supply system.

Implementation/Application
Through cooperative relationships with agencies such as NOAA's National Weather Service, U.S. Department of Agriculture, Natural Resource Conservation Service, and the U.S. Geological Survey (USGS), SPU has secured real-time access to numerous SNOTEL sites[3],

---
[3] The SNOTEL network of weather stations is a snowfall depth monitoring network established by the USGS.

streamflow gages and weather stations in and around Seattle's watersheds. SPU continuously monitors weather and climate data across the maritime Pacific derived from all these above sources. Access to this information has helped to reduce the uncertainty associated with making real-time and seasonal tactical and strategic operational decisions, and enhanced the inherent flexibility of management options available to SPU's water supply managers as they adjust operations for changing weather and hydrologic conditions, including abnormally low levels of snowpack or precipitation.

Among the important consequences of this synthesis of information has been SPU's increasing ability to undertake reservoir operations with higher degrees of confidence than in the past. As an example, SPU was well served by this information infrastructure during the winter of 2005 when the lowest snowpack on record was realized in its watersheds. The consequent reduced probability of spring flooding, coupled with their ongoing understanding of local and regional climate and weather patterns, enabled SPU water managers to safely capture more water in storage earlier in the season than normal. As a result of SPU's ability to continuously adapt its operations, Seattle was provided with enough water to return to normal supply conditions by early summer despite the record low snowpack.

SPU is also using conclusions from a SPU-sponsored University of Washington study that examined potential impacts of climate change on SPU's water supply. To increase the rigor of the study, a set of fixed reservoir operating rules was used and no provisions were made to adjust these to account for changes projected by the study's climate change scenarios. From these conclusions, SPU has created two future climate scenarios, one for 2020 and one for 2040, to examine how the potential impacts of climate change may affect decisions about future supply. While these scenarios indicated a reduction in yield, SPU's existing sources of supply were found to be sufficient to meet official demand forecasts through 2053.

Lessons Learned
SPU has actually incorporated seasonal climate forecasts into their operations and is among the

leaders in considering climate change. SPU is a "receptive audience" for climate tools in that it has a wide range of management and long-term capital investment responsibilities that have clear connections to climate conditions. Further, SPU is receptive to new management approaches due to public pressure and the risk of legal challenges related to the protection of fish populations who need to move upstream to breed.

Specific lessons include: (1) access to skillful seasonal forecasts enhances credibility of using climate information in the Pacific Northwest, even with relatively long lead times; (2) monitoring of snowpack moisture storage and mountain precipitation is essential for effective decision making and for detecting long-term trends that can affect water supply reliability; and (3) while SPU has worked with the research community and other agencies, it also has significant capacity to conduct in-house investigations and assessments. This provides confidence in the use of information.

Experiment 5:
Using Paleoclimate Information to Examine Climate Change Impacts

The Experiment
Can an expanded estimate of the range of natural hydrologic variability from tree ring reconstructions of streamflow, a climate change research tool, be used effectively as a decision-support resource for better understanding SI climate variability and water resource planning? Incorporation of tree ring reconstructions of streamflow into decision making was accomplished through partnerships between researchers and water managers in the intermountain West.

Background and Context
Although water supply forecasts in the intermountain West have become increasingly sophisticated in recent years, water management planning and decision making have generally depended on instrumental gage records of flow, most of which are less than 100 years in length. Drought planning in the Intermountain West has been based on the assumption that the 1950s drought, the most severe drought in the instrumental record, adequately represents the full

range of natural variability and, thus, a likely worst-case scenario.

The recent prolonged drought in the western United States prompted many water managers to consider that the observational gage records of the twentieth century do not contain the full range of natural hydroclimatic variability possible. Gradual shifts in recent decades to more winter precipitation as rain and less as snow, earlier spring runoff, higher temperatures, and unprecedented population growth have resulted in an increase in vulnerability of limited water supplies to a variable and changing climate. The paleoclimate records of streamflow and hydroclimatic variability provide an extended, albeit indirect, record (based on more than 1000 years of record from tree rings in some key watersheds) for assessing the potential impact of a more complete range of natural variability as well as for providing a baseline for detecting possible regional impacts of global climate change.

Implementation/Application
Several years of collaborations between scientists and water resource partners have explored possible applications of tree ring reconstructed flows in water resource management to assess the potential impacts of drought on water systems. Extended records of hydroclimatic variability from tree ring based reconstructions reveal a wider range of natural variability than in gage records alone, but how to apply this information in water management planning has not been obvious. The severe western drought that began in 2000 and peaked in 2002 provided an excellent opportunity to work with water resource providers and agencies on how to incorporate paleoclimate drought information in planning and decision making. These partnerships with water resource managers have led to a range of applications evolving from a basic change in thinking about drought, to the use of tree ring reconstructed flows to run a complex water supply model to assess the impacts of drought on water systems.

The extreme five-year drought that began in 2002 motivated water managers to ask these questions: How unusual was 2002, or the 2000-2004 drought? How often do years or droughts like this occur? What is the likelihood

Several years of collaborations between scientists and water resource partners have explored possible applications of tree ring reconstructed flows in water resource management to assess the potential impacts of drought on water systems.

of it happening again in the future (should we plan for it, or is there too low a risk to justify infrastructure investments)? And, from a long term perspective, is the twentieth and twenty-first century record an adequate baseline for drought planning?

The first three questions could be answered with reconstructed streamflow data for key gages, but to address planning, a critical step is determining how tree ring streamflow reconstruction could be incorporated into water supply modeling efforts. The tree ring streamflow reconstructions have annual resolution, whereas most water system models required weekly or daily time steps, and reconstructions are generated for a few gages, while water supply models typically have multiple input nodes. The challenge has been spatially and temporally disaggregating the reconstructed flow series into the time steps and spatial scales needed as input into models. A variety of analogous approaches have successfully addressed the temporal scale issue, while the spatial challenges have been addressed statistically using nearest neighbor or other approaches.

Another issue addressed has been that the streamflow reconstructions explain only a portion of the variance in the gage record, and the most extreme values are often not fully replicated. Other efforts have focused on characterizing the uncertainty in the reconstructions, the sources of uncertainty, and the sensitivity of the reconstruction to modeling choices. In spite of these many challenges, expanded estimates of the range of natural hydrologic variability from tree ring reconstructions have been integrated into water management decision support and allocation models to evaluate operating policy alternatives for efficient management and sustainability of water resources, particularly during droughts in California and Colorado.

Lessons Learned
Roadblocks to incorporating tree ring reconstructions into water management policy and decision making were overcome through prolonged, sustained partnerships with researchers working to make their scientific findings relevant, useful, and usable to users for planning and management, and water managers willing to take risk and invest time to explore the use

of non-traditional information outside of their comfort zone. The partnerships focused on formulating research questions that led to applications addressing institutional constraints within a decision process addressing multiple timescales.

Workshops requested by water managers have resulted in expansion of application of the tree ring based streamflow reconstructions to drought planning and water management <http://wwa.colorado.edu/resources/paleo/>. In addition, an online resource called TreeFlow <http://wwa.colorado.edu/resources/paleo/data.html> was developed to provide water managers interested in using tree ring streamflow reconstructions access to gage and reconstruction data and information, and a tutorial on reconstruction methods for gages in Colorado and California.

Experiment 6
Climate, Hydrology, and Water Resource Issues in Fire-Prone United States Forests

The Experiment
Improvements in ENSO-based climate forecasting, and research on interactions between climate and wildland fire occurrence, have generated opportunities for improving use of seasonal-to-interannual climate forecasts by fire managers. They can now better anticipate annual fire risk, including potential damage to watersheds over the course of the year. The experiment, consisting of annual workshops to evaluate the utility of climate information for fire management, were initiated in 2000 to inform fire managers about climate forecasting tools and to enlighten climate forecasters about the needs of the fire management community. These workshops have evolved into an annual assessment of conditions and production of pre-season fire-climate forecasts.

Background and Context
Large wildfire activity in the U.S. West and Southeast has increased substantially since the mid-1980s, an increase that has largely been attributed to shifting climate conditions (Westerling *et al.*, 2006). Recent evidence also suggests that global or regional warming trends and a positive phase of the AMO are likely to lead to an even greater increase in risk for ecosystems

Improvements in El Niño–Southern Oscillation-based climate forecasting, and research on interactions between climate and wildland fire occurrence, have generated opportunities for improving use of seasonal-to-interannual climate forecasts by fire managers.

and communities vulnerable to wildfire in the western United States (Kitzberger *et al.*, 2007). Aside from the immediate impacts of a wildfire (e.g., destruction of biomass, substantial altering of ecosystem function), the increased likelihood of high sediment deposition in streams and flash flood events can present post-fire management challenges including impacts to soil stability on slopes and mudslides (e.g., Bisson *et al.*, 2003). While the highly complex nature and substantially different ecologies of fire-prone systems precludes one-size-fits-all fire management approaches (Noss *et al.*, 2006), climate information can help managers plan for fire risk in the context of watershed management and post-fire impacts, including impacts on water resources. One danger is inundation of water storage and treatment facilities with sediment-rich water, creating potential for significant expense for pre-treatment of water or for facilities repair. Post-fire runoff can also raise nitrate concentrations to levels that exceed the federal drinking water standard (Meixner and Wohlgemuth, 2004).

Work by Kuyumjian (2004), suggests that coordination among fire specialists, hydrologists, climate specialists, and municipal water managers may produce useful warnings to downstream water treatment facilities about significant ash- and sediment-laden flows. For example, in the wake of the 2000 Cerro Grande fire in the vicinity of Los Alamos, New Mexico, catastrophic floods were feared, due to the fact that 40 percent of annual precipitation in northern New Mexico is produced by summer monsoon thunderstorms (e.g., Earles *et al.*, 2004). Concern about water quality and about the potential for contaminants carried by flood waters from the grounds of Los Alamos Nuclear Laboratory to enter water supplies prompted a multi-year water quality monitoring effort (Gallaher and Koch, 2004). In the wake of the 2002 Bullock Fire and 2003 Aspen Fire in the Santa Catalina Mountains adjacent to Tucson, Arizona, heavy rainfall produced floods that destroyed homes and caused one death in Canada del Oro Wash in 2003 (Ekwurzel, 2004), destroyed structures in the highly popular Sabino Canyon recreation area and deposited high sediment loads in Sabino Creek in 2003 (Desilets *et al.*, 2006). A flood in 2006 wrought a major transformation to the upper reaches of

the creek (Kreutz, 2006). Residents of Summerhaven, a small community located on Mt. Lemmon, continues to be concerned about the impacts of future fires on their water resources. In all of these situations, climate information can be helpful in assessing vulnerability to both flooding and water quality issues.

Implementation/Application
Little published research specifically targets interactions among climate, fire, and watershed dynamics (OFCM, 2007b). Publications on fire-climate interactions, however, provide a useful entry point for examining needs for and uses of climate information in decision processes involving water resources. A continuing effort to produce fire-climate outlooks was initiated through a workshop held in Tucson, Arizona, in late winter 2000. One of the goals of the workshop was to identify the climate information uses and needs of fire managers, fuel managers, and other decision makers. Another was to actually produce a fire-climate forecast for the coming fire season. The project was initiated through collaboration involving researchers at the University of Arizona, the NOAA-funded Climate Assessment for the Southwest Project (CLIMAS), the Center for Ecological and Fire Applications (CEFA) at the Desert Research Institute in Reno, Nevada and the National Interagency Fire Center (NIFC) located in Boise, Idaho (Morehouse, 2000). Now called the National Seasonal Assessment Workshop (NSAW), the process continues to produce annual fire-climate outlooks (e.g., Crawford *et al.*, 2006). The seasonal fire-climate forecasts produced by NSAW have been published through NIFC since 2004. During this same time period,

Westerling *et al.* (2002) developed a long-lead statistical forecast product for areas burned in western wildfires.

Lessons Learned

The experimental interactions between climate scientists and fire managers clearly demonstrated the utility of climate information for managing watershed problems associated with wildfire. Climate information products used in the most recently published NSAW Proceedings (Crawford *et al.*, 2006), for example, include the following: NOAA Climate Prediction Center (CPC) seasonal temperature and precipitation outlooks, historical temperature and precipitation data, *e.g.*, High Plains Regional Climate Center, National drought conditions, from National Drought Mitigation Center, 12-month standardized precipitation index, spring and summer streamflow forecasts and departure from average greenness.

Based on extensive interactions with fire managers, other products are also used by some fire ecologists and managers, including climate history data from instrumental and paleo (especially tree ring) records and hourly to daily and weekly weather forecasts, (*e.g.*, temperature, precipitation, wind, relative humidity).

Products identified as potentially improving fire management (*e.g.*, Morehouse, 2000; Garfin and Morehouse, 2001) include: improved monsoon forecasts and training in how to use them, annual to decadal (AMO, Pacific Decadal Oscillation) projections, decadal to centennial climate change model outputs, downscaled to regional/finer scales, and dry lightning forecasts.

This experiment is one of the most enduring we have studied. It is now part of accepted practice by agencies, and has produced spin-off activities managed and sustained by the agencies and new participants. The use of climate forecast information in fire management began because decision makers within the wildland fire management community were open to new information, due to legal challenges, public pressure, and a "landmark" wildfire season in 2000. The National Fire Plan (2000) and its associated 10-year Comprehensive Strategy reflected a new receptiveness for new ways

of coping with vulnerabilities, calling for a community-based approach to reducing wildland fires that is proactive and collaborative rather than prior approaches entered on internal agency activities.

Annual workshops became routine forums for bringing scientists and decision makers together to continue to explore new questions and opportunities, as well as involve new participants, new disciplines and specialties, and to make significant progress in important areas (*e.g.*, lightning climatologies, and contextual assessments of specific seasons), quickly enough to fulfill the needs of agency personnel (National Fire Plan, 2000).

Experiment 7:
The CALFED—Bay Delta Program: Implications of Climate Variability

The Experiment

The Sacramento-San Joaquin River Delta, which flows into San Francisco Bay, is the focus of a broad array of environmental issues relating to endangered fish species, land use, flood control and water supply. After decades of debate about how to manage the delta to export water supplies to southern California while managing habitat and water supplies in the region, and maintaining endangered fish species, decision makers are involved in making major long-term decisions about rebuilding flood control levees and rerouting water supply networks through the region. Incorporating the potential for climate change impacts on sea level rise and other regional changes are important to the decision-making process (Hayhoe *et al.*, 2004; Knowles *et al.*, 2006; Lund *et al.*, 2007).

Background and Context

Climate considerations are critical for the managers of the CALFED program, which oversees the 700,000 acres in the Sacramento-San Joaquin Delta. 400,000 acres have been subsiding due to microbial oxidation of peat soils that have been used for agriculture. A significant number of the islands are below sea level, and protected from inundation by dikes that are in relatively poor condition. Continuing sea-level rise and regional climate change are expected to have additional major impacts such as flooding and changes in seasonal precipita-

The use of climate forecast information in fire management began because decision makers within the wildland fire management community were open to new information, due to legal challenges, public pressure, and a "landmark" wildfire season in 2000.

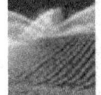

tion patterns. There are concerns that multiple islands would be inundated in a "10-year storm event", which represents extreme local vulnerability to flooding.

In the central delta, there are five county governments in addition to multiple federal and state agencies and non-governmental organizations whose perspectives need to be integrated into the management process, which is one of the purposes of the CALFED program. A key decision being faced is whether delta interests should invest in trying to build up and repair levies to protect subsided soils. What are the implications for other islands when one island floods? Knowing the likelihood of sea-level rise of various magnitudes will significantly constrain the answers to these questions. For example, if the rise is greater than one foot in the next 50 to 100 years, that could end the debate about whether to use levee improvements to further protect these islands. Smaller amounts of sea-level rise will make this decision less clear-cut. Answers are needed in order to support decisions about the delta in the near term.

Implementation/Application
Hundreds of millions of dollars of restoration work has been done in the delta and associated watersheds, and more investment is required. Where should money be invested for effective long-term impact? There is a need to invest in restoring lands at intertidal and higher elevations so that wetlands can evolve uphill while tracking rising sea level (estuarine progression). Protecting only "critical" delta islands (those with major existing infrastructure) to endure a 100-year flood will cost around $2.6 billion.

Another way that climate change-related information is critical to delta management is in estimating volumes and timing of runoff from the Sierra Nevada mountain range (Knowles *et al.*, 2006). To the extent that snowpack will be diminished and snowmelt runoff occurs earlier, there are implications for flood control, water supply and conveyance, and seawater intrusion—all of which affect habitat and land use decisions. One possible approach to water shortages is more recent aggressive management of reservoirs to maximize water supply benefits, thereby possibly increasing flood

risk. The State Water Project is now looking at a ten percent failure rate operating guideline at Oroville rather than a 5 percent failure rate operating guideline; this would provide much more water supply flexibility.

Lessons Learned
Until recently the implications of climate change and sea-level rise were not considered in the context of solutions to the Bay Delta problem—particularly in the context of climate variability. These implications are currently considered to be critical factors in infrastructure planning, and the time horizon for future planning has been extended to to over 100 years (Delta Vision Blue Ribbon Task Force, 2008). The relatively rapid shift in perception of the urgency of climate change impacts was not predicted, but does demand renewed consideration of adaptive management strategies in the context of incremental changes in understanding (as opposed to gradual increases in accumulation of new facts, which is the dominant paradigm in adaptive management).

4.2.2 Organizational and Institutional Dimensions of Decision-Support Experiments

These seven experiments illuminate the need for effective two-way communication among tool developers and users, and the importance of organizational culture in fostering collaboration. An especially important lesson they afford is in underscoring the significance of boundary-spanning entities to enable decision-support transformation. Boundary spanning, discussed in Section 4.3, refers to the activities of special scientific/stakeholder committees, agency coordinating bodies, or task forces that facilitate bringing together tool developers and

users to exchange information, promote communication, propose remedies to problems, foster frequent engagement, and jointly develop decision-support systems to address user needs. In the process, they provide incentives for innovation—frequently noted in the literature—that facilitate the use of climate science information in decisions (e.g., NRC, 2007; Cash and Buizer, 2005; Sarewitz and Pielke, 2007). Before outlining how these seven experiments illuminate boundary spanning, it is important to consider problems identified in recent research.

While there is widespread agreement that decision support involves translating the products of climate science into forms useful for decision makers and disseminating the translated products, there is disagreement over precisely what constitutes translation (NRC, 2008). One view is that climate scientists know which products will be useful to decision makers and that potential users will make appropriate use of decision-relevant information once it is made available. Adherents of this view typically emphasize the importance of developing "decision-support tools", such as models, maps, and other technical products intended to be relevant to certain classes of decisions that, when created, complete the task of decision support. This approach, also called a "translation model", (NRC, 2008) has not proved useful to many decision makers—underscored by the fact that, in our seven cases, greater weight was given to "creating conditions that foster the appropriate use of information" rather than to the information itself (NRC, 2008).

A second view is that decision-support activities should enable climate information producers and users to jointly develop information that addresses users' needs—also called "co-production" of information or reconciling information "supply and demand" (NRC, 1989, 1996, 1999, 2006; McNie, 2007; Sarewitz and Pielke, 2007; Lemos and Morehouse, 2005). Our seven cases clearly delineate the presumed advantages of the second view.

In the SFWMD case, an increase in user trust was a powerful inducement to introduce, and then continue, experiments leading to development of a Water Supply and Environment schedule, employing seasonal and multi-seasonal

climate outlooks as guidance for regulatory releases. As this tool began to help reduce operating system uncertainty, decision-maker confidence in the use of model outputs increased, as did further cooperation between scientists and users—facilitated by SFWMD's communication and agency partnership networks.

In the case of INFORM, participating agencies in California worked in partnership with scientists to design experiments that would allow the state to integrate forecast methods into planning for uncertainties in reservoir regulation. Not only did this set of experiments demonstrate the practical value of such tools, but they built support for adaptive measures to manage risks, and reinforced the use, by decision makers, of tool output in their decisions. Similar to the SFWMD case, through demonstrating how forecast models could reduce operating uncertainties—especially as regards increasing reliability and lead time for crucial decisions—cooperation among partners seems to have been strengthened.

Because the New York City and Seattle cases both demonstrate use of decision-support information in urban settings, they amplify another set of boundary-spanning factors: the need to incorporate public concerns and develop communication outreach methods, particularly about risk, that are clear and coherent. While conscientious efforts to support stakeholder needs for reducing uncertainties associated with sea-level rise and infrastructure relocation are being made, the New York case highlights the need for further efforts to refine communication, tool dissemination, and evaluation efforts to deliver information on potential impacts of climate change more effectively. It also illustrates the need to incorporate new risk-based analysis into existing decision structures related to infrastructure construction and maintenance. The Seattle public utility has had success in conveying the importance of employing SI climate forecasts in operations, and is considered a national model for doing so, in part because of a higher degree of established public support due to: (1) litigation over protection of endangered fish populations and (2) a greater in-house ability to test forecast skill and evaluate decision tools. Both served as incentives for collaboration. Access to highly-skilled

There is a need to incorporate public concerns and develop communication outreach methods, particularly about risk, that are clear and coherent.

forecasts in the region also enhanced prospects for forecast use.

Although not an urban case, the CALFED experiment's focus on climate change, sea-level rise, and infrastructure planning has numerous parallels with the Seattle and New York City cases. In this instance, the public and decision makers were prominent in these cases, and their involvement enhanced the visibility and importance of these issues and probably helped facilitate the incorporation of climate information by water resource managers in generating adaptation policies.

The other cases represent variations of boundary spanning whose lessons are also worth noting. The tree ring reconstruction case documents impediments of a new data source to incorporation into water planning. These impediments were overcome through prolonged and sustained partnerships between researchers and users that helped ensure that scientific findings were relevant, useful, and usable for water resources planning and management, and water managers who were willing to take some risk. Likewise, the case of fire-prone forests represented a different set of impediments that also required novel means of boundary spanning to overcome. In this instance, an initial workshop held among scientists and decision makers itself constituted an experiment on how to: identify topics of mutual interest across the climate and wildland fire management communities at multiple scales; provide a forum for exploring new questions and opportunities; and constitute a vehicle for inviting diverse agency personnel, disciplinary representatives, and operation, planning, and management personnel to facilitate new ways of thinking about an old set of problems. In all cases, the goal is to facilitate successful outcomes in the use of climate information for decisions, including faster adaptation to more rapidly changing conditions.

Before turning to analytical studies on the importance of such factors as the role of key leadership in organizations to empower employees, organizational climate that encourages risk and promote inclusiveness, and the ways organizations encourage boundary innovation (Section 4.3), it is important to reemphasize the distinguishing feature of the above experiments: they

underscore the importance of process as well as product outcomes in developing, disseminating and using information. We return to this issue when we discuss evaluation in Section 4.4.

4.3 APPROACHES TO BUILDING USER KNOWLEDGE AND ENHANCING CAPACITY BUILDING

The previous section demonstrated a variety of contexts where decision-support innovations are occurring. This Section analyzes six factors that are essential for building user knowledge and enhancing capacity in decision-support systems for integration of SI climate variability information, and which are highlighted in the seven cases above: (1) boundary spanning, (2) knowledge-action systems through inclusive organizations, (3) decision-support needs are user driven, (4) proactive leadership that champions change; (5) adequate funding and capacity building, and (6) adaptive management.

4.3.1 Boundary-Spanning Organizations as Intermediaries Between Scientists and Decision Makers

As noted in Section 4.2.2, boundary-spanning organizations link different social and organizational worlds (*e.g.*, science and policy) in order to foster innovation across boundaries, provide two-way communication among multiple sectors, and integrate production of science with user needs. More specifically, these organizations perform translation and mediation functions between producers of information and their users (Guston, 2001; Ingram and Bradley, 2006; Jacobs, *et al.*, 2005). Such activities include convening forums that provide common vehicles for conversations and training, and for tailoring information to specific applications.

Ingram and Bradley (2006) suggest that boundary organizations span not only disciplines, but different conceptual and organizational divides (*e.g.*, science and policy), organizational missions and philosophies, levels of governance, and gaps between experiential and professional ways of knowing. This is important because effective knowledge transfer systems cultivate individuals and/or institutions that serve as

Boundary-spanning organizations perform translation and mediation functions between producers of information and their users.

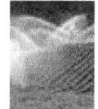

intermediaries between nodes in the system, most notably between scientists and decision makers. In the academic community and within agencies, knowledge, including the knowledge involved in the production of climate forecast information, is often produced in "stove-pipes" isolated from neighboring disciplines or applications.

Evidence for the importance of this proposition—and for the importance of boundary spanning generally—is provided by those cases, particularly in Chapter 3 (e.g., the Apalachicola–Chattahoochee–Flint River basin dispute), where the absence of a boundary spanning entity created a void that made the deliberative consideration of various decision-maker needs all but impossible to negotiate. Because the compact organization charged with managing water allocation among the states of Alabama, Florida, and Georgia would not actually take effect until an allocation formula was agreed upon, the compact could not serve to bridge the divides between decision making and scientific assessment of flow, meteorology, and riverine hydrology in the region.

Boundary spanning organizations are important to decision-support system development in three ways. First, they "mediate" communica-

tion between supply and demand functions for particular areas of societal concern. Sarewitz and Pielke (2007) suggest, for example, that the IPCC serves as a boundary organization for connecting the science of climate change to its use in society—in effect, satisfying a "demand" for science implicitly contained in such international processes for negotiating and implementing climate treaties as the U.N. Framework Convention on Climate Change and Kyoto Protocol. In the United States, local irrigation district managers and county extension agents often serve this role in mediating between scientists (hydrological modelers) and farmers (Cash et al., 2003). In the various cases we explored in Section 4.2.1, and in Chapter 3 (e.g., coordinating committees, post-event "technical sessions" after the Red River floods, and comparable entities), we saw other boundary spanning entities performing mediation functions.

Second, boundary organizations enhance communication among stakeholders. Effective tool development requires that affected stakeholders be included in dialogue, and that data from local resource managers (blended knowledge) be used to ensure credible communication. Successful innovation is characterized by two-way communication between producers and users of

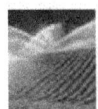

Boundary organizations enhance communication among stakeholders.

Table 4.1 Examples of Boundary Organizations for Decision-Support Tool Development.

Cooperative Extension Services: Housed in land-grant universities in the United States, they provide large networks of people who interact with local stakeholders and decision makers within certain sectors (not limited to agriculture) on a regular basis. In other countries, this agricultural extension work is often done with great effectiveness by local government (e.g., Department of Primary Industries, Queensland, Australia).
Watershed Councils: In some U.S. states, watershed councils and other local planning groups have developed, and many are focused on resolving environmental conflicts and improved land and water management (particularly successful in the State of Oregon).
Natural Resource Conservation Districts: Within the U.S. Department of Agriculture, these districts are highly networked within agriculture, land management, and rural communities.
Non-governmental organizations (NGOs) and public interest groups: Focus on information dissemination and environmental management issues within particular communities. They are good contacts for identifying potential stakeholders, and may be in a position to collaborate on particular projects. Internationally, a number of NGOs have stepped forward and are actively engaged in working with stakeholders to advance use of climate information in decision making (e.g., Asian Disaster Preparedness Center (ADPC), in Bangkok, Thailand).
Federal agency and university research activities: Expanding the types of research conducted within management institutions and local and state governments is an option to be considered—the stakeholders can then have greater influence on ensuring that the research is relevant to their particular concerns.

knowledge, as well as development of networks that allow close and ongoing communication among multiple sectors. Likewise, networks must allow close communication among multiple sectors (Sarewitz and Pielke, 2007).

Third, boundary organizations contribute to tool development by serving the function of translation more effectively than is conceived in the Loading Dock Model of climate products. In relations between experts and decision makers, understanding is often hindered by jargon, language, experiences, and presumptions; e.g., decision makers often want deterministic answers about future climate conditions, while scientists can often only provide probabilistic information, at best. As noted in Chapter 3, decision makers often mistake probabilistic uncertainty as a kind of failure in the utility and scientific merit of forecasts, even though uncertainty is a characteristic of science (Brown, 1997).

One place where boundary spanning can be important with respect to translation is in providing a greater understanding of uncertainty and its source. This includes better information exchange between scientists and decision makers on, for example, the decisional relevance of different aspects of uncertainties, and methods of combining probabilistic estimates of events through simulations, in order to reduce decision-maker distrust, misinterpretation of forecasts, and mistaken interpretation of models (NRC, 2005).

Effective boundary organizations facilitate the co-production of knowledge—generating information or technology through the collaboration of scientists/engineers and nonscientists who incorporate values and criteria from both communities. This is seen, for example, in the collaboration of scientists and users in producing models, maps, and forecast products. Boundary organizations have been observed to work best when accountable to the individuals or interests on both sides of the boundary they bridge, in order to avoid capture by either side and to align incentives such that interests of actors on both sides of the boundary are met.

Jacobs (2003) suggests that universities can be good locations for the development of new ideas and applications, but they may not be ideal for

sustained stakeholder interactions and services, in part because of funding issues and because training cycles for graduate students, who are key resources at universities, do not always allow a long-term commitment of staff. Many user groups and stakeholders either have no contact with universities or may not encourage researchers to participate in or observe decision-making processes. University reward systems rarely recognize interdisciplinary work, outreach efforts, and publications outside of academic journals. This limits incentives for academics to participate in real-world problem solving and collaborative efforts. Despite these limitations, many successful boundary organizations are located within universities.

In short, boundary organizations serve to make information from science useful and to keep information flowing (in both directions) between producers and users of the information. They foster mutual respect and trust between users and producers. Within such organizations there is a need for individuals simultaneously capable of translating scientific results for practical use and framing the research questions from the perspective of the user of the information. These key intermediaries in boundary organizations need to be capable of integrating disciplines and defining the research question beyond the focus of the participating individual disciplines. Table 4.1 depicts a number of boundary organization examples for climate change decision-support tool development. Section 4.3.2 considers the type of organizational leaders who facilitate boundary spanning.

An oft-cited model of the type of boundary-spanning organization needed for the transfer and translation of decision-support information on climate variability is the Regional Integrated Science and Assessment (RISA) teams supported by NOAA. These teams "represent a new collaborative paradigm in which decision makers are actively involved in developing research agendas" (Jacobs, 2003). The eight RISA teams, located within universities and often involving partnerships with NOAA laboratories throughout the United States, are focused on stakeholder-driven research agendas and long-term relationships between scientists and decision makers in specific regions. RISA activities are highlighted in the sidebar below.

Boundary organizations serve to make information from science useful and to keep information flowing (in both directions) between producers and users of the information.

BOX 4.1: Comparative Examples of Boundary Spanning—Australia and the United States

In Australia, forecast information is actively sought both by large agribusiness and government policymakers planning for drought because "the logistics of handling and trading Australia's grain commodities, such as wheat, are confounded by huge swings in production associated with climate variability. Advance information on likely production and its geographical distribution is sought by many industries, particularly in the recently deregulated marketing environment" (Hammer, et al., 2001). Forecast producers have adopted a systems approach to the dissemination of seasonal forecast information that includes close interaction with farmers, use of climate scenarios to discuss the incoming rainfall season and automated dissemination of seasonal forecast information through the RAINMAN interactive software.

In the U.S. Southwest, forecast producers organized stakeholder workshops that refined their understanding of potential users and their needs. Because continuous interaction with stakeholder was well funded and encouraged, producers were able to 'customize' their product—including the design of user friendly and interactive Internet access to climate information—to local stakeholders with significant success (Hartmann, et al., 2002; Pagano, et al., 2002; Lemos and Morehouse, 2005). Such success stories seem to depend largely on the context in which seasonal climate forecasts were deployed—in well-funded policy systems, with adequate resources to customize and use forecasts, benefits can accrue to the local society as a whole. From these limited cases, it is suggested that where income, status, and access to information are more equitably distributed in a society, the introduction of seasonal forecasts may create winners; in contrast, when pre-existing conditions are unequal, the application of seasonal climate forecasts may create more losers by exacerbating those inequities (Lemos and Dilling, 2007). The consequences can be costly both to users and seasonal forecast credibility.

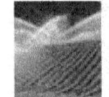

A true dialogue between end users of scientific information and those who generate data and tools is rarely achieved.

This is followed by another sidebar on comparative examples of boundary spanning which emphasizes the "systemic" nature of boundary spanning—that boundary organizations produce reciprocity of benefits to various groups.

One final observation can be made at this juncture concerning boundary spanning and the dissemination of climate information and knowledge. Some suggest a three-pronged process of outreach consisting of "missionary work", "co-discovery", and "persistence". Missionary work is directed toward potential users of climate information who do not fully understand the potential of climate variation and change and the potential of climate information applications. Such non-users may reject science not because they believe it to be invalid, but because they do not envision the strategic threat to their water use, or water rights, through non-application of climate information. Co-discovery, by contrast, is the process of co-production of knowledge aimed at answering questions of concern to both managers and scientists, as we have discussed. Overcoming resistance to using information, in the first case, and ensuring co-production in the second instance—both depend on persistence: the notion that effective introduction of climate applications may require long-term efforts to establish useful relationships, particularly where there is disbelief in the science of climate

change or where there is significant asymmetry of access to information and other resources (i.e., Chambers, 1997; Weiner, 2004).

4.3.2 Regional Integrated Science and Assessment Teams (RISAs) —An Opportunity for Boundary Spanning, and a Challenge

A true dialogue between end users of scientific information and those who generate data and tools is rarely achieved. The eight Regional Integrated Science and Assessment (RISA) teams that are sponsored by NOAA and activities sponsored by the Environmental Protection Agency's Global Change Research Program are among the leaders of this experimental endeavor, and represent a new collaborative paradigm in which decision makers are actively involved in developing research agendas. RISAs explicitly seek to work at the boundary of science and decision making.

There are five principal approaches RISA teams have learned that facilitate engagement with stakeholders and design of climate-related decision-support tools for water managers. First, RISAs employ a "stakeholder-driven research" approach that focuses on performing research on both the supply side (i.e., information development) and demand side (i.e., the user and her/his needs). Such reconciliation efforts require robust communication in which each side informs the other with regard to decisions, needs, and products—this communication cannot be intermittent; it must be robust and ongoing.

Second, some RISAs employ an "information broker" approach. They produce little new scientific information themselves, due to resource limitations or lack of critical mass in a particular scientific discipline. Rather, the RISAs' primary role is providing a conduit for information and facilitating the development of information networks.

Third, RISAs generally utilize a "participant/ advocacy" or "problem-based" approach, which involves focusing on a particular problem or issue and engaging directly in solving that problem. They see themselves as part of a learning system and promote the opportunity for joint learning with a well-defined set of stakeholders who share the RISA's perspective on the problem and desired outcomes.

Fourth, some RISAs utilize a "basic research" approach in which the researchers recognize particular gaps in the fundamental knowledge needed in the production of context sensitive, policy-relevant information. Any RISA may utilize many or most of these approaches at different times depending upon the particular context of the problem. The more well-established RISAs have more formal processes and procedures in place to identify stakeholder needs and design appropriate responses, as well as to evaluate the effectiveness of decision-support tools that are developed.

Finally, a critical lesson for climate science policy from RISAs is that, despite knowing what is needed to produce, package, and disseminate useful climate information—and the well-recognized success of the regional partnerships with stakeholders, RISAs continue to struggle for funding while RISA-generated lessons are widely acclaimed. To a large extent, they have not influenced federal climate science policy community outside of the RISAs themselves, though progress has been made in recent years. Improving feedback between RISA programs and the larger research enterprise need to be enhanced so lessons learned can inform broader climate science policy decisions—not just those

decisions made on the local problem-solving level (McNie *et al.*, 2007).

In April 2002, the House Science Committee held a hearing to explore the connections of climate science and the needs of decision makers. One question it posed was the following: "Are our climate research efforts focused on the right questions"? (<http://www.house.gov/science/hearings/full102/apr17/full_charter_041702.htm>). The Science Committee found that the RISA program is a promising means to connect decision-making needs with the research prioritization process, because "(it) attempts to build a regional-scale picture of the interaction between climate change and the local environment from the ground up. By funding research on climate and environmental science focused on a particular region, [the RISA] program currently supports interdisciplinary research on climate-sensitive issues in five selected regions around the country. Each region has its own distinct set of vulnerabilities to climate change, *e.g.*, water supply, fisheries, agriculture, *etc.*, and RISA's research is focused on questions specific to each region".

4.3.3 Developing Knowledge-Action Systems—a Climate for Inclusive Management

Research suggests that decision makers do not always find seasonal-to-interannual forecast products, and related climate information, to be useful for the management of water resources—this is a theme central to this entire Product (e.g., Weiner, 2004). As our case study experiments suggest, in order to ensure that information is useful, decision makers must be able to affect the substance of climate information production and the method of delivery so that information producers know what are the key questions to respond to in the broad and varied array of decisional needs different constituencies require (Sarewitz and Pielke, 2007; Callahan *et al.*, 1999; NRC, 1999). This is likely the most effective process by which true decision-support activities can be made useful.

Efforts to identify factors that improve the usability of SI climate information have found that effective "knowledge-action" systems focus on promoting broad, user-driven risk management objectives (Cash and Buizer, 2005). These

Decision makers do not always find seasonal-to-interannual forecast products, and related climate information, to be useful for the management of water resources.

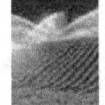

Knowledge systems need to engage a range of participants including those who generate scientific tools and data, those who translate them into predictions for use by decision makers, and the decision makers themselves.

objectives, in turn, are shaped by the decision context, which usually contains multiple stresses and management goals. Research on water resource decision making suggests that goals are defined very differently by agencies or organizations dedicated to managing single-issue problems in particular sectors (e.g., irrigation, public supply) when compared to decision makers working in political jurisdictions or watershed-based entities designed to comprehensively manage and coordinate several management objectives simultaneously (e.g., flood control and irrigation, power generation, and in-stream flow). The latter entities face the unusual challenge of trying to harmonize competing objectives, are commonly accountable to numerous users, and require "regionally and locally tailored solutions" to problems (Water in the West, 1998; Kenney and Lord, 1994; Grigg, 1996).

Effective knowledge-action systems should be designed for learning rather than knowing; the difference being that the former emphasizes the process of exchange between decision makers and scientists, constantly evolving in an iterative fashion, rather than aiming for a one-time-only completed product and structural permanence. Learning requires that knowledge-action systems have sufficient flexibility of processes and institutions to effectively produce and apply climate information (Cash and Buizer, 2005), encourage diffusion of boundary-spanning innovation, be self-innovative and responsive, and develop "operating criteria that measure responsiveness to changing conditions and external advisory processes" (Cash and Buizer, 2005). Often, nontraditional institutions that operate outside of "normal" channels, such as nongovernmental organizations (NGOs) or regional coordinating entities, are less constrained by tradition or legal mandate and thus more able to innovate.

To encourage climate forecast and information producers and end-users to better communicate with one another, they need to be engaged in a long-term dialogue about each others' needs and capabilities. To achieve this, knowledge producers must be committed to establishing opportunities for joint learning. When such communication systems have been established, the result has been the gaining of knowledge by

users. The discovery that climate information must be part of a larger suite of information can help producers understand the decision context, and better appreciate that users manage a broad array of risks. Lead innovators within the user community can lay the groundwork for broader participation of other users and greater connection between producers and users (Cash and Buizer, 2005).

Such tailoring or conversion of information requires organizational settings that foster communication and exchange of ideas between users and scientists. For example, a particular user might require a specific type of precipitation forecast or even a different type of hydrologic model to generate a credible forecast of water supply volume. This producer-user dialogue must be long term, it must allow users to independently verify the utility of forecast information, and finally, must provide opportunities for verification results to "feed back" into new product development (Cash and Buizer, 2005; Jacobs et al., 2005).

Studies of this connection refer to it as an "end-to-end" system to suggest that knowledge systems need to engage a range of participants including those who generate scientific tools and data, those who translate them into predictions for use by decision makers, and the decision makers themselves. A forecast innovation might combine climate factor observations, analyses of climate dynamics, and SI forecasts. In turn, users might be concerned with varying problems and issues such as planting times, instream flows to support endangered species, and reservoir operations.

As Cash and Buizer note, "Often entire systems have failed because of a missing link between the climate forecast and these ultimate user actions. Avoiding the missing link problem varies according to the particular needs of specific users (Cash and Buizer, 2005). Users want useable information more than they want answers—they want an understanding of things that will help them explain, for example, the role of climate in determining underlying variation in the resources they manage. This includes a broad range of information needed for risk management, not just forecasting particular threats.

Organizational measures to hasten, encourage, and sustain these knowledge-action systems must include practices that empower people to use information through providing adequate training and outreach, as well as sufficient professional reward and development opportunities. Three measures are essential. First, organizations must provide incentives to produce boundary objects, such as decisions or products that reflect the input of different perspectives. Second, they must involve participation from actors across boundaries. And finally, they must have lines of accountability to the various organizations spanned (Guston, 2001).

Introspective evaluations of the organizations' ability to learn and adapt to the institutional and knowledge-based changes around them should be combined with mechanisms for feedback and advice from clients, users, and community leaders. However, it is important that a review process not become an end in itself or be so burdensome as to affect the ability of the organization to function efficiently. This orientation is characterized by a mutual recognition on the part of scientists and decision makers of the importance of social learning—that is, learning by doing or by experiment, and refinement of forecast products in light of real-world experiences and previous mistakes or errors—both in forecasts and in their application. This learning environment also fosters an emphasis on adaptation and diffusion of innovation (*i.e.*, social learning, learning from past mistakes, long-term funding).

4.3.4 The Value of User-Driven Decision Support

Studies of what makes climate forecasts useful have identified a number of common characteristics in the process by which forecasts are generated, developed, and taught to—and disseminated among—users (Cash and Buizer, 2005). These characteristics (some previously described) include:

- Ensuring that the problems forecasters address are driven by forecast users;
- Making certain that knowledge-action systems (the process of interaction between scientists and users that produces forecasts) are end-to-end inclusive;
- Employing "boundary organizations" (groups or other entities that bridge the communication void between experts and users) to perform translation and mediation functions between the producers and consumers of forecasts;
- Fostering a social learning environment between producers and users (*i.e.*, emphasizing adaptation); and
- Providing stable funding and other support to keep networks of users and scientists working together.

As noted earlier, "users" encompass a broad array of individuals and organizations, including farmers, water managers, and government agencies; while "producers" include scientists and engineers and those "with relevant expertise derived from practice" (Cash and Buizer, 2005). Complicating matters is that some "users" may, over time, become "producers" as they translate, repackage, or analyze climate information for use by others.

In effective user-driven information environments, the agendas of analysts, forecasters, and scientists who generate forecast information are at least partly set by the users of the information. Moreover, the collaborative process is grounded in appreciation for user perspectives regarding the decision context in which they work, the multiple stresses under which they labor, and their goals so users can integrate climate knowledge into risk management. Most important, this user-driven outlook is reinforced by a systematic effort to link the generation of forecast information with needs of users through soliciting advice and input from the latter at every step in the generation of information process.

Effective knowledge-action systems do not allow particular research or technology capabilities (*e.g.*, ENSO forecasting) to drive the dialogue. Instead, effective systems ground the collaborative process of problem definition in user perspectives regarding the decision context, the multiple stresses bearing on user decisions, and ultimate goals that the knowledge-action system seeks to advance. For climate change information, this means shifting the focus toward "the promotion of broad, user-driven risk-management objectives, rather than advancing the uptake of particular forecasting

There is an emerging consensus that the utility of information intended to make possible sustainable environmental decisions depends on the "dynamics of the decision context and its broader social setting".

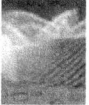

technologies" (Cash and Buizer, 2005; Sarewitz and Pielke, 2007).

In sum, there is an emerging consensus that the utility of information intended to make possible sustainable environmental decisions depends on the "dynamics of the decision context and its broader social setting" (Jasanoff and Wynne, 1998; Pielke *et al.*, 2000; Sarewitz and Pielke, 2007). Usefulness is not inherent in the knowledge generated by forecasters—the information generated must be "socially robust". Robustness is determined by how well it meets three criteria: (1) is it valid outside, as well as inside the laboratory; (2) is validity achieved through involving an extended group of experts, including lay "experts;" and 3) is the information (e.g., forecast models) derived from a process in which society has participated as this ensures that the information is less likely to be contested (Gibbons, 1999).

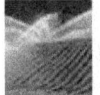

Finally, a user-driven information system relies heavily on two-way communication. Such communication can help bridge gaps between what is produced and what is likely to be used, thus ensuring that scientists produce products that are recognized by the users, and not just the producers, as useful. Effective user-oriented two-way communication can increase users' understanding of how they could use climate information and enable them to ask questions about information that is uncertain or in dispute. It also affords an opportunity to produce "decision-relevant" information that might otherwise not be produced because scientists may not have understood completely what kinds of information would be most useful to water resource decision makers (NRC, 2008).

In conclusion, user-driven information in regard to seasonal-to-interannual climate variability for water resources decision making must be salient (e.g., decision-relevant and timely), credible (viewed as accurate, valid, and of high quality), and legitimate (uninfluenced by pressures or other sources of bias) (see NRC, 2008; NRC, 2005). In the words of a recent National Research Council report, broad involvement of "interested and affected parties" in framing scientific questions helps ensure that the science produced is useful ("getting the right science") by ensuring that decision-support tools are

> User-driven information in regard to seasonal-to-interannual climate variability for water resources decision making must be salient, credible, and legitimate.

explicit about any simplifying assumptions that may be in dispute among the users, and accessible to the end-user (NRC, 2008).

4.3.5 Proactive Leadership— Championing Change

Organizations—public, private, scientific, and political—have leaders: individuals charged with authority, and span of control, over important personnel, budgetary, and strategic planning decisions, among other venues. Boundary organizations require a kind of leadership called inclusive management practice by its principal theorists (Feldman and Khademian, 2004). Inclusive management is defined as management that seeks to incorporate the knowledge, skills, resources, and perspectives of several actors and seeks to avoid creating "winners and losers" among stakeholders.

While there is an enormous literature on organizational leadership, synthetic studies—those that take various theories and models about leaders and try to draw practical, even anecdotal, lessons for organizations—appear to coalesce around the idea that inclusive leaders have context-specific skills that emerge through a combination of tested experience within a variety of organizations, and a knack for judgment (Bennis, 2003; Feldman and Khademan, 2004; Tichy and Bennis, 2007). These skills evolve through trial and error and social learning. Effective "change-agent" leaders have a guiding vision that sustains them through difficult times, a passion for their work and an inherent belief in its importance, and a basic integrity toward the way in which they interact with people and approach their jobs (Bennis, 2003).

While it is difficult to discuss leadership without focusing on individual leaders (and difficult to disagree with claims about virtuous leadership), inclusive management also embraces the notion of "process accountability": that leadership is embodied in the methods by which organizations make decisions, and not in charismatic personality alone. Process accountability comes not from some external elected political principle or body that is hierarchically superior, but instead infuses through processes of deliberation and transparency. All of these elements make boundary organizations capable of being solution focused and integrative and, thus, able to span the domains of climate knowledge production and climate knowledge for water management use.

Adaptive and inclusive management practices are essential to fulfilling these objectives. These practices must empower people to use information through providing adequate training and outreach, as well as sufficient professional reward and development opportunities; and they must overcome capacity-building problems within organizations to ensure that these objectives are met, including adequate user support. The cases discussed below—on the California Department of Water Resources' role in adopting climate variability and change into regional water management, and the efforts of the Southeast consortium and its satellite efforts—are examples of inclusive leadership which illustrate how scientists as well agency managers can be proactive leaders. In the former case, decision makers consciously decided to develop relationships with other western states' water agencies and partnership (through a Memorandum of Understanding [MOU]) with NOAA. In the latter, scientists ventured into collaborative efforts—across universities, agencies, and states—because they shared a commitment to exchanging information in order to build institutional capacity among the users of the information themselves.

Case Study A:
Leadership in the California Department of Water Resources

The deep drought in the Colorado River Basin that began with the onset of a La Niña episode in 1998 has awakened regional water resources managers to the need to incorporate climate variability and change into their plans and reservoir forecast models. Paleohydrologic estimates of streamflow, which document extended periods of low flow and demonstrate greater streamflow variability than the information found in the gage record, have been particularly persuasive examples of the non-stationary behavior of the hydroclimate system (Woodhouse *et al.*, 2006; Meko *et al.*, 2007). Following a 2005 scientist-stakeholder workshop on the use of paleohydrologic data in water resource management <http://www.climas.arizona.edu/ calendar/details.asp?event_id=21>, NOAA

RISA and California Department of Water Resources (CDWR) scientists developed strong relationships oriented toward improving the usefulness and usability of science in water management. Since the 2005 workshop, CDWR, whose mission in recent years includes preparation for potential impacts of climate change on California's water resources, has led western states' efforts in partnering with climate scientists to co-produce hydroclimatic science to inform decision making. CDWR led the charge to clarify scientific understanding of Colorado River Basin climatology and hydrology, past variations, projections for the future, and impacts on water resources, by calling upon the National Academy of Sciences to convene a panel to study the aforementioned issues (NRC, 2007). This occurred, and in 2007, CDWR developed a Memorandum of Agreement with NOAA, in order to better facilitate cooperation with scientists in NOAA's RISA program and research laboratories (CDWR, 2007a).

Case Study B:
Cooperative Extension Services, Watershed Stewardship: The Southeast Consortium

Developing the capacity to use climate information in resource management decision making requires both outreach and education, frequently in an iterative fashion that leads to two-way communication and builds partnerships. The Cooperative Extension Program has long been a leader in facilitating the integration of scientific information into decision maker of practice in the agricultural sector. Cash (2001) documents an example of successful

Cooperative Extension leadership in providing useful water resources information to decision makers confronting policy changes in response to depletion of groundwater in the High Plains aquifer. Cash notes the Cooperative Extension's history of facilitating dialogue between scientists and farmers, encouraging the development of university and agency research agendas that reflect farmers' needs, translating scientific findings into site-specific guidance, and managing demonstration projects that integrate farmers into researchers' field experiments.

In the High Plains aquifer example, the Cooperative Extension's boundary-spanning work was motivated from a bottom-up need of stakeholders for credible information on whether water management policy changes would affect their operations. By acting as a liaison between the agriculture and water management decision making communities, and building bridges between many levels of decision makers, Kansas Cooperative Extension was able to effectively coordinate information flows between university and USGS modelers, and decision makers. The result of their effort was collaborative development of a model with characteristics needed by agriculturalists (at a sufficient spatial resolution) and that provided credible scientific information to all parties. Kansas Cooperative Extension effectiveness in addressing groundwater depletion and its impact on farmers sharply contrasted with the Cooperative Extension efforts in other states where no effort was made to establish multi-level linkages between water management and agricultural stakeholders.

The Southeast Climate Consortium RISA (SECC), a confederation of researchers at six universities in Alabama, Georgia, and Florida, has used more of a top-down approach to developing stakeholder capacity to use climate information in the Southeast's $33 billion agricultural sector (Jagtap et al., 2002). Early in its existence, SECC researchers recognized the potential to use knowledge of the impact of the El Niño-Southern Oscillation on local climate to provide guidance to farmers, ranchers, and forestry sector stakeholders on yields and changes to risk (e.g., frost occurrence). Through a series of needs and vulnerability assessments (Hildebrand et al., 1999, Jagtap et al., 2002), SECC

researchers determined that the potential for producers to benefit from seasonal forecasts depends on factors that include the flexibility and willingness to adapt farming operations to the forecast, and the effectiveness of the communication process—and not merely documenting the effects of climate variability and providing better forecasts (Jones et al., 2000). Moreover, Fraisse et al. (2006) explain that climate information is only valuable when both the potential response and benefits of using the information are clearly defined. SECC's success in championing integration of new information is built upon a foundation of sustained interactions with agricultural producers in collaboration with extension agents. Extension specialists and faculty are integrated as members of the SECC research team. SECC engages agricultural stakeholders through planned communication and outreach, such as monthly video conferences, one-on-one meetings with extension agents and producers, training workshops designed for extension agents and resource managers to gain confidence in climate decision tool use and to identify opportunities for their application, and by attending traditional extension activities (e.g., commodity meetings, field days) (Fraisse et al., 2005). SECC is able to leverage the trust engendered by Cooperative Extension's long service to the agricultural community and Extension's access to local knowledge and experience, in order to build support for its AgClimate online decision-support tool <http://www.agclimate.org> (Fraisse et al., 2006). This direct engagement with stakeholders provides feedback to improve the design of the tool and to enhance climate forecast communication (Breuer et al., 2007).

Yet another Cooperative Extension approach to integrating scientific information into decision making is the Extension's Master Watershed Steward (MWS) programs. MWS was first developed at Oregon State University <http://seagrant.oregonstate.edu/wsep/index.html>. In exchange for 40 hours of training on aspects of watersheds that range from ecology to water management, interested citizen volunteers provide service to their local community through projects, such as drought and water quality monitoring, developing property management plans, and conducting riparian habitat restoration. Arizona's MWS program includes training

Climate information is only valuable when both the potential response and benefits of using the information are clearly defined.

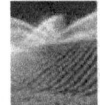

in climate and weather (Garfin and Emanuel, 2006); stewards are encouraged to participate in drought impact monitoring through Arizona's Local Drought Impact Groups (GDTF, 2004; Garfin, 2006). MWS enhances the capacity for communities to deploy new climate information and to build expertise for assimilating scientific information into a range of watershed management decisions.

4.3.6 Funding and Long-Term Capacity Investments Must Be Stable and Predictable

Provision of a stable funding base, as well as other investments, can help to ensure effective knowledge-action systems for climate change. Stable funding promotes long-term stability and trust among stakeholders because it allows researchers to focus on user needs over a period of time, rather than having to train new participants in the process. Given that these knowledge-action systems produce benefits for entire societies, as well as for particular stakeholders in a society, it is not uncommon for these systems to be thought of as producing both public and private goods, and thus, needing both public and private sources of support (Cash and Buizer, 2005). Private funders could include, for example, farmers whose risks are reduced by the provision of climate information (as is done in Queensland, Australia, where the individual benefits of more profitable production are captured by farmers who partly support drought-warning systems). In less developed societies, by contrast, it would not be surprising for these systems to be virtually entirely supported by public sources of revenue (Cash and Buizer, 2005).

Experience suggests that a public-private funding balance should be shaped on the basis of user needs and capacities to self-tailor knowledge-action systems. More generic systems that could afterwards be tailored to users' needs might be most suitable for public support, while co-funding with particular users can then be pursued for developing a collaborative system that more effectively meets users' needs. Funding continuity is essential to foster long-term relationship building between users and producers. The key point here is that—regardless of who pays for these systems, continued funding of the social and economic investigations of the

use of scientific information is essential to ensure that these systems are used and are useful (Jacobs *et al.*, 2005).

Other long-term capacity investments relate to user training—an important component that requires drawing upon the expertise of "integrators". Integrators are commonly self-selected managers and decision makers with particular aptitude or training in science, or scientists who are particularly good at communication and applications. Training may entail curriculum development, career and training development for users as well as science integrators, and continued mid-career in-stream retraining and re-education. Many current integrators have evolved as a result of doing interdisciplinary and applied research in collaborative projects, and some have been encouraged by funding provided by NOAA's Climate Programs Office (formerly Office of Global Programs) (Jacobs, *et al.*, 2005).

4.3.7 Adaptive Management for Water Resources Planning— Implications for Decision Support

Since the 1970s, an "adaptive management paradigm" has emerged that is characterized by: greater public and stakeholder participation in decision making; an explicit commitment to environmentally sound, socially just outcomes; greater reliance upon drainage basins as planning units; program management via spatial and managerial flexibility, collaboration, participation, and sound, peer-reviewed science; and finally, embracing of ecological, economic, and equity considerations (Hartig *et al.*, 1992; Landre and Knuth, 1993; Cortner and Moote, 1994; Water in the West, 1998; May *et al.*, 1996; McGinnis, 1995; Miller *et al.*, 1996; Cody, 1999; Bormann *et al.*, 1993; Lee, 1993). Adaptive management traces its roots to a convergence of intellectual trends and disciplines, including industrial relations theory, ecosystems management, ecological science, economics, and engineering. It also embraces a constellation of concepts such as social learning, operations research, environmental monitoring, precautionary risk avoidance, and many others (NRC, 2004).

Adaptive management can be viewed as an alternative decision-making paradigm that seeks

Regardless of who pays for these systems, continued funding of the social and economic investigations of the use of scientific information is essential to ensure that these systems are used and are useful.

> An adaptive management approach is one that is flexible and subject to adjustment in an iterative, social learning process.

insights into the behavior of ecosystems utilized by humans. In regard to climate variability and water resources, adaptive management compels consideration of questions such as the following: What are the decision-support needs related to managing in-stream flows/low flows? How does climate variability affect runoff? What is the impact of increased temperatures on water quality or on cold-water fisheries' (e.g., lower dissolved oxygen levels)? What other environmental quality parameters does a changing climate impact related to endangered or threatened species? And, what changes to runoff and flow will occur in the future, and how will these changes affect water uses among future generations unable to influence the causes of these changes today? What makes these questions particularly challenging is that they are interdisciplinary in nature[4].

While a potentially important concept, applying adaptive management to improving decision support requires that we deftly avoid a number of false and sometimes uncritically accepted suppositions. For example, adaptive management does not postpone actions until "enough" is known about a managed ecosystem, but supports actions that acknowledge the limits of scientific knowledge, "the complexities and stochastic behavior of large ecosystems", and the uncertainties in natural systems, economic demands, political institutions, and ever-changing societal social values (NRC, 2004; Lee, 1999). In short, an adaptive management approach is one that is flexible and subject to adjustment in an iterative, social learning process (Lee, 1999). If treated in such a manner, adaptive management can encourage timely responses by: encouraging protagonists involved in water management to bound disputes; investigating

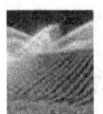

---

[4] Underscored by the fact that scholars concur, adaptive management entails a broad range of processes to avoid environmental harm by imposing modest changes on the environment, acknowledging uncertainties in predicting impacts of human activities on natural processes, and embracing social learning (i.e., learning by experiment). In general, it is characterized by managing resources by learning, especially about mistakes, in an effort to make policy improvements using four major strategies that include: (1) modifying policies in the light of experience, (2) permitting such modifications to be introduced in "mid-course, (3) allowing revelation of critical knowledge heretofore missing and analysis of management outcomes, and (4) incorporating outcomes in future decisions through a consensus-based approach that allows government agencies and NGOs to conjointly agree on solutions (Bormann, et al., 1993; Lee, 1993; Definitions of Adaptive Management, 2000).

environmental uncertainties; continuing to constantly learn and improve the management and operation of environmental control systems; learning from error; and "reduc(ing) decision-making gridlock by making it clear...that there is often no 'right' or 'wrong' management decision, and that modifications are expected" (NRC, 2004).

The four cases discussed below illustrate varying applications, and context specific problems, of adaptive management. The discussion of Integrated Water Resource Planning stresses the use of adaptive management in a variety of local political contexts where the emphasis is on reducing water use and dependence on engineered solutions to provide water supply. The key variables are the economic goals of cost savings coupled with the ability to flexibly meet water demands. The Arizona Water Institute case illustrates the use of a dynamic organizational training setting to provide "social learning" and decisional responsiveness to changing environmental and societal conditions. A key trait is the use of a boundary-spanning entity to bridge various disciplines.

The Glen Canyon and Murray–Darling Basin cases illustrate operations-level decision making aimed at addressing a number of water management problems that, over time, have become exacerbated by climate variability, namely: drought, streamflow, salinity, and regional water demand. On one hand, adaptive management has been applied to "re-engineer" a large reservoir system. On the other, a management authority that links various stakeholders together has attempted to instill a new set of principles into regional river basin management. It should be borne in mind that transferability of lessons from these cases depends not on some assumed "randomness" in their character (they are not random; they were chosen because they are amply studied), but on the similarity between their context and that of other cases. This is a problem also taken up in Section 4.5.2.

4.3.8 Integrated Water Resources Planning—Local Water Supply and Adaptive Management

A significant innovation in water resources management in the United States that affects climate information use is occurring in the local water supply sector: the growing use of integrated water resource planning (or IWRP) as an alternative to conventional supply-side approaches for meeting future demands. IWRP is gaining acceptance in chronically water-short regions such as the Southwest and portions of the Midwest, including Southern California, Kansas, Southern Nevada, and New Mexico (*e.g.*, Beecher, 1995; Warren *et al.*, 1995; Fiske and Dong, 1995; Wade, 2001).

IWRP's goal is to "balanc(e) water supply and demand management considerations by identifying feasible planning alternatives that meet the test of least cost without sacrificing other policy goals" (Beecher, 1995). This can be variously achieved through depleted aquifer recharge, seasonal groundwater recharge, conservation incentives, adopting growth management strategies, wastewater reuse, and/or applying least cost planning principles to large investor-owned water utilities. The latter may encourage IWRP by demonstrating the relative efficiency of efforts to reduce demand as opposed to building more supply infrastructure. A particularly challenging alternative is the need to enhance regional planning among water utilities in order to capitalize on the resources of every water user, eliminate unnecessary duplication of effort, and avoid the cost of building new facilities for water supply (Atwater and Blomquist, 2002).

In some cases, short-term applications of least cost planning may increase long-term project costs, especially when environmental impacts, resource depletion, and energy and maintenance costs are included. The significance of least cost planning is that it underscores the importance of long- and short-term costs (in this case, of water) as an influence on the value of certain kinds of information for decisions. Models and forecasts that predict water availability under different climate scenarios can be especially useful to least cost planning and make more credible efforts to reducing demand. Specific questions IWRP raises for decision support given a changing climate include: How precise must climate information be to enhance long-term planning? How might predicted climate change provide an incentive for IWRP strategies? and, What climate information is needed to optimize decisions on water pricing, re-use,

shifting from surface to groundwater use, and conservation?

Case Study C:
Approaches to Building User Knowledge and Enhancing Capacity Building—the Arizona Water Institute

The Arizona Water Institute was initiated in 2006 to focus the resources of the State of Arizona's university system on the issue of water sustainability. Because there are 400 faculty and staff members in the three Arizona universities who work on water-related topics, it is clear that asking them and their students to assist the state in addressing the major water quantity and quality issues should make a significant contribution to water sustainability. This is particularly relevant given that the state budget for supporting water resources related work is exceedingly small by comparison to many other states, and the fact that Arizona is one of the fastest-growing states in the United States. In addition to working towards water sustainability, the Institute's mission includes water-related technology transfer from the universities to the private sector to create and develop economic opportunities, as well as build capacity, to enhance the use of scientific information in decision making.

The Institute was designed from the beginning as a "boundary organization" to build pathways for innovation between the universities and state agencies, communities, Native American tribal representatives, and the private sector. In addi-

In some cases, short-term applications of least cost planning may increase long-term project costs, especially when environmental impacts, resource depletion, and energy and maintenance costs are included.

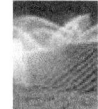

tion, the Institute is specifically designed as an experiment in how to remove barriers between groups of researchers in different disciplines and across the universities. The Institute's projects involve faculty members from more than one of the universities, and all involve true engagement with stakeholders. The faculty is provided incentives to engage both through small grants for collaborative projects and through the visibility of the work that the Institute supports. Further, the Institute's structure is unique, in that there are high level Associate Directors of the Institute whose assignment is to build bridges between the universities and the three state agencies that are the Institute's partners: Water Resources, Environmental Quality, and Commerce. These Associate Directors are physically located inside the state agencies that they serve. The intent is to build trust between university researchers (who may be viewed as "out of touch with reality" by agency employees), and agency or state employees (whom researchers may believe are not interested in innovative ideas). Physical proximity of workspaces and daily engagement has been shown to be an ingredient of trust building.

A significant component of the Institute's effort is focused on: capacity building, training students through engagement in real-world water policy issues, providing better access to hydrologic data for decision makers, assisting them in visualizing the implications of the decisions that they make, workshops and training programs for tribal entities, joint definition of research agendas between stakeholders and researchers, and building employment pathways to train students for specific job categories where there is an insufficient supply of trained workers, such as water and wastewater treatment plant operators. Capacity-building in interdisciplinary planning applications such as combining land use planning and water supply planning to focus on sustainable water supplies for future development is emerging as a key need for many communities in the state.

The Institute is designed as a "learning organization" in that it will regularly revisit its structure and function, and redesign itself as needed to maintain effectiveness in the context of changing institutional and financial conditions.

Case Study D:
Murray–Darling Basin—Sustainable Development and Adaptive Management

The Murray-Darling Basin Agreement (MDBA), formed in 1985 by New South Wales, Victoria, South Australia and the Commonwealth, is an effort to provide for the integrated and conjoint management of the water and related land resources of the world's largest catchment system. The problems initially giving rise to the agreement included rising salinity and irrigation-induced land salinization that extended across state boundaries (SSCSE, 1979; Wells, 1994). However, embedded in its charter was a concern with using climate variability information to more effectively manage drought, runoff, riverine flow and other factors in order to meet the goal of "effective planning and management for the equitable, efficient and sustainable use of the water, land and environmental resources (of the basin)" (MDBC, 2002).

Some of the more notable achievements of the MDBA include programs to promote the management of point and non-point source pollution; balancing consumptive and in-stream uses (a decision to place a cap on water diversions was adopted by the commission in 1995); the ability to increase water allocations—and rates of water flow—in order to mitigate pollution and protect threatened species (applicable in all states except Queensland); and an explicit program for "sustainable management". The latter hinges on implementation of several strategies, including a novel human dimension strategy adopted in 1999 that assesses the social, institutional and cultural factors impeding sustainability; as well as adoption of specific policies to deal with salinity, better manage wetlands, reduce the frequency and intensity of algal blooms by better managing the inflow of nutrients, reverse declines in native fisheries populations (a plan which, like that of many river basins in the United States, institutes changes in dam operations to permit fish passage), and preparing floodplain management plans.

Moreover, a large-scale environmental monitoring program is underway to collect and analyze basic data on pressures upon the basin's resources as well as a "framework for evaluating and reporting on government and community

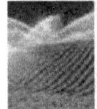

investment" efforts and their effectiveness. This self-evaluation program is a unique adaptive management innovation rarely found in other basin initiatives. To support these activities, the Commission funds its own research program and engages in biophysical and social science investigations. It also establishes priorities for investigations based, in part, on the severity of problems, and the knowledge acquired is integrated directly into commission policies through a formal review process designed to assure that best management practices are adopted.

From the standpoint of adaptive management, the Murray–Darling Basin Agreement seeks to integrate quality and quantity concerns in a single management framework; has a broad mandate to embrace social, economic, environmental and cultural issues in decisions; and has considerable authority to supplant, and supplement, the authority of established jurisdictions in implementing environmental and water development policies. While water quality policies adopted by the Basin Authority are recommended to states and the federal government for approval, generally, the latter defer to the commission and its executive arm. The MDBA also promotes an integrated approach to water resources management. Not only does the Commission have responsibility for functions as widely varied as floodplain management, drought protection, and water allocation, but for coordinating them as well. For example, efforts to reduce salinity are linked to strategies to prevent waterlogging of floodplains and land salinization on the Murray and Murrumbidgee Valleys (MDBC, 2002). Also, the Basin commission's environmental policy aims to utilize water allocations not only to control pollution and benefit water users, but to integrate its water allocation policy with other strategies for capping diversions, governing in-stream flow, and balancing in-stream needs and consumptive (*i.e.*, agricultural irrigation) uses. Among the most notable of MDBC's innovations is its community advisory effort.

In 1990, the ministerial council for the MDBC adopted a Natural Resources Management Strategy that provides specific guidance for a community-government partnership to develop plans for integrated management of the Basin's

water, land and other environmental resources on a catchment basis. In 1996, the ministerial council put in place a Basin Sustainability Plan that provides a planning, evaluation and reporting framework for the Strategy, and covers all government and community investment for sustainable resources management in the basin.

According to Newson (1997), while the policy of integrated management has "received wide endorsement", progress towards effective implementation has fallen short—especially in the area of floodplain management. This has been attributed to a "reactive and supportive" attitude as opposed to a proactive one. Despite such criticism, it is hard to find another initiative of this scale and sophistication that has attempted adaptive management based on community involvement.

Case Study E:
Adaptive Management in Glen Canyon,
Arizona and Utah

Glen Canyon Dam was constructed in 1963 to provide hydropower, water for irrigation, flood control, and public water supply—and to ensure adequate storage for the upper basin states of the Colorado River Compact (*i.e.*, Utah, Wyoming, New Mexico, and Colorado). Lake Powell, the reservoir created by Glen Canyon Dam, has a storage capacity equal to approximately two years flow of the Colorado River. Critics of Glen Canyon Dam have insisted that its impacts on the upper basin have been injurious almost from the moment it was completed. The flooding of one of the West's most beautiful canyons under the waters of Lake Powell increased rates of evapotranspiration and other forms of water loss (*e.g.*, seepage of water into canyon walls) and eradicated historical flow regimes. The latter has been the focus of recent debate. Prior to Glen Canyon's closure, the Colorado River, at this location, was highly variable with flows ranging from 120,000 cubic feet per second (cfs) to less than 1,000 cfs.

When the dam's gates were closed in 1963, the Colorado River above and below Glen Canyon was altered by changes in seasonal variability. Once characterized by muddy, raging floods, the river became transformed into a clear, cold stream. Annual flows were stabilized and

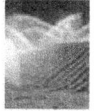

replaced by daily fluctuations by as much as 15 feet. A band of exotic vegetation colonized a river corridor no longer scoured by spring floods; five of eight native fish species disappeared; and the broad sand beaches of the pre-dam river eroded away. Utilities and cities within the region came to rely on the dam's low cost power and water, and in-stream values were ignored (Carothers and Brown, 1991).

Attempts to abate or even reverse these impacts came about in two ways. First, in 1992, under pressure from environmental organizations, Congress passed the Grand Canyon Protection Act that mandated Glen Canyon Dam's operations coincide with protection, migration, and improvement of the natural and cultural resources of the Colorado River. Second, in 1996, the Bureau of Reclamation undertook an experimental flood to restore disturbance and dynamics to the river ecosystem. Planners hoped that additional sand would be deposited on canyon beaches and that backwaters (important rearing areas for native fish) would be revitalized. They also hoped the new sand deposits would stabilize eroding cultural sites while high flows would flush some exotic fish species out of the system (Moody, 1997; Restoring the Waters, 1997). The 1996 flood created over 50 new sandbars, enhanced existing ones, stabilized cultural sites, and helped to restore some downstream sport fisheries. What made these changes possible was a consensus developed through a six-year process led by the Bureau that brought together diverse stakeholders on a regular basis. This process developed a new operational plan for Lake Powell, produced an environmental impact statement for the

project, and compelled the Bureau (working with the National Park Service) to implement an adaptive management approach that encouraged wide discussion over all management decisions.

While some environmental restoration has occurred, improvement to backwaters has been less successful. Despite efforts to restore native fisheries, the long-term impact of exotic fish populations on the native biological community, as well as potential for long-term recovery of native species, remains uncertain (Restoring the Waters, 1997). The relevance for climate variability decision support in the Glen Canyon case is that continued drought in the Southwest is placing increasing stress on the land and water resources of the region, including agriculture lands. Efforts to restore the river to conditions more nearly approximating the era before the dam was built will require changes in the dam's operating regime that will force a greater balance between instream flow considerations and power generation and offstream water supply. This will also require imaginative uses of forecast information to ensure that these various needs can be optimized.

4.3.9 Measurable Indicators of Progress to Promote Information Access and Use

These cases, and our previous discussion about capacity building, point to four basic measures that can be used to evaluate progress in providing equitable access to decision-support-generated information. First, the overall process of tool development should be inclusive. This could be measured and documented over time by the interest of groups to continue to participate and to be consulted and involved. Participants should view the process of collaboration as fair and effective—this could be gauged by elicitation of feedback from process participants.

Second, there should be progress in developing an interdisciplinary and interagency environment of collaboration, documented by the presence of dialogue, discussion, and exchange of ideas and data among different professions—in other words, documented boundary-spanning progress and building of trusted relationships. One documentable measure of interdisciplinary, boundary-spanning collaboration is the growth, over time, of professional reward systems within organizations that reward and recognize people who develop, use, and translate such systems for use by others.

Third, the collaborative process must be viewed by participants as credible. This means that participants feel it is believable and trustworthy and that there are benefits to all who engage in it. Again, this can be documented by elicitation of feedback from participants. Finally, outcomes of decision-support tools must be implementable in the short term, as well as longer-term. It is necessary to see progress in assimilating and using such systems in a short period of time in order to sustain the interest, effort, and participatory conviction of decision makers in the process. Table 4.2 suggests some specific,

Outcomes of decision-support tools must be implementable in the short term, as well as longer-term.

Table 4.2 Promoting Access to Information and its Use Between Scientists and Decision Makers—A Checklist (adopted from: Jacobs, 2003).

Information Integration
• Was information received by stakeholders and integrated into decision makers' management framework or world view?
• Was capacity built? Did the process lead to a result where institutions, organizations, agencies, officials can use information generated by decision-support experts? Did experts who developed these systems rely upon the knowledge and experience of decision makers—and respond to their needs in a manner that was useful?
• Will stakeholders continue to be invested in the program and participate in it over the long term?
Stakeholder Interaction/Collaboration
• Were contacts/relationships sustained over time and did they extend beyond individuals to institutions?
• Did stakeholders invest staff time or money in the activity?
• Was staff performance evaluated on the basis of quality or quantity of interaction?
• Did the project take on a life of its own, become at least partially self-supporting after the end of the project?
• Did the project result in building capacity and resilience to future events/conditions rather than focus on mitigation?
• Was quality of life or economic conditions improved due to use of information generated or accessed through the project?
• Did the stakeholders claim or accept partial ownership of final product?
Tool Salience/Utility
• Are the tools actually used to make decisions; are they used by high-valued uses and users?
• Is the information generated/provided by these tools accurate/valid?
• Are important decisions made on the basis of the tool?
• Does the use of these tools reduce vulnerabilities, risks, and hazards?
Collaborative Process Efficacy
• Was the process representative (all interests have a voice at the table)?
• Was the process credible (based on facts as the participants knew them)?
• Were the outcomes implementable in a reasonable time frame (political and economic support)?
• Were the outcomes disciplined from a cost perspective (i.e., there is some relationship between total costs and total benefits)?
• Were the costs and benefits equitably distributed, meaning there was a relationship between those who paid and those who benefited?

discrete measures that can be used to assess progress toward effective information use.

4.3.10 Monitoring Progress

An important element in the evaluation of process outcomes is the ability to monitor progress. A recent National Academy report (NRC, 2008) on NOAA's Sectoral Applications Research Program (SARP), focusing on climate-related information to inform decisions, encourages the identification of process measures that can be recorded on a regular basis, and of outcome measures tied to impacts of interest to NOAA and others that can also be recorded on a comparable basis.

These metrics can be refined and improved on the basis of research and experience, while consistency is maintained to permit time-series comparisons of progress (NRC, 2008). An advantage of such an approach includes the ability to document learning (e.g., Is there progress on the part of investigators in better project designs? Should there be a redirection of funding toward projects that show a large payoff in benefits to decision makers?).

Finally, the ability to consult with agencies, water resource decision makers, and a host of other potential forecast user communities can be an invaluable means of providing "mid-course" or interim indicators of progress in integrating forecast use in decisions. The Transition of Research Applications to Climate Services Program (TRACS), also within the NOAA Climate Program Office, has a mandate to support users of climate information and forecasts at multiple spatial and geographical scales—the transitioning of "experimentally mature climate information tools, methods, and processes, including computer-related applications (e.g. web interfaces, visualization tools), from research mode into settings where they may be applied in an operational and sustained manner" (TRACS, 2008). While TRACS primary goal is to deliver useful climate information products and services to local, regional, national, and even international policy makers, it is also charged with learning from its partners how to better accomplish technology transition processes. NOAA's focus is to infer how effectively transitions of research applications (i.e. experimentally developed and tested, end-user-

friendly information to support decision making), and climate services (i.e. the routine and timely delivery of that information, including via partnerships) are actually occurring.

While it is far too early to conclude how effectively this process of consultation has advanced, NOAA has established criteria for assessing this learning process, including clearly identifying decision makers, research, operations and extension partners, and providing for post-audit evaluation (e.g., validation, verification, refinement, maintenance) to determine at the end of the project if the transition of information has been achieved and is sustainable. Effectiveness will be judged in large part by the partners, and will focus on the developing means of communication and feedback, and on the deep engagement with the operational and end-user communities (TRACS, 2008).

The Southeast Climate Consortium case discussed below illustrates how a successful process of ongoing stakeholder engagement can be developed through the entire cycle (from development, introduction, and use) of decision-support tools. This experiment affords insights into how to elicit user community responses in order to refine and improve climate information products, and how to develop a sense of decision-support ownership through participatory research and modeling. The Potomac River case focuses on efforts to resolve a long-simmering water dispute and the way collaborative processes can themselves lead to improved decisions. Finally, the Upper San Pedro Partnership exemplifies the kind of sustained partnering efforts that are possible when adequate funding is made available, politicization of water management questions is prevalent, and climate variability has become an important issue on decision-makers' agenda, while the series of fire prediction workshops illustrate the importance of a highly-focused problem—one that requires improvements to information processes, as well as outcomes, to foster sustained collaboration.

The ability to consult with agencies, water resource decision makers, and a host of other potential forecast user communities can be an invaluable means of providing "mid-course" or interim indicators of progress in integrating forecast use in decisions.

Case Study F:
Southeast Climate Consortium Capacity Building, Tool Development

The Southeast Climate Consortium is a multidisciplinary, multi-institutional team, with members from Florida State University, University of Florida, University of Miami, University of Georgia, University of Auburn and the University of Alabama-Huntsville. A major part of the Southeast Climate Consortium's (SECC) effort is directed toward developing and providing climate and resource management information through AgClimate <http://www.agclimate.org/>, a decision-support system (DSS) introduced for use by Agricultural Extension, agricultural producers, and resource managers in the management of agriculture, forests, and water resources. Two keys to SECC's progress in promoting the effective use of climate information in agricultural sector decision making are (1) iterative ongoing engagement with stakeholders, from project initiation to decision-support system completion and beyond (further product refinement, development of ancillary products, *etc.*) (Breuer *et al.*, 2007; Cabrera *et al.*, 2007), and (2) co-developing a stakeholder sense of decision-support ownership through participatory research and modeling (Meinke and Stone, 2005; Breuer *et al.*, 2007; Cabrera *et al.*, 2007).

The SECC process has begun to build capacity for the use of climate information with a rapid assessment to understand stakeholder perceptions and needs regarding application of climate information that may have benefits (*e.g.*, crop yields, nitrogen pollution in water) (Cabrera *et al.*, 2006). Through a series of engagements, such as focus groups, individual interviews, research team meetings (including stakeholder advisors), and prototype demonstrations, the research team assesses which stakeholders are most likely adopt the decision-support system and communicate their experience with other stakeholders (Roncoli *et al.*, 2006), as well as stakeholder requirements for decision support (Cabrera *et al.*, 2007). Among the stakeholder requirements gleaned from more than six years of stakeholder engagements, are: present information in an uncomplicated way (often deterministic), but allow the option to view probabilistic information; provide information

timed to allow users to take revised or preventative actions; include an economic component (because farmer survival, *i.e.* cost of practice adoption, takes precedence over stewardship concerns); and allow for confidential comparison of model results with proprietary data.

The participatory modeling approach used in the development of DyNoFlo, a whole-farm decision-support system to decrease nitrogen leaching while maintaining profitability under variable climate conditions (Cabrera *et al.*, 2007), engaged federal agencies, individual producers, cooperative extension specialists, and consultants (who provided confidential data for model verification). Cabrera *et al.* (2007) report that the dialogue between these players, as equals, was as important as the scientific underpinning and accuracy of the model in improving adoption. They emphasize that the process, including validation (defined as occurring when researchers and stakeholders agree the model fits real or measured conditions adequately) is a key factor in developing stakeholder sense of ownership and desire for further engagement and decision-support system enhancement. These findings concur with recent examples of the adoption of climate data, predictions and information to improve water supply model performance by Colorado River Basin water managers (Woodhouse and Lukas, 2006).

Case Study G:
The Potomac River Basin

Water wars, traditionally seen in the West, are spreading to the Midwest, East, and South. The

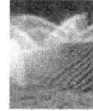

> Water wars, traditionally seen in the West, are spreading to the Midwest, East, and South.

"Water Wars" report (Council of State Governments, 2003) underlines the stress a growing resident population is imposing on a limited natural resource, and how this stress is triggering water wars in areas formerly with plentiful water. An additional source of concern would be the effect on supply and the increase in demand due to climate variability and change. Although the study by Hurd *et al.* (1999) indicated that the Northeastern water supply would be less vulnerable to the effect of climate change, the Interstate Commission on the Potomac River Basin (ICPRB) periodically studies the impact of climate change on the supply reliability to the Washington metropolitan area (WMA). (See also: *Restoring the Waters*. 1997, Boulder, CO, Natural Resources Law Center, the University of Colorado School of Law, May.)

The ICPRB was created in 1940 by the States of Maryland and West Virginia, the Commonwealths of Virginia and Pennsylvania, and the District of Columbia. The ICPRB was recognized by the United States Congress, which also provided a presence in the Commission. The ICPRB's purpose is "regulating, controlling, preventing, or otherwise rendering unobjectionable and harmless the pollution of the waters of said Potomac drainage area by sewage and industrial and other wastes".

The Potomac River constitutes the primary source of water for the WMA. Out of the five reservoirs in the WMA, three are in the Potomac River Basin. Every five years, beginning in April, 1990, the Commission evaluates the adequacy of the different sources of water supply to the Metropolitan Washington area. The latest report, (Kame'enui *et al.*, 2005), includes a report of a study by Steiner and Boland (1997) of the potential effects of climate variability and change on the reliability of water supply for that area.

The ICPRB inputs temperature, precipitation from five general circulation models (GCMs), and soil moisture capacity and retention, to a water balance model, to produce monthly average runoff records. The computed Potential Evapotranspiration (PET) is also used to estimate seasonal water use in residential areas.

The results of the 2005 study indicated that, depending on the climate change scenario, the demand in the Washington metropolitan area in 2030 could be 74 to 138 percent greater than that of 1990. According to the report, "resources were significantly stressed or deficient" at that point. The water management component of the model helped determine that, with aggressive plans in conservation and operation policies, existing resources would be sufficient through 2030. In consequence, the study recommended "that water management consider the need to plan for mitigation of potential climate change impacts" (Kame'enui *et al.*, 2005; Steiner and Boland, 1997).

Case Study H:
Fire Prediction Workshops as a Model for a Climate Science-Water Management Process to Improve Water Resources Decision Support

Fire suppression costs the United States about $1 billion each year. Almost two decades of research into the associations between climate and fire (*e.g.*, Swetnam and Betancourt, 1998), demonstrate a high potential to predict various measures of fire activity, based on direct influences, such as drought, and indirect influences, such as growth of fire fuels such as grasses and shrubs (*e.g.*, Westerling *et al.*, 2002; Roads *et al.*, 2005; Preisler and Westerling, 2007). Given strong mutual interests in improving the range of tools available to fire management, with the goals of reducing fire related damage and loss of life, fire managers and climate scientists have developed a long-term process to improve fire potential prediction (Garfin *et al.*, 2001; Wordell and Ochoa, 2006) and to better estimate the costs and most efficient deployment of fire fighting resources. The strength of collaborations between climate scientists, fire ecologists, fire managers, and operational fire weather forecasters, is based upon mutual learning and meshing of both complementary knowledge (*e.g.*, atmospheric science and forestry science) and expertise (*e.g.*, dynamical modeling and command and control operations management) (Garfin, 2005). The emphasis on process, as well as product, may be a model for climate science in support of water resources management decision making. Another key facet in maintaining this collaboration and di-

Almost two decades of research into the associations between climate and fire demonstrate a high potential to predict various measures of fire activity, based on direct influences, such as drought, and indirect influences, such as growth of fire fuels such as grasses and shrubs.

rect application of climate science to operational decision-making has been the development of strong professional relationships between the academic and operational partners. Aspects of developing these relationships that are germane to adoption of this model in the water management sector include:

- Inclusion of climate scientists as partners in annual fire management strategic planning meetings;
- Development of knowledge and learning networks in the operational fire management community;
- Inclusion of fire managers and operational meteorologists in academic research projects and development of verification procedures (Corringham *et al.*, 2008)
- Co-location of fire managers at academic institutions (Schlobohm *et al.*, 2003).

Case Study I:
Incentives to Innovate—Climate Variability and Water Management along the San Pedro River

The San Pedro River, though small in size, supports one of the few intact riparian systems remaining in the Southwest. Originating in Sonora, Mexico, the stream flows northward into rapidly urbanizing southeastern Arizona, eventually joining with the Gila River, a tributary of the Lower Colorado River. On the American side of the international boundary, persistent conflict plagues efforts to manage local water resources in a manner that supports demands generated at Fort Huachuca Army Base and the nearby city of Sierra Vista, while at the same time preserving the riparian area. Located along a major flyway for migratory birds and providing habitat for a wide range of avian and other species, the river has attracted major interest from an array of environmental groups that seek its preservation. Studies carried out over the past decade highlight the vulnerability of the river system to climate variability. Recent data indicate that flows in the San Pedro have declined significantly due, in part, to ongoing drought. More controversial is the extent to which intensified groundwater use is depleting water that would otherwise find its way to the river.

The highly politicized issue of water management in the upper San Pedro River Basin has led to establishment of the Upper San Pedro Partnership, whose primary goal is balancing water demands with water supply in a manner that does not compromise the region's economic viability, much of which is directly or indirectly tied to Fort Huachuca Army base. Funding from several sources, including, among others, several NOAA programs and the Netherlands-based Dialogue on Climate and Water, has supported ongoing efforts to assess vulnerability of local water resources to climate variability on both sides of the border. These studies, together with experience from recent drought, point toward escalating vulnerability to climatic impacts, given projected increases in demand and likely diminution of effective precipitation over time in the face of rising temperatures and changing patterns of winter *versus* summer rainfall (IPCC, 2007). Whether recent efforts to reinforce growth dynamics by enhancing the available supply through water reuse or water importation from outside the basin will buffer impacts on the riparian corridor remains to be seen. In the meantime, climatologists, hydrologists, social scientists, and engineers continue to work with members of the Partnership and others in the area to strengthen capacity and interest in using climate forecast products. A relatively recent decision to include climate variability and change in a decision-support model being developed by a University of Arizona engineer in collaboration with members of the Partnership constitutes a significant step forward in integrating climate into local decision processes.

The incentives for engagement in solving the problems in the San Pedro include both a "carrot" in the form of federal and state funding for the San Pedro Partnership, and a newly formed water management district, and a "stick" in the form of threats to the future of Fort Huachuca. Fort Huachuca represents a significant component of the economy of southern Arizona, and its existence is somewhat dependent on showing that endangered species in the river, and the water rights of the San Pedro Riparian Conservation Area, are protected.

Effective integration of climate information in decisions requires identifying topics of mutual interest to sustain long-term collaborative research and application of decision-support outcomes.

4.4 SUMMARY FINDINGS AND CONCLUSIONS

The decision-support experiments discussed here and in Chapter 3, together with the analytical discussion, have depicted several barriers to use of decision-support experiment information on SI climate conditions by water resource managers. The discussion has also pinpointed a number of ways to overcome these barriers and ensure effective communication, transfer, dissemination, and use of information. Our major findings are as follows.

Effective integration of climate information in decisions requires identifying topics of mutual interest to sustain long-term collaborative research and application of decision-support outcomes: Identifying topics of mutual interest, through forums and other means of formal collaboration, can lead to information penetration into agency (and stakeholder group) activities, and produce self-sustaining, participant-managed spin-off activities. Long-term engagement also allows time for the evolution of scientist/decision-maker collaborations, ranging from understanding the roles of various players to connecting climate to a range of decisions, issues, and adaptation strategies—and building trust.

Tools must engage a range of participants, including those who generate them, those who translate them into predictions for decision-maker use, and the decision makers who apply the products. Forecast innovations might combine climate factor observations, analyses of climate dynamics, and SI forecasts. In turn, users are concerned with varying problems and issues such as planting times, instream flows to support endangered species, and reservoir operations. While forecasts vary in their skill, multiple forecasts that examine various factors (e.g., snow pack, precipitation, temperature variability) are most useful because they provide decision makers more access to data that they can manipulate themselves.

A critical mass of scientists and decision makers is needed for collaboration to succeed: Development of successful collaborations requires representation of multiple perspectives, including diversity of disciplinary and agency-group affiliation. For example, operations, planning, and management personnel should all be involved in activities related to integrating climate information into decision systems; and there should be sound institutional pathways for information flow from researchers to decision makers, including explicit responsibility for information use. Cooperative relationships that foster learning and capacity building within and across organizations, including restructuring organizational dynamics, are important, as is training of "integrators" who can assist stakeholders with using complex data and tools.

What makes a "critical mass" critical? Research on water resource decision making suggests that agencies and other organizations define problems differently depending on whether they are dedicated to managing single-issue problems in particular sectors (e.g., irrigation, public supply) or working in political jurisdictions or watershed-based entities designed to comprehensively manage and coordinate several management objectives simultaneously (e.g., flood control and irrigation, power generation, and in-stream flow). The latter entities face the unusual challenge of trying to harmonize competing objectives, are commonly accountable to numerous users, and require "regionally and locally tailored solutions" to problems (Water in the West, 1998; also, Kenney and Lord, 1994; Grigg, 1996). A lesson that appears to resonate in our cases is that decision makers representing the affected organizations should be incorporated into collaborative efforts.

Forums and other means of engagement must be adequately funded and supported. Discussions that are sponsored by boundary organizations and other collaborative institutions allow for co-production of knowledge, legitimate pathways for climate information to enter assessment processes, and a platform for building trust. Collaborative products also give each community something tangible that can be used within its own system (i.e., information to support decision making, climate service, or academic research products). Experiments that effectively incorporate seasonal forecasts into operations generally have long-term financial support, facilitated, in turn, by high public concern over potential adverse environmental and/or economic impacts. Such concern helps generate

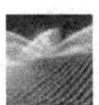

> While forecasts vary in their skill, multiple forecasts that examine various factors are most useful because they provide decision makers more access to data that they can manipulate themselves.

a receptive audience for new tools and ideas. Flexible and appropriate sources of funding must be found that recognize benefits received by various constituencies on the one hand, and ability to pay on the other. A combination of privately-funded, as well as publicly-supported revenue sources may be appropriate in many cases—both because of the growing demands on all sources of decision-support development, and because such a balance better satisfies demands that support for these experiments be equitably borne by all who benefit from them (Cash and Buizer, 2005). Federal agencies within CCSP can help in this effort by developing a database of possible funding sources from all sectors, public and private (CDWR, 2007b).

There is a need to balance national decision-support tool production against customizable, locally specific conditions. Given the diversity of challenges facing decision makers, the diverse needs and aspirations of stakeholders, and the diversity of decision-making authorities, there is little likelihood of providing comprehensive climate services or "one-stop-shop" information systems to support all decision making or risk assessment. Support for tools to help communities and other self-organizing groups develop their own capacity and conduct their own assessments within a regional context is essential.

There is a growing push for smaller scale products that are tailored to specific users, as well as private sector tailored products (e.g., "Weatherbug"). However, private sector products are generally available only to specific paying clients, and may not be equitable to those who lack access to publicly-funded information sources. Private observing systems also generate issues related to trustworthiness of information and quality control. What are the implications of this push for proprietary vs. public domain controls and access? This problem is well-documented in policy studies of risk-based information in the fields of food labeling, toxic pollutants, medical and pharmaceutical information, and other forms of public disclosure programs (Graham, 2002).

4.5 FUTURE RESEARCH NEEDS AND PRIORITIES

Six major research needs are at the top of our list of priorities for investigations by government agencies, private sector organizations, universities, and independent researchers. These are:

1. Better understanding the decision context within which decision support tools are used,
2. Understanding decision-maker perceptions of climate risk and vulnerability;
3. Improving the generalizability/transferability of case studies on decision-support experiments,
4. Understanding the role of public pressures and networks in generating demands for climate information,
5. Improving the communication of uncertainties, and
6. Sharing lessons for collaboration and partnering with other natural resource areas.

Better understanding of the decision-maker context for tool use is needed. While we know that the institutional, political and economic context has a powerful influence on the use of tools, we need to learn more about how to promote user interactions with researchers at all junctures within the tool development process.

The institutional and cultural circumstances of decision makers and scientists are important to determining the level of collaboration. Among the topics that need to be addressed are the following:

- understanding how organizations engage in transferring and developing climate variability information,
- defining the decision space occupied by decision makers,
- determining ways to encourage innovation within institutions, and
- understanding the role of economics and chain-of-command in the use of tools.

Access to information is an equity issue: large water management agencies may be able to afford sophisticated modeling efforts, consultants to provide specialized information, and a higher quality of data management and analysis, while smaller or less wealthy stakeholders generally

Those most likely to use weather and climate information are individuals who have experienced weather and climate problems in the past.

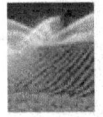

Much more needs
to be known about
how to make
decision makers
aware of their
possible vulnerability
from climate
variability impacts
to water resources.

do not have the same access or the consequent ability to respond (Hartmann, 2001). This is especially true where there are no alternatives to private competitive markets where asymmetries of economic buying power may affect information access. Scientific information that is not properly disseminated can inadvertently result in windfall profits for some and disadvantage others (Pfaff et al., 1999; Broad and Agrawalla, 2000; Broad et al., 2002). Access and equity issues also need to be explored in more detail.

4.5.1 Understanding Decision-Makers' Perceptions of Climate Vulnerability

Much more needs to be known about how to make decision makers aware of their possible vulnerability from climate variability impacts to water resources. Research on the influence of climate science on water management in western Australia, for example, (Power et al., 2005) suggests that water resource decision makers can be persuaded to act on climate variability information if a strategic program of research in support of specific decisions (e.g., extended drought) can be wedded to a dedicated, timely risk communication program.

While we know, based on research in specific applications, that managers who find climate forecasts and projections to be reliable may be more likely to use them, those most likely to use weather and climate information are individu-

als who have experienced weather and climate problems in the recent past. The implication of this finding is that simply delivering weather and climate information to potential users may be insufficient in those cases in which the manager does not perceive climate to be a hazard—at least in humid, water-rich regions of the United States that we have studied[5].

We also need to know more about how the financial, regulatory, and management contexts influence perceptions of usefulness (Yarnal et al., 2006; Dow et al., 2007). Experience suggests that individual responses, in the aggregate, may have important impacts on one's capacity to use, access, and interpret information. Achieving a better understanding of these factors and of the informational needs of resource managers will require more investigation of their working environments and intimate understanding of their organizational constraints, motivations, and institutional rewards.

4.5.2 Possible Research Methodologies

Case studies increase understanding of how decisions are made by giving specific examples of decisions and lessons learned. A unique

---

[5] Additional research on water system manager perceptions is needed, in regions with varying hydrometeorological conditions, to discern if this finding is universally true.

strength offered by the case study approach is that "...only when we confront specific facts, the raw material on the basis of which decisions are reached—not general theories or hypotheses—do the limits of public policy become apparent (Starling, 1989)". In short, case studies put a human face on environmental decision making by capturing, even if only in a temporal "snapshot", the institutional, ethical, economic, scientific, and other constraints and factors that influence decisions.

4.5.3 Public Pressures, Social Movements and Innovation

The extent to which public pressures can compel innovation in decision-support development and use is an important area of prospective research. As has been discussed elsewhere in this Product, knowledge networks—which provide linkages between various individuals and interest groups that allow close, ongoing communication and information dissemination among multiple sectors of society involved in technological and policy innovations—can be sources of non-hierarchical movement to impel innovation (Sarewitz and Pielke, 2007; Jacobs, 2005). Such networks can allow continuous feedback between academics, scientists, policy-makers, and NGOs in at least two ways: (1) by cooperating in seeking ways to foster new initiatives, and (2) providing means of encouraging common evaluative and other assessment criteria to advance the effectiveness of such initiatives.

Since the late 1980s, there has arisen an extensive collection of local, state (in the case of the United States) and regional/sub-national climate change-related activities in an array of developed and developing nations. These activities are wide-ranging and embrace activities inspired by various policy goals, some of which are only indirectly related to climate variability. These activities include energy efficiency and conservation programs; land use and transportation planning; and regional assessment. In some instances, these activities have been enshrined in the "climate action plans" of so-called Annex I nations to the UN Framework Convention on Climate Change (UNCED, 1992; Rabe, 2004).

An excellent example of an important network initiative is the International Council of Local Environmental Initiatives, or ICLEI is a Toronto, Canada-based NGO representing local governments engaged in sustainable development efforts worldwide. Formed in 1990 at the conclusion of the World Congress of Local Governments involving 160 local governments, it has completed studies of urban energy use useful for gauging growth in energy production and consumption in large cities in developing countries (e.g., Dickinson, 2007; ICLEI, 2007). ICLEI is helping to provide a framework of cooperation to evaluate energy, transportation, and related policies and, in the process, may be fostering a form of "bottom-up" diffusion of innovation processes that function across jurisdictions—and even entire nation-states (Feldman and Wilt, 1996; 1999). More research is needed on how, and how effectively networks actually function and whether their efforts can shed light on the means by which the diffusion of innovation can be improved and evaluated.

Another source of public pressure is social movements for change—hardly unknown in water policy (e.g., Donahue and Johnston, 1998). Can public pressures through such movements actually change the way decision makers look at available sources of information? Given the anecdotal evidence, much more research is warranted. One of the most compelling recent accounts of how public pressures can change such perceptions is that by the historian Norris Hundley on the gradual evolution on the part of city leaders in Los Angeles, California, as well as members of the public, water agencies, and state and federal officials—toward diversion of water from the Owens Valley.

After decades of efforts and pressures from interested parties to, at first prevent and then later, roll back, the amount of water taken from the Owens River, the city of Los Angeles sought an out-of-court settlement over diversion; in so doing, they were able to study the reports of environmental degradation caused by the volumes of water transferred, and question whether to compensate the Valley for associated damages (Hundley, 2001). While Hundley's chronicling of resistance has a familiar ring to students of water policy, remarkably little research has been done to draw lessons using the grounded theory

> While uncertainty is an inevitable factor in regards to climate variability and weather information, the communication of uncertainty—as our discussion has shown—can be significantly improved.

approach discussed earlier—about the impacts of such social movements.

While uncertainty is an inevitable factor in regards to climate variability and weather information, the communication of uncertainty—as our discussion has shown—can be significantly improved. Better understanding of innovative ways to communicate uncertainty to users should draw on additional literatures from the engineering, behavioral and social, and natural science communities (e.g., NRC 2005; NRC 2006). Research efforts are needed by various professional communities involved in the generation and dissemination of climate information to better establish how to define and communicate climate variability risks clearly and coherently and in ways that are meaningful to water managers. Additional research is needed to determine the most effective communication, dissemination and evaluation tools to deliver information on potential impacts of climate variability, especially with regards to such factors as further reducing uncertainties associated with future sea-level rise, more reliable predictions of changes in frequency and intensity of tropical and extra-tropical storms, and how saltwater intrusion will impact freshwater resources, and the frequency of drought. Much can be learned from the growing experience of RISAs and other decision-support partnerships and networks.

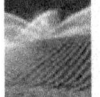

Research on lessons from other resource management sectors on decision-support use and decision maker/researcher collaboration would be useful. While water issues are ubiquitous and connect to many other resource areas, a great deal of research has been done on the impediments to, and opportunities for, collaboration in other resource areas such as energy, forests, coastal zone and hydropower. This research suggests that there is much that water managers and those who generate SI information on climate variability could learn from this literature. Among the questions that need further investigation are issues surrounding the following subject areas: (1) innovation (Are there resource areas in which tool development and use is proceeding at a faster pace than in water management?); (2) organizational culture and leadership (Are some organizations and agencies more resistant to change, more hierarchical

in their decision making, more formalized in their decisional protocols than is the case in water management?); and (3) collaborative style (Are some organizations in certain resource areas or science endeavors better at collaborating with stakeholder groups in the generation of information tools, or other activities? [e.g., Kaufman, 1967; Bromberg, 2000]). Much can also be learned about public expectations and the expectations of user groups from their collaborations with such agencies that could be valuable to the water sector.

CHAPTER 5

Looking Toward The Future

Convening Lead Authors: Helen Ingram, Univ. of Arizona; David
Feldman, Univ. of California, Irvine; Katharine L. Jacobs, Arizona
Water Institute; Nathan Mantua, Climate Impacts Group, Univ. of
Washington

Lead Authors: Maria Carmen Lemos, Univ. of Michigan; Barbara
Morehouse, Univ. of Arizona

Contributing Authors: Nancy Beller-Simms, NOAA; Anne M.
Waple, STG, Inc.

5.1 INTRODUCTION

The future context for decision support for
seasonal-to-interannual (SI) climate forecast-
ing-related decisions in water resources and
other sectors will evolve in response to future
climate trends and events, advances in monitoring, predicting and communicating information about hydrologically-significant aspects
of climate, and social action. Climate-related
issues have a much higher profile among the
public, media, and policy makers than they
did even a few years ago. In water resources
and other sectors, climate is likely to be only
one of a number of factors affecting decision
making, and the extent to which it is given
priority will depend both on the experiences
associated with "focusing events" such as major
droughts, floods, hurricanes and heat waves,
and on how strong knowledge networks have
become (Pulwarty and Melis, 2001). The utility
of climate information will depend largely on
how salient, credible, valuable and legitimate it
is perceived to be. These qualities are imparted
through knowledge networks that can be fostered and strengthened using decision-support
tools. Increasingly, climate forecasting and data
have become integrated with water resources
decisions at multiple levels, and some of the
lessons learned in the water sector can improve
the application of SI climate forecasts in other
climate sensitive sectors. Better integration of

climate forecasting science into water resources
and other sectors will likely save and improve
lives, reduce damages from weather extremes,
and lower economic cost related to adapting to
continued climate variability.

Section 5.2 of this Chapter highlights a number
of overarching themes that need to be emphasized as important to understanding the overall
challenges facing decision support and its use.
Section 5.3 addresses research priorities that are
critical to progress. Section 5.4 discusses other
sectors that are likely to be affected by climate
variation that could profit from lessons in the
water resources sector.

5.2 OVERARCHING THEMES AND FINDINGS

5.2.1 The "Loading Dock Model" of Information Transfer is Unworkable

Only recently have climate scientists come to
realize that improving the skill and accuracy
of climate forecasting products does not necessarily make them more useful or more likely
to be adopted (e.g., see Chapter 2, Box 2.4).
Skill is a necessary ingredient in perceived
forecast value, yet more forecast skill by itself
does not imply more forecast value. Lack of
forecast skill and/or accuracy may be one of
the impediments to forecast use, but there are
many other barriers to be overcome. Better

*Only recently
have climate
scientists come
to realize that
improving the skill
and accuracy of
climate forecasting
products does not
necessarily make
them more useful
or more likely
to be adopted.*

technical skill must be accompanied by better communication and stronger linkages between forecasters and potential users. In this Product, we have stressed that forecasts flow through knowledge networks and across disciplinary and occupational boundaries. Thus, forecasts need to support a range of activities including research and applications, and be "end-to-end useful". End-to-end useful implies a broad fabric of utility, created by multiple entities that adopt forecasts for their own reasons and adapt them to their own purposes by blending forecast knowledge with local know-how, practices, and other sources of information more familiar to those participants. These network participants then pass the blended information to other participants who, in turn, engage in the same process. By the end of the process of transfer, translation and transformation of information, forecast information may look very different from what scientists initially envisioned.

Skill and accuracy are only two of the values important to the use of climate knowledge; others might include relevance, timeliness, and credibility. Using climate information and decision tools can have obvious economic benefits, and these advantages can extend into the political, organizational, and professional realms as well. Salience is a product of framing in the larger political community and the professional circles in which different decision makers travel. Novel ideas are difficult for organizations to adopt, and therefore, such ideas become more credible if they are consistent with, and tempered by, already existing information channels and organizational routines.

5.2.2 Decision Support is a Process Rather Than a Product
As knowledge systems have become better understood, providing decision support has evolved into a communications process that links scientists with users rather than a one-time exchange of information products. While decision tools such as models, scenarios, and other boundary objects that connect scientific forecasters to various stakeholder groups can be helpful, the notion of tools insufficiently conveys the relational aspects of networks. Relevance, credibility, and legitimacy are human perceptions built through repeated interactions. For this reason, decision support does not result

in a product that can be shelved until needed or reproduced for different audiences. Clearly, lessons from decision-support experience are portable from one area to another but only as the differences in context are interpreted, understood, and taken into account.

Governments are not the only producers of climate variability forecasts. Non-governmental actors, including private businesses, play a critical role in knowledge networks, particularly in tailoring climate forecast products to fit the needs of particular sectors and user groups. Nothing in this Product should suggest that knowledge networks must be wholly or even primarily developed in the public sector. Just as numerous entrepreneurs have taken National Weather Service forecasts and applied them to different sectors and user-group needs, SI climate information transfer, translation and transformation may become functions largely provided by the private sector. However, as argued in the following section, there is clearly a role for the public sector because information access is related to economic and social outcomes that must be acknowledged.

Ensuring that information is accessible and relevant will require paying greater attention to the role of institutions in furthering the process of decision support; particularly *boundary spanning* activities that bring together tool developers and users to exchange information, promote communication, propose remedies to problems, foster stakeholder engagement, and conjointly develop decision-support systems to address user needs. An important facet of boundary spanning is that the exchange (including coproduction, transference, communication and dissemination) of climate information to water decision makers requires partnerships among public and private sector entities. In short, to avoid the Loading Dock Model previously discussed, efforts to further boundary-spanning partnerships is essential to fostering a process of decision support (NRC, 2007; Cash and Buizer, 2005; Sarewitz and Pielke, 2007).

5.2.3 Equity May Not Be Served
Information is power in global society and, unless it is widely shared, the gaps between the advantaged and the disadvantaged may widen. Lack of resources is one of the causes of

As knowledge systems have become better understood, providing decision support has evolved into a communications process that links scientists with users rather than a one-time exchange of information products.

poverty, and resources are required to tap into knowledge networks. Unequal distribution of knowledge can insulate decision making, facilitate elite capture of resources, and alienate disenfranchised groups. In contrast, an approach that is open, interactive and inclusive can go a long way in supporting informed decisions that, in turn, can yield better outcomes from the perspective of fairness.

While United Nations Millennium Development Goals attract attention to equity in poor countries, the unequal availability of and access to knowledge and technology, including SI forecast products, exacerbates inequalities within the United States. The case of agriculture is especially important because of the high impacts the agricultural sector has upon the long-term quality of the general environment. The dust bowl of the 1930s and its broad national impact stand as a reminder of the consequences of poorly informed and unsustainable practices. Avoiding repetition of such top soil losses, desertification increases, and social dislocations is more likely if early warning of variations in seasonal precipitation and run off are available, trusted, and credible. To build and maintain networks in the agricultural sector, particularly among smaller, less-advantaged farmers will require greater efforts (Wiener, 2007).

The emergence of seasonal climate forecasting initially raised great expectations of its potential role to decrease the vulnerability of poor farmers around the world to climate variability and the development and dissemination of forecasts have been justified in equity terms (Glantz, 1996; McPhaden *et al.*, 2006). However, ten years of empirical research on seasonal forecasting application and effect on agriculture, disaster response and water management have tempered these expectations (Klopper, 1999; Vogel, 2000; Valdivia *et al.*, 2000; Letson *et al.*, 2001; Hammer *et al.*, 2001; Lemos *et al.*, 2002; Patt and Gwata, 2002; Broad *et al.*, 2002; Archer, 2003; Lusenso *et al.*, 2003; Roncoli *et al.*, 2006; Bharwani *et al.*, 2005; Meinke *et al.*, 2006; Klopper *et al.*, 2006). Examples of SI climate forecast applications show that not only are the most vulnerable often unable to benefit, but in some

situations may even be harmed (Broad *et al.*, 2002; Lemos *et al.*, 2002; Patt and Gwata, 2002; Roncoli *et al.*, 2004). However, some users have been able to benefit significantly from this new information. For example, many Pacific island nations respond to El Niño forecasts and avoid potential disasters from water shortages. Similarly, agricultural producers in Australia have been better able to cope with swings in their commodity production associated with drought and water managers. In the Southwest United States, managers have been able to incorporate seasonal-to-interannual climate forecasts into their decision-making processes in order to respond to crises—and this is also beginning to occur in more water-rich regions such as the Southeast United States that are currently facing prolonged drought (Hammer *et al.*, 2001; Hartmann *et al.*, 2002; Pagano *et al.*, 2002; Georgia DNR, 2003). But, unless greater effort is expended to rectify the differential impacts of climate information in contexts where the poor lack resources, SI climate forecasts will not contribute to global equity.

There are several factors that help to explain when and where equity goals are served in SI climate forecasting and when they are not (Lemos and Dilling, 2007). Understanding existing levels of underlying inequities and differential vulnerabilities is critical (Agrawala *et al.*, 2001). Forecasts are useful only when recipients of information have sufficient decision space or options to be able to respond to lower vulnerability and risk. Differential levels in the ability to respond can create winners and losers within the same policy context.

In the Southwest United States, managers have been able to incorporate seasonal-to-interannual climate forecasts into their decision-making processes in order to respond to crises—and this is also beginning to occur in more water-rich regions such as the Southeast United States that are currently facing prolonged drought.

For example, in Zimbabwe and northeastern Brazil, news of poor rainfall forecasts for the planting season influence bank managers who systematically deny credit, especially to poor farmers they perceive as high risk (Hammer *et al.*, 2001; Lemos *et al.*, 2002). In Peru, a forecast of El Niño and the prospect of a weak season gives fishing companies incentives to accelerate seasonal layoffs of workers (Broad *et al.*, 2002). Some users (bankers, businesses) who were able to act based on forecasted outcomes (positive or negative) benefited while those who could not (farmers, fishermen), were harmed. Financial, social and human resources to engage forecast producers are often out of reach of the poor (Lemos and Dilling, 2007). Even when the information is available, differences in resources, social status, and empowerment limit hazard management options. As demonstrated by Hurricane Katrina, for example, the poor and minorities were reluctant to leave their homes for fear of becoming victims of crime and looting, and were simply not welcome as immigrants fleeing from disaster (Hartmann *et al.*, 2002; Carbone and Dow, 2005; Subcommittee on Disaster Reduction, 2005; Leatherman and White, 2005).

Native American farmers who are unable to move their farming enterprises as do agribusinesses, and cannot lease their water rights strategically to avoid planting during droughts, are disadvantaged because of their small decision space or lack of alternatives. Moreover, poorer groups often distrust experts who are in possession of risk information because the latter are often viewed as elitist; focused more on probabilities rather than on the consequences of disaster; or unable to communicate in terms comprehensible to the average person (Jasanoff, 1987; Covello *et al.*, 1990). However, other research has found that resources, while desirable, are not an absolute constraint to poor people's ability to benefit from seasonal forecast use. In these cases, farmers have been able to successfully use seasonal climate forecasts by making small adjustments to their decision-making process (Eakin, 2000; Patt *et al.*, 2005; Roncoli *et al.*, 2006).

A more positive future in terms of redressing inequity and reducing poverty can take place if application policies and programs create al-

ternative types of resources, such as sustained relationships with information providers and web-based tools that can be easily tailored to specific applications; promotion of inclusionary dissemination practices; and paying attention to the context of information applications (Valdivia *et al.*, 2000; Archer, 2003; Ziervogel and Calder, 2003; Roncoli *et al.*, 2006). Examples in the literature show that those who benefit from SI climate forecasts usually have the means to attend meetings or to access information through the media (at least through the radio). For example, small farmers in Tamil Nadu, India (Huda *et al.*, 2004) and Zimbabwe (Patt and Gwata, 2002) benefited from climate information through a close relationship with forecast "brokers"[1] who spent considerable effort in sustaining communication and providing expert knowledge to farmers. However, the number of farmers targeted in these projects was very limited. For any real impact, such efforts will need to be scaled up and sustained beyond research projects.

Equitable communication and access are critical to fairness with respect to potential benefit from forecast information, but such qualities often do not exist. Factors such as levels of education, access to electronic media such as the Internet, and expert knowledge critically affect the ability of different groups to take advantage of seasonal forecasts (Lemos and Dilling, 2007). While the adoption of participatory processes of communication and dissemination can defray some of these constraints, the number of positive cases documented is small (*e.g.*, Patt *et al.*, 2005; Roncoli *et al.*, 2006; O'Brien and Vogel, 2003). Also, because forecasts are mostly disseminated in the language of probabilities, they may be difficult to assimilate by those who do not generally think probabilistically nor interpret probabilities easily, or those whose framing of environmental issues is formed through experience with extreme events (Nicholls, 1999; Yarnal *et al.*, 2006; Dow *et al.*, 2007; Weingert *et al.*, 2000). In a situation where private enterprise is important for participants in knowledge networks, serving the poor may not be profitable, and for that reason they become marginalized.

---

[1]   Researchers in the India case and researchers and extension agents in the Zimbabwe case.

Factors such as levels of education, access to electronic media such as the Internet, and expert knowledge critically affect the ability of different groups to take advantage of seasonal forecasts.

Fostering inclusive, equitable access, therefore, will require a combination of organizational practices that empower employees, and engage agency clients, outside stakeholder groups, and the general public through providing training and outreach in tool use, and the infusion of trust in communication of risks. The latter will require use of public forums and other vehicles that provide opportunities for open, clear, jargon-free information as well as opportunity for discussion and public reaction (Freudenburg and Rursch, 1994; Papadakis, 1996; Jasanoff, 1987; Covello *et al.*, 1990; NRC, 1989). If climate science applications are to more clearly put vulnerable poor people on an equal footing or to go further toward reducing inequality, decision support must target the vulnerable poor specifically. Specific training and a concerted effort to "fit" the available information to local decision-making patterns and culture can be a first step to enhance its relevance. Seasonal forecast producers and policy makers need to be aware of the broader sociopolitical context and the institutional opportunities and constraints presented by seasonal forecast use and understand potential users and their decision environment. A better fit between product and client can avoid situations in which forecast use may harm those it could help. Finally, as some of the most successful examples show, seasonal forecasting applications should strive to be more transparent, inclusionary, and interactive as a means to counter power imbalances.

5.2.4 Science Citizenship Plays an Important Role in Developing Appropriate Solutions

Some scholars observe that a new paradigm in science is emerging, one that emphasizes science-society collaboration and production of knowledge tailored more closely to society's decision-making needs (Gibbons, 1999; Nowotny *et al.*, 2001; Jasanoff, 2004a). The philosophy is that, through mobilizing both academic and pragmatic knowledge and experience, better solutions may be produced for pressing problems. Concerns about climate impacts on water resource management are among the most pressing problems that require close collaboration between scientists and decision makers. Examples of projects that are actively pursuing collaborative science to address climate-related water resource problems include the Sustain-

ability of Semi-Arid Hydrology and Riparian Area (SAHRA) project <http://www.sahra.arizona.edu>, funded by the National Science Foundation (NSF) and located at the University of Arizona and the NSF-funded Decision Center for a Desert City, located at Arizona State University <http://dcdc.asu.edu>. The regional focus of NOAA's Regional Integrated Sciences and Assessments (RISA) program is likewise providing opportunities for collaborations between scientists and citizens to address climate impacts and information needs in different sectors, including water resource management. An examination of the Climate Assessment for the Southwest (CLIMAS), one of the RISA projects, provided insight into some of the ways in which co-production of science and policy is being pursued in a structured research setting (Lemos and Morehouse, 2005).

Collaborative efforts to produce knowledge for policy applications not only expand the envelope of the scientific enterprise, but also change the terms of the relationship between scientists and citizens. This emergence of new forms of science/society interactions has been documented from various perspectives, including the place of local, counter-scientific, and non-scientific knowledge (Eden, 1996; Fischer, 2000), links with democracy and democratic ideals (Jasanoff, 1996; Harding, 2000; Durodié, 2003), and environmental governance and decision making (Jasanoff and Wynne, 1998; Bäckstrand, 2003; Brunner *et al.*, 2005). These types of collaboration present opportunities to bridge the gaps between abstract scientific conceptualizations and knowledge needs generated by a grounded understanding of the nature and intensity of actual and potential risks, and the specific vulnerabilities experienced by different populations at different times and in different places. As we are coming to understand, seasonal and interannual variations of past climate may be misleading about future variation, and a heightened awareness and increased observation on the part of citizens in particular contexts is warranted. Moreover, engaged citizens may well come to think more deeply about the longer-term environmental impacts of both human activities and the variable climate.

Unlike the more traditional "pipeline" structure of knowledge transfer uni-directionally

> A new paradigm in science is emerging, one that emphasizes science-society collaboration and production of knowledge tailored more closely to society's decision-making needs.

from scientists to citizens, multi-directional processes involving coproduction of science and policy may take a more circuitous form, one that requires experimentation and iteration (Lemos and Morehouse, 2005; Jasanoff and Wynne, 1998). This model of science-society interaction has a close affinity to concepts of adaptive management and adaptive governance (Pulwarty and Melis, 2001; Gunderson, 1999; Holling, 1978; Brunner *et al.*, 2005), for both of these concepts are founded on notions that institutional and organizational learning can be facilitated through careful experimentation with different decision and policy options. Such experimentation is ideally based on best available knowledge but allows for changes based on lessons learned, emergence of new knowledge, and/or changing conditions in the physical or social realms. The experiments described in this Product offer examples of adaptive management and adaptive governance in practice.

Less extensively documented, but no less essential to bringing science to bear effectively on climate-related water resource management challenges is the notion of science citizenship (Jasanoff, 2004b), whereby the fruits of collaboration between scientists and citizens produces capacity to bring science-informed knowledge into processes of democratic deliberation, including network building, participation in policy-making, influencing policy interpretation and implementation processes, and even voting in elections. Science citizenship might, for example, involve participating in deliberations about how best to avert or mitigate the impacts of climate variability and change on populations, economic sectors, and natural systems vulnerable to reduced access to water. Indeed, water is fundamental to life and livelihood, and, as noted above, climate impacts research has revealed that deleterious effects of water shortages are unequally experienced; poorer and more marginalized segments of populations often suffer the most (Lemos, 2008). Innovative drought planning processes require precisely these kinds of input, as does planning for long-term reductions in water availability due to reduced snowpack. Issues such as these require substantial evaluation of how alternative solutions are likely to affect different entities at different times and in different places. For example, substantial reduction in snowpack,

together with earlier snowmelt and longer periods before the onset of the following winter, will likely require serious examination of social values and practices as well as of economic activities throughout a given watershed and water delivery area. As these examples demonstrate, science citizenship clearly has a crucial role to play in building bridges between science and societal values in water resource management. It is likely that this will occur primarily through the types of knowledge networks and knowledge-to-action networks discussed earlier in this Chapter.

5.2.5 Trends and Reforms in Water Resources Provide New Perspectives

As noted in Chapters 1 and 4, since the 1980s a "new paradigm" or frame for federal water planning has developed that appears to reflect the ascendancy of an environmental protection ethic among the general public. The new paradigm emphasizes greater stakeholder participation in decision making; explicit commitment to environmentally-sound, socially-just outcomes; greater reliance upon drainage basins as planning units; program management via spatial and managerial flexibility, collaboration, participation, and sound, peer-reviewed science; and an embrace of ecological, economic, and equity considerations (Hartig *et al.*, 1992; Landre and Knuth, 1993; Cortner and Moote, 1994; Water in the West, 1998; McGinnis, 1995; Miller *et al.*, 1996; Cody, 1999; Bormann *et al.*, 1994; Lee, 1993).

This "adaptive management" paradigm results in a number of climate-related SI climate information needs, including questions pertaining to the following: what are the decision-support needs related to managing in-stream flows/ low flows? and, what changes to water quality, runoff and streamflow will occur in the future, and how will these changes affect water uses among future generations unable to influence the current causes of these changes? The most dramatic change in decision support that emerges from the adaptive management paradigm is the need for real-time monitoring and ongoing assessment of the effectiveness of management practices, and the possibility that outcomes recommended by decision-support tools be iterative, incremental and reversible if they prove unresponsive to critical groups, in-

Science citizenship clearly has a crucial role to play in building bridges between science and societal values in water resource management.

effective in managing problems, or both. What makes these questions particularly challenging is that they are interdisciplinary in nature[2].

Because so many of the actions necessary to implement either adaptive management or integrated water resources management rest with private actors who own either land or property rights, the importance of public involvement can not be overemphasized. At the same time, the difficulties of implementing these new paradigm approaches should not be overlooked. The fragmented patchwork of jurisdictions involved and the inflexibility of laws and other institutions present formidable obstacles that will require both greater efforts and investments if they are to be overcome.

Another significant innovation in U.S. water resources management that affects climate information use is occurring in the *local* water supply sector, as discussed in Chapter 4, the growing use of integrated water resource planning (or IWRP) as an alternative to conventional supply-side approaches for meeting future demands. IWRP is gaining acceptance in chronically water-short regions such as the Southwest and portions of the Midwest—including Southern California, Kansas, Southern Nevada, and New Mexico (Beecher, 1995; Warren *et al.*, 1995; Fiske and Dong, 1995; Wade, 2001). IWRP supports the use of multiple sources of water integration of quality and quantity issues and information like that of SI climate and water supply forecasts as well as feedback from experience and experiments.

IWRP's goal is to "balance water supply and demand management considerations by

identifying feasible planning alternatives that meet the test of least cost without sacrificing other policy goals (Beecher, 1995)". This can be variously achieved through depleted aquifer recharge, seasonal groundwater recharge, conservation incentives, adopting growth management strategies, wastewater reuse, and applying least-cost planning principles to large investor-owned water utilities. The latter may encourage IWRP by demonstrating the relative efficiency of efforts to reduce demand as opposed to building more supply infrastructure. A particularly challenging alternative is the need to enhance regional planning among water utilities in order to capitalize on the resources of every water user, eliminate unnecessary duplication of effort, and avoid the cost of building new facilities for water supply (Atwater and Blomquist, 2002).

In some cases, short-term, least-cost planning may *increase* long-term project costs, especially when environmental impacts, resource depletion, and energy and maintenance costs are included. The significance of least-cost planning is that it underscores the importance of long- and short-term costs (in this case, of water) as an influence on the value of certain kinds of information for decisions. The most dramatic change in decision support that emerges from the adaptive management paradigm is the need for real-time monitoring and ongoing assessment of the effectiveness of management practices, and the possibility that outcomes recommended by decision-support tools be iterative, incremental and reversible if they prove unresponsive to critical groups, ineffective in managing problems, or both. Models and forecasts that predict water availability under different climate scenarios can be especially useful to least-cost planning and make more credible efforts to reducing demand. Specific questions IWRP raises for decision-support-generated climate information include: how precise must climate information be to enhance long-term planning? How might predicted climate change provide an incentive for IWRP strategies? And, what climate information is needed to optimize decisions on water pricing, re-use, shifting from surface to groundwater use, and conservation?

In some cases, short-term, least-cost planning may *increase* long-term project costs, especially when environmental impacts, resource depletion, and energy and maintenance costs are included.

---

[2] Underscored by the fact that scholars concur adaptive management entails a broad range of processes to avoid environmental harm by imposing modest changes on the environment, acknowledging uncertainties in predicting impacts of human activities on natural processes, and embracing social learning (*i.e.*, learning by experiment). In general, it is characterized by four major strategies: (1) managing resources by learning, especially about mistakes, in an effort to make policy improvements, (2) modifying policies in the light of experience—and permitting such modifications to be introduced in "mid-course", (3) allowing revelation of critical knowledge heretofore missing, as feedback to improve decisions, and (4) incorporating outcomes in future decisions through a consensus-based approach that allows government agencies and non-governmental organizations (NGOs) to conjointly agree on solutions (Bormann *et. al.*, 1993; Lee, 1993; Definitions of Adaptive Management, 2000).

5.2.6 Useful Evaluation of Applications of Climate Variation Forecasts Requires Innovative Approaches

There can be little argument that SI climate and hydrologic forecast applications must be evaluated just as are most other programs that involve substantial public expenditures. This Product has evidenced many of the difficulties in using standard evaluation techniques. While there have been some program evaluations, mostly from the vantage point of assessing the influence of RISAs on federal climate science policy (*e.g.*, McNie *et al.*, 2007; Cash *et al.*, 2006), there has been little formal, systematic, standardized evaluation as to whether seasonal-to-interannual climate and hydrologic forecast applications are optimally designed to learn from experience and incorporate user feedback. Evaluation works best on programs with a substantial history so that it is possible to compare present conditions with those that existed some years ago. The effort to promote the use of SI climate forecasts is relatively new and has been a moving target, with new elements being regularly introduced, making it difficult to determine what features of those federal programs charged with collaborating with decision makers in the development, use, application, and evaluation of climate forecasts have which consequences. As the effort to promote greater use of SI climate and hydrologic forecasts accelerates in the future, it is important to foster developments that facilitate evaluation. It is imperative that those promoting forecast use have a clear implementation chain with credible rationales or incentives for participants to take desired actions. Setting clear goals and priorities for allocation of resources among different elements is essential to any evaluation of program accomplishments (NRC, 2007). It is especially difficult to measure the accomplishment of some types of goals that are important to adaptive management, such as organizational learning. For this reason, we believe that consistent monitoring and regular evaluation of processes and tools at different time and spatial scales will be required in order to assess progress.

> As the effort to promote greater use of seasonal-to-interannual climate and hydrologic forecasts accelerates in the future, it is important to foster developments that facilitate evaluation.

An NRC panel addressing a closely related challenge for standard evaluation recommended that the need for evaluation should be addressed primarily through monitoring (NRC, 2007). The language of that report seems entirely applicable here:

"Monitoring requires the identification of process measures that could be recorded on a regular (for instance, annual) basis and of useful output or outcome measures that are plausibly related to the eventual effects of interest and can be feasibly and reliably recorded on a similar regular basis. Over time, the metrics can be refined and improved on the basis of research, although it is important to maintain some consistency over extended periods with regard to at least some of the key metrics that are developed and used".

There are signals of network building and collaborative forecaster/user interaction and collaboration that can be monitored. Meetings and workshops held, new contacts made, new organizations involved in information diffusion, websites, list serves, newsletters and reports targeted to new audiences are but a few of the many activities that are indicative of network creation activity.

5.3 RESEARCH PRIORITIES

As a result of the findings in this Product, we suggest that a number of research priorities should constitute the focus of attention for the foreseeable future: (1) improved vulnerability assessment, (2) improved climate and hydrologic forecasts, (3) enhanced monitoring and modeling to better link climate and hydrologic forecasts, (4) identification of pathways for better integration of SI climate science into decision making, (5) better balance between physical science and social science research related to the use of scientific information in decision making, (6) better understanding and support for small-scale, specially-tailored tools, and (7) significant funding for sustained long-term scientist/decision-maker interactions and collaborations. The following discussion identifies each priority in detail, and recommends ways to implement them.

5.3.1 A Better Understanding of Vulnerability is Essential

Case studies of the use of decision-support tools in water resources planning and management suggest that the research and policy-making communities need a far more comprehensive picture of the vulnerability of water and related resources to climate variability. This assessment must account for vulnerability along several dimensions.

As we have seen, there are many forms of climate vulnerability—ranging from social and physical vulnerability to ecological fragmentation, economic dislocation, and even organizational change and turmoil. Vulnerability may also range across numerous temporal and spatial scales. Spatially, it can affect highly localized resources or spread over large regions. Temporally, vulnerability can be manifested as an extreme and/or rapid onset problem that lasts briefly, but imposes considerable impact on society (e.g., intense tropical storms) or as a prolonged or slow-onset event, such as drought, which may produce numerous impacts for longer time periods.

In order to encompass these widely varying dimensions of vulnerability, we also need more research on how decision makers perceive the risks from climate variability and, thus, what variables incline them to respond proactively to threats and potential hazards. As in so many other aspects of decision-support information use, previous research indicates that merely delivering weather and climate information to potential users may be insufficient in those cases in which the manager does not perceive climate variability to be a hazard—for example, in humid, water rich regions of the United States that we have studied (Yarnal *et al.*, 2006; Dow *et al.*, 2007). Are there institutional incentives to using risk information, or—conversely—not using it? In what decisional contexts (e.g., protracted drought, sudden onset flooding hazards) are water managers most likely—or least likely—to be susceptible to employing climate variability hazard potential information?

More research is needed on the relationship of perceived vulnerability and the credibility of different sources of information including disinformation. What is the relationship of sources of funding, and locus of researchers such as government or private enterprise, and discounting of information?

5.3.2 Improving Hydrologic and Climate Forecasts

Within the hydrologic systems, accurate measures and assimilation of the initial state are crucial for making skillful hydrologic forecasts; therefore, a sustained high-quality monitoring system tracking stream flow, soil moisture, snowpack, and evaporation, together with tools for real-time data assimilation, are fundamental to the hydrologic forecasting effort. In addition, watersheds with sparse monitoring networks, or relatively short historical data series, are also prone to large forecast errors due to a lack of historical and real-time data and information about its hydrologic state.

Monitoring and assimilation are also essential for climate forecasting, as well as exercises of hindcasting to compare present experience with the historical record. Moreover, monitoring is critical for adaptive and integrated water resources management, and for the more effective adoption of strategies currently widely embraced by natural resources planners and managers.

On going improvements in the skill of climate forecasting will continue to provide another important avenue for improving the skill in SI hydrologic and water supply forecasts. For many river basins and in many seasons, the single greatest source of hydrologic forecast error is unknown precipitation after the forecast issue date. Thus, improvements in hydrologic

> We also need more research on how decision makers perceive the risks from climate variability and, thus, what variables incline them to respond proactively to threats and potential hazards.

forecasting are directly linked with improvements in forecasts for precipitation and temperature.

In addition, support for coordinated efforts to standardize and quantify the skill in hydrologic forecasts is needed. While there is a strong culture and tradition of forecast evaluation in meteorology and climatology, this sort of retrospective analysis of the skill of seasonal hydrologic forecasts has historically not been commonly disseminated. Hydrologic forecasts have historically tended to be more often deterministic than probabilistic with products focused on water supplies (e.g., stream flow, reservoir inflows). In operational settings, seasonal hydrologic forecasts have generally been taken with a grain of salt, in part because of limited quantitative assurance of how accurate they can be expected to be. In contrast, operational climate forecasts and many of today's experimental and newer operational hydrologic forecasts are probabilistic, and contain quantitative estimates for the forecast uncertainty.

New efforts are needed to extend "forecasts of opportunity" beyond those years when anomalous El Niño-Southern Oscillation (ENSO) conditions are underway. At present, the skill available from combining SI climate forecasts with hydrologic models is limited when all years are considered, but can provide useful guidance in years having anomalous ENSO conditions. During years with substantial ENSO effects, the climate forecasts have high enough skill for temperatures, and mixed skill for precipitation, so that hydrologic forecasts for some seasons and some basins provide measurable improvements over approaches that do not take advantage of ENSO information. In contrast, in years where the state of ENSO is near neutral, most of the skill in U.S. climate forecasts is due to decadal temperature trends, and this situation leads to substantially more limited skill in hydrologic forecasts. In order to improve this situation, additional sources of climate and hydrologic predictability must be exploited; these sources likely include other patterns of ocean temperature change, sea ice, land cover, and soil moisture conditions.

Linkages between climate and hydrologic scientists are getting stronger as they collaboratively

create forecast products. A great many complex factors influence the rate at which seasonal water supply forecasts and climate forecast-driven hydrologic forecasts are improving in terms of skill level. Mismatches between needs and information resources continue to occur at multiple levels and scales. There is currently substantial tension between providing tools at the space and time scales useful for water resources decisions and ensuring that they are also scientifically defensible, accurate, reliable, and timely. Further research is needed to identify ways to resolve this tension.

5.3.3 Better Integration of Climate Information into Decision Making

It cannot be expected that information that promises to lower costs or improve benefits for organizations or groups will simply be incorporated into decisions. Scholarly research on collaboration among organizations indicates that straightforward models of information transfer are not operative in situations where a common language between organizations has not been adopted, or more challenging, when organizations must transform their own perspectives and information channels to adjust to new information. It is often the case that organizations are path dependent, and will continue with decision routines even when they are suboptimal. The many case examples provided in this Product indicate the importance of framing issues; framing climate dependent natural resources issues that emphasize the sources of uncertainty and variability of climate and the need for adaptive action helps in integrating forecasting information. What is needed are not more case studies, however, but better case investigations employing grounded theory approaches to discerning general characteristics of decision-making contexts and their factors that impede, or provide better opportunities for collaboration with scientists and other tool developers. The construction of knowledge networks in which information is viewed as relevant, credible, and trusted is essential, and much can be learned from emerging experiences in climate-information networks being formed among local governments, environmental organizations, scientists, and others worldwide to exchange information and experiences, influence national policy-making agendas, and leverage international organization resources

Linkages between climate and hydrologic scientists are getting stronger as they collaboratively create forecast products.

on climate variability and water resources—as well as other resource—vulnerability.

Potential barriers to information use that must be further explored include: the cultural and organizational context and circumstances of scientists and decision makers; the decision space allowed to decision makers and their real range of choice; opportunities to develop—and capacity to exercise—science citizenship; impediments to innovation within institutions; and solutions to information overload and the numerous conflicting sources of already available information. As our case studies have shown, there is often a relatively narrow range of realistic options open to decision makers given their roles, responsibilities, and the expectations placed upon them.

There are also vast differences in water laws and state-level scientific and regulatory institutions designed to manage aquifers and stream-flows in the United States and information can be both transparent and yet opaque simultaneously. While scientific products can be precise, accurate, and lucid, they may still be inaccessible to those who most need them because of proprietary issues restricting access except to those who can pay, or due to agency size or resource base. Larger agencies and organizations, and wealthier users, can better access information in part because scientific information that is restricted in its dissemination tends to drive up information costs (Pfaff *et al.*, 1999; Broad and Agrawalla, 2000; Broad *et al.*, 2002; Hartmann, 2001). Access and equity issues also need to be explored in more detail. Every facet of tool use juncture needs to be explored.

Priority in research should be toward focused, solution-oriented, interdisciplinary projects that involve sufficient numbers and varieties of kinds of knowledge. To this end, NOAA's Sectoral Applications Research Program is designed to support these types of interactions between research and development of decision-support tools. Although this program is small, it is vital for providing knowledge on impacts, adaptation, and vulnerability and should be supported especially as federal agencies are contemplating a larger role in adaptation and vulnerability assessments and in light of pending legislation by Congress.

Regional Integrated Science Assessments are regarded as a successful model of effective knowledge-to-action networks because they have developed interdisciplinary teams of scientists working as (and/or between) forecasts producers while being actively engaged with resource managers. The RISAs have been proposed as a potentially important component of a National Climate Service (NCS), wherein the NCS engages in observations, modeling, and research nested in global, national, and regional scales with a user-centric orientation (Figure 1 of Miles *et al.*, 2006). The potential for further development of the RISAs and other boundary spanning organizations that facilitate knowledge-to-action networks deserves study. While these programs are small in size, they are the most successful long-term efforts by the federal government to integrate climate science in sectors and regions across the United States.

5.3.4 Better Balance Between Physical Science and Social Science

Throughout this Product, the absence of systematic research on applications of climate variation forecasting information has required analysis to be based on numerous case study materials often written for a different purpose, upon the accumulated knowledge and wisdom of authors, and logical inference. The dearth of hard data in this area attests to the very small research effort afforded the study of use-inspired social science questions. Five years ago a social science review panel recommended that NOAA should readjust its research priorities by additional investment in a wide variety of use-inspired social science projects (Anderson *et al.*, 2003). What was once the Human Dimensions of Climate Change Program within NOAA now exists only in the Sectoral Applications Research Program. Managers whose responsibilities may be affected by climate variability need detailed understanding of relevant social, economic, organizational and behavioral systems—as well as the ethical dilemmas faced in using, or not using information; including public trust, perceived competence, social stability and community well-being, and perceived social equity in information access, provision, and benefit. Much more needs to be known about the economic and other factors that shape demands for water, roads, and land conversion for residential and commercial development,

Priority in research should be toward focused, solution-oriented, interdisciplinary projects that involve sufficient numbers and varieties of kinds of knowledge.

Future progress
in making climatic
forecasts useful
depends upon
advancing our
understanding of
the incorporation of
available knowledge
into decisions in water
related sectors, since
there are already many
useful applications of
climate variation and
change forecasts at
present skill levels.

and shape social and economic resilience in face of climate variability.

A recent NRC Report (2007) set out five research topics that have direct relevance to making climate science information better serve the needs of various sectors: human influences on vulnerability to climate; communications processes; science produced in partnership with users; information overload; and innovations at the individual and organizational level necessary to make use of climate information. The last research topic is the particular charge of NOAA's Sectoral Applications Research Program and is of great relevance to the subject of this Product. However, the lack of use of theoretically-infused social science research is a clear impediment to making investments in physical sciences useful and used. Committed leadership that is poised to take advantage of opportunities is fundamental to future innovation, yet not nearly enough research has been done on the necessary conditions for recruitment, promotion and rewarding leadership in public organizations, particularly as that leadership serves in networks involving multiple agencies, both public and private, at different organizational levels.

5.3.5 Better Understanding of the Implications of Small-Scale, Tailored Decision-Support Tools is Needed

While there is almost universal agreement that specially tailored, small scale forecast tools are needed, concern is growing that the implications of such tools for trustworthiness, quality control, and ensuring an appropriate balance between proprietary *versus* public domain controls have not been sufficiently explored.

There is a growing push for smaller scale products that are tailored to specific users but are expensive, as well as private sector tailored products (e.g., "Weatherbug" and many reservoir operations proprietary forecasts have restrictions on how they share data with NOAA); this also generates issues related to trustworthiness of information and quality control. What are the implications of this push for proprietary *versus* public domain controls and access? This problem is well-documented in policy studies of risk-based information in the fields of food labeling, toxic pollutants, medical and

pharmaceutical information, and other public disclosure or "right-to-know" programs, but has not been sufficiently explored in the context of climate forecasting tool development.

Related to this issue of custom-tailoring forecast information is the fact that future progress in making climatic forecasts useful depends upon advancing our understanding of the incorporation of available knowledge into decisions in water related sectors, since there are already many useful applications of climate variation and change forecasts at present skill levels. Here, the issue is tailoring information to the *type* of user. Research related to specific river systems, and/or sectors such as energy production, flood plain and estuary planning and urban areas is important. Customizable products rather than generic services are the most needed by decision makers. The uptake of information is more likely when the form of information provided is compatible with existing practice. It makes sense to identify decision-support experiments where concerted efforts are made to incorporate climate information into decision making. Such experimentation feeds into a culture of innovation within agencies that is important to foster at a time when historically conservative institutions are evolving more slowly than the pace of change in the natural and social systems, and where, in those instances when evolution is taking place relatively quickly—there are few analogues that can be used as reference points for how to accommodate these changes and ensure that organizations can adapt to stress—an important role of visionary leadership (Bennis, 2003; Tichy and Bennis, 2007)

Given the diversity of challenges facing decision makers, the varied needs and aspirations of stakeholders, and the diverse array of decision-making authorities, there is little hope of providing comprehensive climate services or a "one-stop-shop" information system to support the decision-making or risk-assessment needs of a wide audience of users. Development of products to help nongovernmental communities and groups develop their own capacity and conduct their own assessments is essential for future applications of climate information.

A seasonal hydrologic forecasting and applications testbed program would facilitate the rapid development of better decision-support tools for water resources planning. Testbeds, as described in Chapter 2, are intermediate activities, a hybrid mix of research and operations, serving as a conduit between the operational, academic and research communities. A testbed activity may have its own resources to develop a realistic operational environment. However, the testbed would not have real-time operational responsibilities and instead, would be focused on introducing new ideas and data to the existing system and analyzing the results through experimentation and demonstration. The old and new system may be run in parallel and the differences quantified (a good example of this concept is the INFORM program tested in various reservoir operations in California described in Chapter 4). Other cases that demonstrate aspects of this same parallelism are the use of paleoclimate data in the Southwest (tree ring data being compared to current hydrology) and the South Florida WMD (using decade-scale data together with current flow and precipitation information). The operational system may even be deconstructed to identify the greatest sources of error, and these findings can serve as the motivation to drive new research to find solutions to operations-relevant problems. The solutions are designed to be directly integrated into the mock-operational system and therefore should be much easier to directly transfer to actual production. While NOAA has many testbeds currently in operation, including testbeds focused on: Hydrometeorology (floods), Hazardous Weather (thunderstorms and tornadoes), Aviation Weather (turbulence and icing for airplanes), Climate (El Niño, seasonal precipitation and temperature) and Hurricanes, a testbed for seasonal stream flow forecasting does not exist. Generally, satisfaction with testbeds has been high, with the experience rewarding for operational and research participants alike.

5.4 THE APPLICATION OF LESSONS LEARNED FROM THIS REPORT TO OTHER SECTORS

Research shows the close interrelationships among climate change, deep sustained drought, beetle infestations, high fuel load levels, forest fire activity, and the secondary impacts of fire activity including soil erosion, decreases in recharge, and increases in water pollution. Serious concern about the risks faced by communities in wildland-urban interface areas as well as about the long-term viability of the nation's forests is warranted. It is important to know more about climate-influenced changes in marine environments that have significant implications for the health of fisheries and for saltwater ecosystems. Potential changes in the frequency and severity of extreme events such as tropical storms, floods, droughts, and strong wind episodes threaten urban and rural areas alike and need to be better understood. Rising temperatures, especially at night, are already driving up energy use and contributing to urban heat island effects. They also pose alarming potential for heat wave-related deaths such as those experienced in Europe a few years ago. The poor and the elderly suffer most from such stresses. Clearly, climate conditions affect everyone's daily life.

Some of the lessons learned and described in this Product from the water sector are directly transferable to other sectors. The experiments described in Chapters 2, 3, and 4 are just as relevant to water resource managers as they are to farmers, energy planners or city planners. Of the overarching lessons described in this Chapter, perhaps the most important to all sectors is that the climate forecast delivery system in the past, where climatologists and meteorologists produced forecasts and other data in a vacuum, can be improved. This Product reiterates in each chapter that the Loading Dock Model of information transfer (see Chapter 2, Box 2.4) is unworkable. Fortunately, this Product highlights experiments where interaction

Research shows the close interrelationships among climate change, deep sustained drought, beetle infestations, high fuel load levels, forest fire activity, and the secondary impacts of fire activity including soil erosion, decreases in recharge, and increases in water pollution.

between producers and users is successful. A note of caution is warranted, however, against supposing that lessons from one sector are directly transferable to others. Contexts vary widely in the severity of problems, the level of forecasting skill available, and the extent to which networks do not exist or are already built and only need to be engaged. Rather than diffusion of model practices, we suggest judicious attention to a wide variety of insights suggested in the case studies and continued support for experimentation.

This Report has emphasized that decision support is a process rather than a product. Accordingly, we have learned that communication is key to delivering and using climate products. One example where communication techniques are being used to relay relevant climate forecast and other relevant information can be found in the Climate Assessment for the Southwest (RISA) project where RISA staff are working with the University of Arizona Cooperative Extension to produce a newsletter that contains official and non-official forecasts and other information useful to a variety of decision makers in that area, particularly farmers <http://www.climas.arizona.edu/forecasts/swoutlook.html>.

Equity is an issue that arises in other sectors as well. Emergency managers preparing for an ENSO-influenced season already understand that while some have access to information and evacuation routes, others, notably the elderly and those with financial difficulties, might not have the same access. To compound this problem, information may also not be in a language understood by all citizens. While these managers already realize the importance of climate forecast information, improved climate forecast and data delivery and/or understanding will certainly help in assuring that the response to a potential climate disaster is performed equitably for all of their residents (Beller-Simms, 2004).

Finally, science citizenship is and will be increasingly important in all sectors. Science citizenship clearly has a crucial role to play in building bridges between science and societal values in all resource management arenas and increased collaboration and production of knowledge between scientists and decision makers. The use of SI and climate forecasts and observational data will continue to be increasingly important in assuring that resource-management decisions bridge the gap between climate science, and the implementation of scientific understanding in our management of critical resources.

> Decision support is a process rather than a product. Accordingly, communication is key to delivering and using climate products.

APPENDIX A

Transitioning the National Weather Service Hydrologic Research Into Operations

Convening Lead Author: Nathan Mantua, Climate Impacts Group, Univ. of Washington

Lead Authors: Michael D. Dettinger, U.S.G.S., Scripps Institution of Oceanography; Thomas C. Pagano, National Water and Climate Center, NRCS/USDA; Andrew W. Wood, 3TIER™, Inc./ Dept. of Civil and Environmental Engineering, Univ. of Washington; Kelly Redmond, Western Regional Climate Center, Desert Research Institute

Contributing Author: Pedro Restrepo, NOAA

(Adapted from the National Weather Service Instruction 10-103, June, 2007, available at:
<http://www.weather.gov/directives/sym/pd01001003curr.pdf>).

Because of the operational nature of the National Weather Service's mission, transition of research into operations is of particular importance. Transition of all major NOAA research into operations is monitored by the NOAA Transition Board. Within the National Weather Service (NWS), two structured processes are followed to transition research into operations, in coordination with the NOAA Transition Board. The Operations and Service Improvement Process (OSIP) is used to guide all projects, including non-hydrology projects, through field deployment within the Advanced Weather Interactive System (AWIPS). A similar process called Hydrologic Operations and Service Improvement Process (HOSIP), with nearly identical stages and processes as OSIP, is used exclusively for the hydrology projects. For those hydrology projects that will be part of AWIPS, HOSIP manages the first two stages of hydrologic projects, and, upon approval, they are moved to OSIP. The OSIP process is described below.

The Operations and Service Improvement Process consists of five stages (Table A.1). In order for a project to advance from one stage to the next, it must pass a review process (a "gate") which determines that the requirements for each gate are met and that the typical gate questions are satisfactorily answered.

Each gate requires that the project be properly documented up to that point. The first stage, *Collection and Validation of Need or Opportunity*, allows people who have a need, an idea, or an opportunity (including people external to the NWS) to hold discussions with an OSIP Submitting Authority to explore the merits of that idea, and to have that idea evaluated. For this evaluation, the working team prepares two documents:

1. A Statement of Need or Opportunity Form, which describes the Need or Opportunity for consideration, and

2. The OSIP Project Plan, which identifies what is to be done next and what resources will be needed. For

Table A.1 National Weather Service Transition of Research to Operations: Operational and Service Improvement Process, OSIP.

Stage	Major Activity	Typical Decision Point (Gate) Questions
1	Collection and Validation of Need or Opportunity	Is this valid for the Weather Service? What is to be done next and who will do it?
2	Concept Exploration and Definition	Are the concept and high level requirements adequately defined or is research needed? What is to be done next and who will do it?
3	Applied Research and Analysis	What solutions are feasible, which is best? What is to be done next and who will do it?
4	Operational Development	Does developed solution meet requirements? Is there funding for deployment and subsequent activities? What is to be done next and who will do it?
5	Deploy, Maintain, and Assess	Survey—How well did the solution meet the requirements?

Hydrology projects, the Statement of Need requires the endorsement of a field office.

The *Concept Exploration and Definition* stage requires the preparation of the following documents:
1. The Exploratory Research Results Document which, as required for research projects, documents the results from exploratory research to determine effectiveness, use, or concept for associated need or opportunity, and documents the availability of already-developed solutions that will meet the Statement of Need;
2. The Concept of Operations and Operational Requirements Document, which describes how the system operates from the perspective of the user in terms that define the system capabilities required to satisfy the need; and
3. An updated OSIP Project Plan.

During the *Applied Research and Analysis* stage, the team conducts applied research, development, and analysis; identifies possible solutions; defines and documents the technical requirements; prepares a Business Case Analysis (BCA) to present a detailed comparison of the potential alternative solutions, with the recommendation of the working team as to which alternative is preferred. The BCA is a critical element in demonstrating to NWS, NOAA, and Department of Commerce management that a program is a prudent investment and will support and enhance the ability of the NWS to meet current and planned demand for its products and services. This stage requires the preparation of four documents:
1. The Applied Research Evaluation, which documents how the research was carried out, how the processes were validated, and the algorithm description for operational implementation;
2. The Technical Requirements document, which states what the operational system must explicitly address;
3. The Business case, which collects the business case analysis that describes how the system will be used; and
4. An updated OSIP Project Plan.

During the *Operational Development* stage, the team performs the operational development activities summarized in the approved Project Plan as described in the Operational Development Plan. The purpose of this stage is to fully implement the previously selected solution, to verify that the solution meets the operational and technical requirements, to conduct preparations to deploy the solution to operations, and to carry out the actions stated in the Training Plan. During this stage, the team prepares:
1. The Deployment Decision Document, which summarizes the results of the development and verification activities and presents the results of preparations for deployment, support, and training;
2. The Deployment, Maintenance and Assessment Plan, which is the plan for the final OSIP stage; and
3. An updated OSIP Project Plan and other documentation as needed.

During the final stage, *Deploy, Maintain and Assess*, the team performs the deployment activities summarized in the approved Project Plan as described in the Deployment, Assessment, and Lifecycle Support Plan. The primary purpose of this stage is to fully deploy the developed and verified solution.

How the National Weather Service Prioritizes the Development of Improved Hydrologic Forecasts

Convening Lead Author: Nathan Mantua, Climate Impacts Group, Univ. of Washington

Lead Authors: Michael D. Dettinger, U.S. Geological Survey, Scripps Institution of Oceanography; Thomas C. Pagano, National Water and Climate Center, NRCS/USDA; Andrew W. Wood, 3TIER™, Inc/Dept. of Civil and Environmental Engineering, Univ. of Washington; Kelly Redmond, Western Regional Climate Center, Desert Research Institute

Contributing Author: Pedro Restrepo, NOAA

(Adapted from Mary Mulluski's *Hydrologic Services Division [HSD] Requirements Process: How to Solicit, Collect, Refine, and Integrate Formal Ideas into Funded Projects*, NWS internal presentation, 2008.)

There are three sources of requirements toward the development of improved hydrologic forecasts at the National Weather Service: internal and external forecast improvements, and Web page information improvement. All improvements are coordinated by the National Weather Service Hydrologic Services Division (HSD).

The internal hydrologic forecast improvement requirements at the National Weather Service are a result of one of more of these sources:

- HSD routine support
- Proposed research and research-to-operations projects by annual planning teams, with the participation of HSD, the Office of Hydrologic Development (OHD), River Forecast Center and Weather Forecast Offices employees
- Teams chartered to address specific topics
- The result of service assessments
- Solicitation by the National Weather Service (NWS) Regions of improved forecast requirements to services leaders
- Semi-annual Hydrologists-in-charge (HIC), Advanced Hydrologic Prediction Service (AHPS) Review Committee (ARC), and HSD Chiefs coordination meetings
- Monthly hydro program leader calls
- Monthly ARC calls

- Biennial National Hydrologic Program Manager's Conference (HPM)
- Training classes, workshops, and customer satisfaction surveys

A flow diagram of the internal hydrologic forecast process is shown in Figure B.1.

The external requirements for hydrologic forecast improvements are the results of:

- Congressional mandates
- Office of Inspector General (OIG) requirements
- National Research Council (NRC) recommendations
- NOAA Coordination
- Biennial customer satisfaction surveys

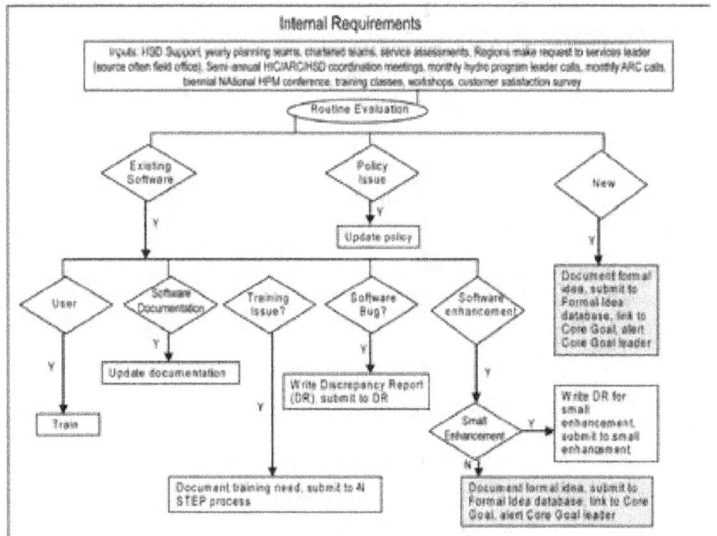

Figure B.1 Hydrologic forecast improvement: internal requirements process.

- Annual meetings, quarterly meetings on the subcommittee on hydrology, quarterly meetings of the Satellite Telemetry Information Working Group of the Advisory Committee on Water Information (ACWI)
- NOAA/USGS quarterly meetings (consistently for over 30 years)
- Local, regional and national outreach such as the National Safety Council, National Association of Flood Plain Managers, (NASFPM), National Hydrologic Warning Council (NHWC) and associated ALERT (Automated Local Evaluation in Real Time) user group conferences, International Association of Emergency Managers, (IAEM), American Geophysical Union (AGU), American Meteorological Society (AMS)
- Local and regional user forums (e.g., briefing to the Delaware River Basin Commission (DRBC), and Susquehanna River Basin Commission (SRBC))
- Federal Emergency Management Agency (FEMA) National Flood conference and coordination meetings with FEMA and regional headquarters
- Hurricane conferences, annual NWS partners meeting, NOAA constituent meetings

A flow diagram of the external hydrologic forecast process is shown in Figure B.2.

A fundamental part of the overall service of issuing hydrologic forecasts is the communication of those forecasts to the users, and the Web is an important part of that communication process. The requirement process for Web page improvements would arise from:
- Requests arising from user feedback on the web
- User calls
- Direct contact with national partners/customers
- Local NWS offices and NWS regions input
- Customer satisfaction survey

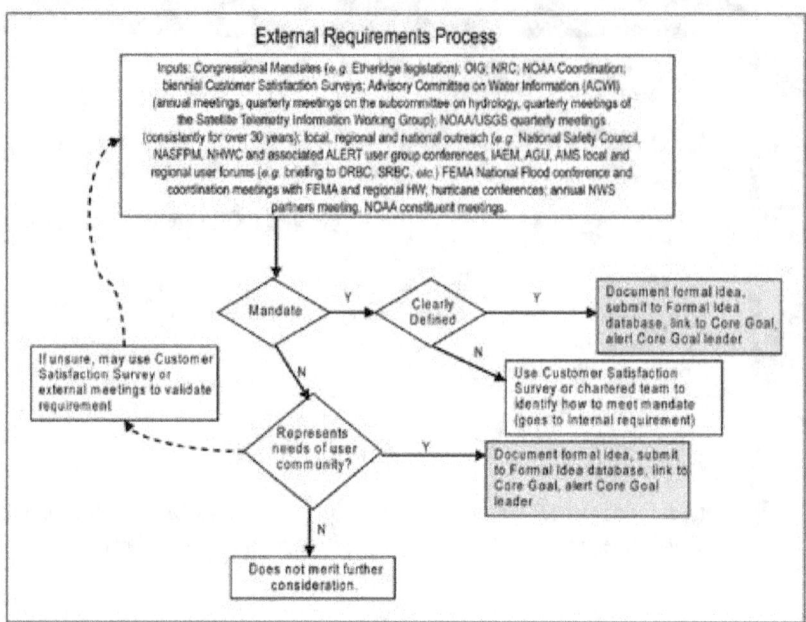

Figure B.2 Hydrologic forecast improvement: external requirements process.

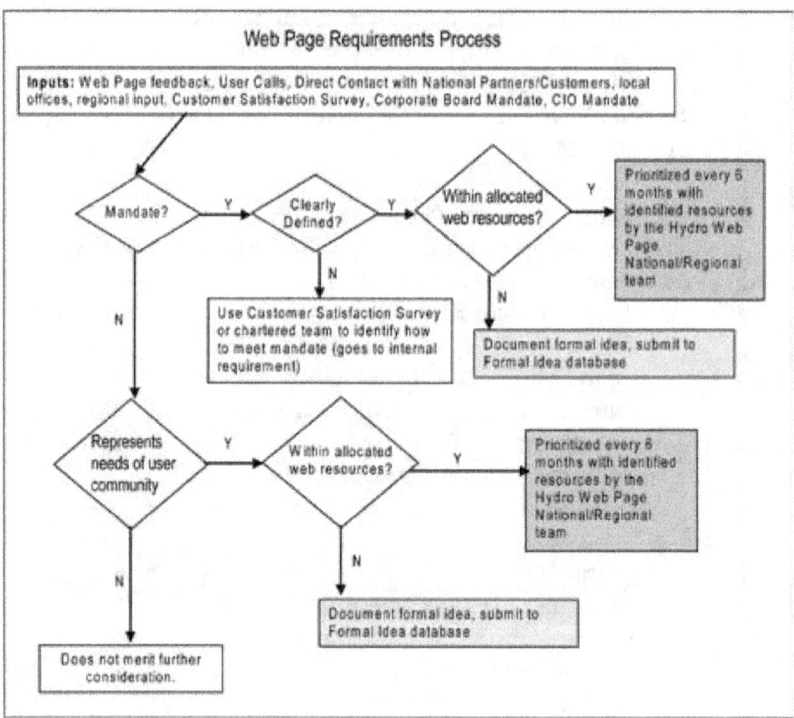

Figure B.3 Web-page improvement process.

- Corporate Board Mandate
- Chief Information Office Mandate

Figure B.3 shows the flow diagram for the web-page improvement requirement process.

GLOSSARY AND ACRONYMS

adaptive capacity
an ability of people to mitigate or reduce the potential for harm, or their vulnerability to various hazards that can cause them harm, by taking action to reduce exposure or sensitivity, both before and after the hazardous event

adaptive management
approach to water resource management that emphasizes stakeholder participation in decisions; commitment to environmentally sound, socially just outcomes; reliance upon drainage basins as planning units; program management via spatial and managerial flexibility, collaboration, participation, and sound, peer-reviewed science; and embracing ecological, economic, and equity considerations

boundary object
a prototype, model or other artifact through which collaboration can occur across different kinds of boundaries

boundary organizations
entities that perform translation and mediation functions between producers (*i.e.*, scientists) and users (*i.e.*, policy makers) of information which include: convening forums to discuss information needs, providing training, assessing problems in communication, and tailoring information for specific applications; individuals within these organizations who lead these activities are often termed "integrators"

boundary spanning
the effort to translate tools to a variety of audiences –it is usually an organization or group of people that translates scientific or difficult language to audiences so that they can use it in the future (for planning, *etc.*)

conjunctive use
the conjoint use of surface and groundwater supplies within a region to supply various uses and permit comprehensive management of both sources; this requires co-management of a stream or system of streams and an aquifer system to meet several objectives such as conserving water supplies, preventing saltwater intrusion into aquifers, and preventing contamination resulting from one supply source polluting another

decision maker
a vast assortment of elected and appointed local, state, and national agency officials, as well as public and private sector managers with policy-making responsibilities in various water management areas

decision-support experiments
practical exercises where scientists and decision makers explicitly set out to use decision–support tools–such as climate forecasts, hydrological forecasts, *etc.*–to aid in making decisions in order to address the impacts of climate variability and change upon various water issues

deterministic forecast
a single-valued prediction for a weather phenomenon

disaggregation
similar to downscaling, but in the temporal dimension; *e.g.*, seasonal climate forecasts may need to be translated into daily or subdaily temperature and precipitation inputs for a given application

downscaling
the process of bridging the spatial scale gap between the climate forecast resolution and the application's climate input resolution, if they are not the same; if the climate forecasts are from climate models, for instance, they are likely to be at a grid resolution of several hundred km, whereas the application may require climate information at a point (*e.g.*, station location)

dynamical forecasts
physics-based forecasts that are developed from conservation equations

ensemble streamflow prediction (ESP)
a method for prediction that uses an ensemble of historical meteorological sequences as model inputs (*e.g.*, temperature and precipitation) to simulate hydrology in the future (or forecast) period

hindcasts
the simulated forecasts for periods in the past using present day tools and monitoring systems; hindcasts are often used to evaluate the potential skill of present day forecast systems

integrated water resource planning
efforts to manage water by balancing supply and demand considerations through identifying feasible alternatives that meet the test of least cost without sacrificing other policy goals—such as depleted aquifer recharge, seasonal groundwater recharge, conservation, growth management strategies, and wastewater reuse

knowledge-to-action networks
the interaction among scientists and decision makers that results in decision-support system development; it begins with basic research, continues through development of information products, and concludes with end use application of information products; what makes this process a "system" is that scientists and users discuss what is needed as well as what can be provided; learn from one another's perspectives; and try to understand one another's roles and professional constraints

Loading Dock model
issuing forecasts with little notion of whether they will be used by other organizations—they are organizations that provide information to the public—but provide what they think are relevant for decision making without consulting the user to see if the information is useful

objective hybrid forecasts
forecast that uses some combination of objective forecast tools (typically, a combination of dynamical and statistical approaches)

physical vulnerability
the hazard posed to, for example, water resources and water resource systems by exposure to harmful natural or harmful technological events such as pollution, flooding, sea-level rise, or temperature change

predictand
a target variable used in statistics-based methods of forecasting

probabilistic forecast
a forecast that summarizes the results in terms of statistics of the forecast ensemble and presents the probabilistic forecast in terms of selected statistics, like probabilities of being more or less than normal

sensitivity
the degree to which people and the things they value can be harmed by exposure to a hazardous event; all other factors being equal, a water system with old infrastructure will be more sensitive to a flood or drought than one with state-of-the-art infrastructure

social vulnerability
the social factors (*e.g.*, level of income, knowledge, institutional capacity, disaster experience) that affect a system's sensitivity to exposure to a hazardous event, and that also influences its capacity to respond and adapt to exposure

statistical forecasts
objective forecasts based on empirically determined relationships between observed predictors and predictands

subjective consensus forecasts
forecasts in which expert judgment is subjectively applied to modify or combine outputs from other forecast approaches

water year or hydrologic year
October 1st through September 30th; this reflects the natural cycle in many hydrologic parameters such as the seasonal cycle of evaporative demand, and of the snow accumulation, melt, and runoff periods in many parts of the United States

ACRONYMS AND ABBREVIATIONS

ACCAP	Alaska Center for Climate Assessment and Policy	**NGOs**	non-governmental organizations
ACF	Apalachicola–Chattahoochee– Flint river basin compact	**NIFC**	National Interagency Fire Center, Boise, Idaho
AHPS	Advanced Hydrologic Prediction System	**NRC**	National Research Council
AMO	Atlantic Multidecadal Oscillation	**NSAW**	National Seasonal Assessment Workshop
CALFED	California Bay–Delta Program	**NWS**	National Weather Service
CDWR	California Department of Water Resources	**NYCDEP**	New York City Department of Environmental Protection
CEFA	Center for Ecological and Fire Applications	**OASIS**	A systems model used for reconstructing daily river flows
CFS	Climate Forecast System (see NCEP)	**ORNL**	Oak Ridge National Laboratory
CLIMAS	Climate Assessment for the Southwest Project	**PDO**	Pacific Decadal Oscillation
		PET	potential evapotranspiration
CVP	Central Valley (California) Project	**RGWM**	Regional Groundwater Model
DO	dissolved oxygen	**RISAs**	Regional Integrated Science Assessment teams
DOE	U.S. Department of Energy		
DOI	U.S. Department of the Interior	**SARP**	Sectoral Applications Research Program
DRBC	Delaware River Basin Commission		
DSS	decision support system	**SECC**	Southeast Climate Consortium
ENSO	El Niño–Southern Oscillation	**SFWMD**	South Florida Water Management District
ESA	Endangered Species Act		
ESP	Ensemble Streamflow Prediction	**SI**	Seasonal to Interannual
FEMA	Federal Emergency Management Agency	**SPU**	Seattle Public Utilities
		SRBC	Susquehanna River Basin Commission
FERC	Federal Energy Regulatory Commission		
GCM	General Circulation Model	**SST**	sea surface temperature
ICLEI	International Council of Local Environmental Initiatives	**SWE**	snow water equivalent
		SWP	State Water Project (California)
ICPRB	Interstate Commission on the Potomac River Basin	**TOGA**	Tropical Ocean–Global Atmosphere
		TRACS	Transition of Research Applications to Climate Services program
INFORM	Integrated Forecast and Reservoir Management project		
		TVA	Tennessee Valley Authority
IJC	International Joint Commission	**USACE**	U.S. Army Corps of Engineers
IPCC	United Nations Intergovernmental Panel on Climate Change	**USGS**	U.S. Geological Survey
		WMA	Washington (D.C.) Metropolitan Area
IWRP	integrated water resource planning	**WRC**	U.S. Water Resources Council
KAF	thousand acre feet	**WSE**	Water Supply and Environment —a regulation schedule for Lake Okeechobee
NCEP	National Center for Environmental Predictions		
GFS	Global Forecast System (see NCEP)		
MDBA	Murray–Darling Basin Agreement		
MLR	Multiple Linear Regression		
MOS	Model Output Statistics		
NCRFC	North Central River Forecast Center		

REFERENCES

References marked with (*) are
non peer-reviewed literature and
are available upon request.

CHAPTER I REFERENCES

Blatter, J. and H. Ingram, 2001: *Reflections on Water: New Approaches to Transboundary Conflicts and Cooperation.* MIT Press, Cambridge, MA, 358 pp.

Carbone, G.J. and K. Dow, 2005: Water resource management and drought forecasts in South Carolina. *Journal of the American Water Resources Association,* **41(1)**, 145-155.

Carlile, P., 2002: A pragmatic view of knowledge and boundaries: boundary objects in new product development. *Organization Science,* **13(4)**, 442-455.

Carlile, P.R., 2004: Transferring, translating and transforming: an integrative framework for managing knowledge across boundaries. *Organization Science,* **15(5)**, 555-568.

Cash, D.W., 2001: In order to aid in diffusing useful and practical information: agricultural extension and boundary organizations. *Science, Technology, & Human Values,* **26(4)**, 431-453.

Cash, D.W. and J. Buizer, 2005: *Knowledge-Action Systems for Seasonal to Interannual Climate Forecasting: Summary of a Workshop.* National Academies Press, Washington, DC, 44 pp.

Cash, D.W., W.C. Clark, F. Alcock, N.M. Dickson, N. Eckley, D.H. Guston, J. Jäger, and R.B.H. Mitchell, 2003: Knowledge systems for sustainable development. *Proceedings of the National Academy of Sciences,* **100(14)**, 8086-8091.

CCSP (Climate Change Science Program), 2008: *Our Changing Planet: The U.S. Climate Change Science Program for Fiscal Year 2008.* A Report by the Climate Change Science Program and the Subcommittee on Global Change Research, A Supplement to the President's Fiscal Year 2008 Budget. U.S. Climate Change Science Program, Washington, DC, 212 pp.

Clarkson, J.M. and E.T. Smerdon, 1989: Effects of climate change on water resources. *Phi Kappa Phi Journal,* **Winter,** 29-31.

de Villiers, M., 2003: *Water: The Fate of Our Most Precious Resource.* McClelland and Stewart, Toronto, 453 pp.

Dewulf, A., B. Gray, L. Putnam, N. Aarts, R. Lewicki, R. Bouwen, and C. Van Woerkum, 2005: Disentangling approaches to framing: mapping the terrain. Presented at: *18th Annual Conference of the International Association for Conflict Management,* June 12-15, 2005, Seville, Spain.

Feldman, M.S., A.M. Khadamian, H. Ingram, and A.S. Schneider, 2006: Ways of knowing and inclusive management practices. *Public Administration Review,* **66(s1)**, 89-99.

Gleick, P.H. (ed.), 2000: *Water: The Potential Consequences of Climate Variability and Change for the Water Resources of the United States.* Pacific Institute for Studies in Development, Environment, and Security, Oakland, CA, 151 pp. <http://www.gcrio.org/NationalAssessment/water/water.pdf>

Guston, D.H., 2001: Boundary organizations in environmental science and policy: an introduction. *Science, Technology, and Human Values,* **26(4)**, 399-408.

Hamlet, A.F., D. Huppert, and D.P. Lettenmaier, 2002: Economic value of long-lead streamflow forecasts for Columbia River hydropower. *Journal of Water Resources Planning and Management,* **128(2)**, 91-101.

Hartmann, H.C., T.C. Pagano, S. Sorooshian, and R. Bales, 2002: Confidence builders: evaluating seasonal climate forecasts from user perspectives. *Bulletin of the American Meteorological Society,* **83(5)**, 683-698.

IPCC (Intergovernmental Panel on Climate Change), 2007a: *Climate Change 2007: The Physical Science Basis.* Contribution of Working Group I to the Fourth Assessment Report of the Intergovernmental Panel on Climate Change [Solomon, S., D. Qin, M. Manning, Z. Chen, M. Marquis, K.B. Averyt, M. Tignor, and H.L. Miller (eds.)]. Cambridge University Press, Cambridge, UK, and New York, 996 pp.

IPCC (Intergovernmental Panel on Climate Change), 2007b: Summary for policymakers. In: *Climate Change 2007: Impacts, Adaptation and Vulnerability.* Contribution of Working Group II to the Fourth Assessment Report of the Intergovernmental Panel on Climate Change [Parry, M.L., O.F. Canziani, J.P. Palutikof, P.J. van der Linden, and C.E. Hanson (eds.)]. Cambridge University Press, Cambridge, UK, and New York, pp. 7-22.

Jacobs, K.L. and J.M. Holway, 2004: Managing for sustainability in an arid climate: lessons learned from 20 years of groundwater management in Arizona, USA. *Hydrogeology Journal,* **12(1)**, 52-65.

Jacobs, K.L., G.M. Garfin, and M. Lenart, 2005: More than just talk: connecting science and decisionmaking. *Environment*, **47(9)**, 6-22.

Keenleyside, N.S., M. Latif, J. Jungclaus, L. Kornblueh, and E. Roeckner, 2008: Advancing decadal-scale climate prediction in the North Atlantic sector. *Nature*, **453(7191)**, 84-88.

Lemos, M.C. and B. Morehouse, 2005: The co-production of science and policy in integrated climate assessments. *Global Environmental Change: Human and Policy Dimensions*, **15(1)**, 57-68.

Martin, W.E., 1984: Saving Water in a Desert City. Resources for the Future, Washington, DC, 111 pp.

Milly, P.C.D., K.A. Dunne, and A.V. Vecchia, 2005: Global pattern of trends in streamflow and water availability in a changing climate. *Nature*, **438(7066)**, 347-350.

Milly, P.C.D., J. Betancourt, M. Falkenmark, R.M. Hirsch, Z.W. Kundzewicz, D.P. Lettenmaier, and R.J. Stouffer, 2008: Stationarity is dead: Whither water management? *Science*, **319(5863)**, 573-574.

Minsky, M., 1980: A framework for representing knowledge. In: *Frame Conceptions and Text Understandings* [Metzig, D. (ed.)]. Walter de Gruter, Berlin and New York, pp. 96-119.

NRC (National Research Council), 1989: *Improving Risk Communication*. National Academy Press, Washington, DC, 332 pp.

NRC (National Research Council), 1996: *Understanding Risk: Informing Decisions in a Democratic Society*. [Stern, P.C. and H.V. Fineberg (eds.)]. National Academy Press, Washington, DC, 249 pp.

NRC (National Research Council), 1999a: *Making Climate Forecasts Matter*. National Academy Press, Washington, DC, 175 pp.

NRC (National Research Council), 1999b: *Our Common Journey: A Transition Toward Sustainability*. National Academy Press, Washington, DC, 363 pp.

NRC (National Research Council), 2004: *Adaptive Management for Water Resources Project Planning*. National Academies Press, Washington, DC, 123 pp.

NRC (National Research Council), 2005: *Decision Making for the Environment: Social and Behavioral Science Research Priorities*. [Brewer, G.D. and P.C. Stern (eds.)]. National Academies Press, Washington, DC, 281 pp.

NRC (National Research Council), 2006: *Toward a New Advanced Hydrologic Prediction Service (AHPS)*. National Academies Press, Washington, DC, 84 pp.

NRC (National Research Council), 2007: *Colorado River Basin Water Management: Evaluating and Adjusting to Hydroclimatic Variability*. National Academies Press, Washington, DC, 222 pp.

NRC (National Research Council), 2008: *Research and Networks for Decision Support in the NOAA Sectoral Applications Research Program*. [Ingram, H.M. and P.C. Stern (eds.)]. National Academies Press, Washington DC, 85 pp.

O'Connor, R.E., B. Yarnal, K. Dow, C.L. Jocoy, and G.J. Carbone, 2005: Feeling at-risk matters: water managers and the decision to use forecasts. *Risk Analysis*, **25(5)**, 1265-1275.

Rayner, S., D. Lach, and H. Ingram, 2005: Weather forecasts are for wimps: why water resource managers do not use climate forecasts. *Climatic Change*, **69(2-3)**, 197-227.

Star, S.L. and J. Griesemer, 1989: Institutional ecology, translations and boundary objects: amateurs and professionals in Berkeley's Museum of Vertebrate Zoology. *Social Studies of Science*, **19(3)**, 387-420.

Weber, M., 1947: *The Theory of Social and Economic Organization*. [Henderson, A.M. and T. Parsons, translators]. Oxford University Press, New York, 436 pp.

Weick, K., 1995: *Sensemaking in Organizations*. Sage, Thousand Oaks, CA, 231 pp.

Whiteley, J., H.M. Ingram, and R. Perry, 2008: *Water, Place and Equity*. MIT Press, Cambridge, MA, 312 pp.

Worster, D. 1985: *Rivers of Empire: Water, Aridity and the Growth of the American West*. Pantheon Books, New York, 402 pp.

Yarnal, B., A.L. Heasley, R.E. O'Connor, K. Dow, and C.L. Jocoy, 2006: The potential use of climate forecasts by community water system managers. *Land Use and Water Resources Research*, **6,** 3.1-3.8. <http://www.luwrr.com>

CHAPTER 2 REFERENCES

Anderson, E.A., 1972: *NWSRFS Forecast Procedures*. NOAA technical memorandum NWS HYDRO-14. Office of Hydrologic Development, Hydrology Laboratory, NWS/NOAA, Silver Spring, MD.

Anderson, E.A., 1973: *National Weather Service River Forecast System-Snow Accumulation and Ablation Model*. NOAA technical memorandum NWS HYDRO-17. Office of Hydrologic Development, Hydrology Laboratory, NWS/NOAA, Silver Spring, MD.

Andreadis, K.M. and D.P. Lettenmaier, 2006: Assimilating remotely sensed snow observations into a macroscale hydrology model. *Advances in Water Resources*, **29(6)**, 872-886.

Atger, F., 2003: Spatial and interannual variability of the reliability of ensemble-based probabilistic forecasts: consequences for calibration. *Monthly Weather Review*, **131(8)**, 1509-1523.

Baldwin, M.P. and T.J. Dunkerton, 1999: Propagation of the Arctic Oscillation from the stratosphere to the troposphere. *Journal of Geophysical Research*, **104(D24)**, 30937-30946.

Bales, R., N. Molotch, T. Painter, M. Dettinger, R. Rice, and J. Dozier, 2006: Mountain hydrology of the western US. *Water Resources Research*, **42**, W08432, doi:10.1029/2005WR004387.

Barnett, T.P., D.W. Pierce, H.G. Hidalgo, C. Bonfils, B.D. Santer, T. Das, G. Bala, A.W. Wood, T. Nazawa, A.A. Mirin, D.R. Cayan, and M.D. Dettinger, 2008: Human-induced changes in the hydrology of the western United States. *Science*, **319(5886)**, 1080-1083.

Barnston, A.G., A. Leetmaa, V.E. Kousky, R.E. Livezey, E. O'Lenic, H. Van den Dool, A.J. Wagner, and D.A. Unger, 1999: NCEP forecasts of the El Niño of 1997-98 and its U.S. impacts. *Bulletin of the American Meteorological Society*, **80(9)**, 1829-1852.

Beljaars, A.C.M., P.V. Viterbo, M.J. Miller, and A.K. Betts 1996: The anomalous rainfall over the United States during July 1993: sensitivity to land surface parameterization and soil moisture anomalies. *Monthly Weather Review*, **124(3)**, 362-383.

Boyle, D.P., H.V. Gupta, and S. Sorooshian, 2000: Toward improved calibration of hydrologic models: combining the strengths of manual and automatic methods. *Water Resources Research*, **36(12)**, 3663-3674.

Brier, G.W. and R.A. Allen, 1951: Verification of weather forecasts. In: *Compendium of Meteorology* [Malone, T. (ed.)]. American Meteorological Society, Boston, MA, pp. 841-848.

Burnash, R.J.C., R.L. Ferral, and R.A. McGuire, 1973: *A Generalized Streamflow Simulation System: Conceptual Modeling for Digital Computers*. Joint Federal and State River Forecast Center and State of California Department of Water Resources, Sacramento, 204 pp.

Callahan, B., E. Miles, and D. Fluharty, 1999: Policy implications of climate forecasts for water resources management in the Pacific Northwest. *Policy Sciences*, **32(3)**, 269-293.

Cash, D.W. and J. Buizer, 2005: *Knowledge-Action Systems for Seasonal to Interannual Climate Forecasting: Summary of a Workshop*. National Academies Press, Washington, DC, 44 pp.

Cash, D.W., J.D. Borck, and A.G. Pratt, 2006: Countering the loading-dock approach to linking science and decision making: comparative analysis of El Niño/Southern Oscillation (ENSO) forecasting systems. *Science, Technology and Human Values*, **31(4)**, 465-494.

Cayan, D.R. and R.H. Webb, 1992: El Niño/Southern Oscillation and streamflow in the western United States. In: *El Niño: Historical and Paleoclimatic Aspects of the Southern Oscillation* [Diaz, H.F. and V. Markgraf (eds.)]. Cambridge University Press, Cambridge, UK, and New York, pp. 29-68.

CCSP (Climate Change Science Program), 2008: *The Effects of Climate Change on Agriculture, Land Resources, Water Resources, and Biodiversity in the United States.* [Backlund, P., A. Janetos, D.S. Schimel, J. Hatfield, M. Ryan, S. Archer, and D. Lettenmaier (eds.)]. Synthesis and assessment product 4.3. U.S. Climate Change Science Program, Washington, DC, 193 pp.

Chen, W.Y. and H.M. Van den Dool, 1997: Atmospheric predictability of seasonal, annual, and decadal climate means and the role of the ENSO cycle: a model study. *Journal of Climate*, **10(6)**, 1236-1254.

Clark, M.P. and M.C. Serreze, 2000: Effects of variations in East Asian snow cover on modulating atmospheric circulation over the North Pacific Ocean. *Journal of Climate*, **13(20)**, 3700-3710.

Clark, M., S. Gangopadhyay, L. Hay, B. Rajagopalan, and R. Wilby, 2004: The Schaake Shuffle: a method for reconstructing space-time variability in forecasted precipitation and temperature fields. *Journal of Hydrometeorology*, **5(1)**, 243-262.

Coelho, C.A.S., S. Pezzuli, M. Balmaseda, F.J. Doblas-Reyes, and D.B. Stephenson, 2004: Forecast calibration and combination: a simple Bayesian approach for ENSO. *Journal of Climate*, **17(7)**, 1504-1516.

Cohen, J. and D. Entekhabi, 1999: Eurasian snow cover variability and northern hemisphere climate predictability. *Geophysical Research Letters*, **26(3)**, 345-348.

Collins, M., 2002: Climate predictability on interannual to decadal time scales: the initial value problem. *Climate Dynamics*, **19(8)**, 671-692.

Cook, E.R., D.M. Meko, D.W. Stahle, and M.K. Cleaveland, 1999: Drought reconstructions for the continental United States. *Journal of Climate*, **12(4)**, 1145-1162.

Crawford, N.H. and R.K. Linsley, 1966: *Digital Simulation in Hydrology: Stanford Watershed Model IV.* Technical report 39. Department of Civil Engineering, Stanford University, Stanford, CA, 210 pp.

Croley, T.E., II, 1996: Using NOAA's new climate outlooks in operational hydrology. *Journal of Hydrologic Engineering*, **1(3)**, 93-102.

Dallavalle, J.P. and B. Glahn, 2005: Toward a gridded MOS system. In: *21st Conference on Weather Analysis and Forecasting/17th*

Conference on Numerical Weather Prediction. American Meteorological Society, Boston, Paper 13B2. <http://ams.confex. com/ams/WAFNWP34BC/techprogram/paper_94998.htm>

Davidson, P., R. Hedrich, T. Leavy, W. Sharp, and N. Wilson, 2002: Information systems development techniques and their application to the hydrologic database derivation application. In: *Proceedings of the Second Federal Interagency Hydrologic Modeling Conference,* Las Vegas, Nevada, July 28-August 1, 2002. [11 pp.]

Day, G.N., 1985: Extended streamflow forecasting using NWS-RFS. *Journal of Water Resources Planning and Management,* **111(2)**, 157-170.

Dettinger, M.D., 2007: Sources of seasonal water-supply forecast skill in the western U.S. *EOS Transactions of the American Geophysical Union,* **88(52)**, Fall Meeting Supplement, Abstract H43A-0956.

Dettinger, M.D., D.R. Cayan, and K.T. Redmond, 1999: United States streamflow probabilities based on forecasted La Niña, winter-spring 2000: *Experimental Long-Lead Forecast Bulletin,* **8(4)**, 57-61.

Dirmeyer, P.A., R.D. Koster, and Z. Guo, 2006: Do global models properly represent the feedback between land and atmosphere? *Journal of Hydrometeorology,* **7(6)**, 1177-1198.

Doblas-Reyes, F.J., R. Hagedorn, and T.N. Palmer, 2005: The rationale behind the success of multi-model ensembles in seasonal forecasting. Part II: calibration and combination. *Tellus A,* **57(3)**, 234-252.

Earman, S., A.R. Campbell, F.M. Phillips, and B.D. Newman, 2006: Isotopic exchange between snow and atmospheric water vapor: estimation of the snowmelt component of greenwater recharge in the southwestern United States. *Journal of Geophysical Research,* **111(D9)**, D09302, doi:10.1029/2005JD006470.

Enfield, D.B., A.M. Mestas-Nuñez, and P.J. Trimble, 2001: The Atlantic Multidecadal Oscillation and its relation to rainfall and river flows in the continental US. *Geophysical Research Letters,* **28(10)**, 2077-2080.

Famiglietti, J.S., J.A. Devereaux, C.A. Laymon, T. Tsegaye, P.R. Houser, T.J. Jackson, S.T. Graham, M. Rodell, and P.J. van Oevelen, 1999: Ground-based investigation of soil moisture variability within remote sensing footprints during the Southern Great Plains 1997 (SGP97) hydrology experiment. *Water Resources Research,* **35(6)**, 1839-1851.

Ferranti, L. and P. Viterbo, 2006: The European summer of 2003: sensitivity to soil water initial conditions. *Journal of Climate,* **19(15)**, 3659-3680.

Franz, K.J., H.C. Hartmann, S. Sorooshian, and R. Bales, 2003: Verification of National Weather Service ensemble streamflow

predictions for water supply forecasting in the Colorado River basin. *Journal of Hydrometeorology,* **4(6)**, 1105-1118.

Garen, D.C., 1992: Improved techniques in regression-based streamflow volume forecasting. *Journal of Water Resources Planning and Management,* **118(6)**, 654-670.

Georgakakos, K.P., N.E. Graham, T.M. Carpenter, A.P. Georgakakos, and H. Yao, 2005: Integrating climate-hydrology forecasts and multi-objective reservoir management for Northern California. *EOS Transactions of the American Geophysical Union,* **86(12)**, 122, 127.

Glahn, H.R. and D.A. Lowry, 1972: The use of Model Output Statistics (MOS) in objective weather forecasting. *Journal of Applied Meteorology,* **11(8)**, 1203-1211.

Gleick, P.H. (ed.), 2000: *Water: The Potential Consequences of Climate Variability and Change for the Water Resources of the United States.* Pacific Institute for Studies in Development, Environment, and Security, Oakland, CA, 151 pp. <http://www. gcrio.org/NationalAssessment/water/water.pdf>

Goddard, L. and M. Dilley, 2005: El Niño: catastrophe or opportunity. *Journal of Climate,* **18(5)**, 651-665.

Hamill, T.M., J.S. Whitaker, and S.L. Mullen, 2006: Reforecasts: an important dataset for improving weather predictions. *Bulletin of the American Meteorological Society,* **87(1)**, 33-46.

Hamlet, A.F. and D.P. Lettenmaier, 1999: Columbia River streamflow forecasting based on ENSO and PDO climate signals. *Journal of Water Resources Planning and Management,* **125(6)**, 333-341.

Hartmann, H.C., T.C. Pagano, S. Sorooshian, and R. Bales, 2002: Confidence builders: evaluating seasonal climate forecasts from user perspectives. *Bulletin of the American Meteorological Society,* **83(5)**, 683-698.

Hashino, T., A.A. Bradley, and S.S. Schwartz, 2007: Evaluation of bias-correction methods for ensemble streamflow volume forecasts. *Hydrology and Earth System Sciences,* **11(2)**, 939-950.

Hermann, R. (ed.), 1999: *Managing water resources during global change.* AWRA 28th Annual Conference and Symposium, Reno, NV, November 1-5, 1992. Technical publication series TPS-92-4. American Water Resources Association, Bethesda, MD, 860 pp.

Higgins, W., H. Pan, M. Gelman, and M. Ji, 2006: *NOAA Climate Test Bed (CTB): Part I. Science Plan and Implementation Strategy.* Climate Prediction Center, Camp Springs, MD, 33 pp. <http://www.cpc.ncep.noaa.gov/products/ctb/CTB-Science-Plan-Mar06.pdf>

Hoerling, M.P. and A. Kumar, 2003: The perfect ocean for drought. *Science,* **299(5607)**, 691-694.

Hughes, M.K. and P.M. Brown, 1992: Drought frequency in central California since 101 BC recorded in giant sequoia tree rings. *Climate Dynamics*, **6(3-4)**, 161-197.

HVSRT (Hydrologic Verification System Requirements Team), 2006: *National Weather Service River Forecast Verification Plan*. National Weather Service, Silver Spring MD, 44 pp. <http://www.weather.gov/oh/rfcdev/docs/Final_Verification_Report.pdf>

Kang, H. and C.K. Park, 2007: Error analysis of dynamical seasonal predictions of summer precipitation over the East Asian-Western Pacific region. *Geophysical Research Letters*, **34**, L13706, doi:10.1029/2007GL029392.

Kim, J., N.L. Miller, J.D. Farrara, and S.Y. Hong, 2000: A seasonal precipitation and stream flow hindcast and prediction study in the western United States during the 1997-98 winter season using a dynamic downscaling system. *Journal of Hydrometeorology*, **1(4)**, 311-329.

Koster, R.D. and M.J. Suarez, 1995: Relative contributions of land and ocean processes to precipitation variability. *Journal of Geophysical Research*, **100(D7)**, 13775-13790.

Koster, R.D. and M.J. Suarez, 2001: Soil moisture memory in climate models. *Journal of Hydrometeorology*, **2(6)**, 558-570.

Krishnamurti, T.N., C.M. Kishtawal, Z. Zhang, T. LaRow, D. Bachiochi, E. Williford, S. Gadgil, and S. Surendran, 2000: Multimodel ensemble forecasts for weather and seasonal climate. *Journal of Climate*, **13(23)**, 4196-4216.

Krzysztofowicz, R., 1999: Bayesian theory of probabilistic forecasting via deterministic hydrologic model. *Water Resources Research*, **35(9)**, 2739-2750.

Kumar, A., 2007: On the interpretation and utility of skill information for seasonal climate predictions. *Monthly Weather Review*, **135(5)**, 1974-1984.

Kumar, S.V., C.D. Peters-Lidard, Y. Tian, P.R. Houser, J. Geiger, S. Olden, L. Lighty, J.L. Eastman, B. Doty, P. Dirmeyer, J. Adams, K. Mitchell, E.F. Wood, and J. Sheffield, 2006: Land information system: an interoperable framework for high resolution land surface modeling. *Environmental Modelling & Software*, **21(10)**, 1402-1415.

Kwon, H.-H., U. Lall, and A.F. Khalil, 2007: Stochastic simulation model for nonstationary time series using an autoregressive wavelet decomposition: applications to rainfall and temperature. *Water Resources Research*, **43**, W05407, doi:10.1029/2006WR005258.

Kyriakidis, P.C., N.L. Miller, and J. Kim, 2001: Uncertainty propagation of regional climate model precipitation forecasts to hydrologic impact assessment. *Journal of Hydrometeorology*, **2(2)**, 140-160.

Landman, W.A. and L. Goddard, 2002: Statistical recalibration of GCM forecasts over southern Africa using model output statistics. *Journal of Climate*, **15(15)**, 2038-2055.

Liang, X., D.P. Lettenmaier, E.F. Wood, and S.J. Burges, 1994: A simple hydrologically based model of land surface water and energy fluxes for general circulation models. *Journal of Geophysical Research*, **99(D7)**, 14415-14428.

Liu, J., X. Yuan, D.G. Martinson, and D. Rind, 2004: Re-evaluating Antarctic sea-ice variability and its teleconnections in a GISS global climate model with improved sea ice and ocean processes. *International Journal of Climatology*, **24(7)**, 841-852.

Lo, F. and M.P. Clark, 2002: Relationships between spring snow mass and summer precipitation in the southwestern United States associated with the North American monsoon system. *Journal of Climate*, **15(11)**, 1378-1385.

Lu, C.-H. and K. Mitchell, 2004: Impact of land-surface initial conditions spin-up on warm season predictability simulated by the NCEP GFS coupled with the NOAH LSM. In: *18th Conference on Hydrology*, American Meteorological Society, Boston, MA, J3.3-1 to J3.3-5. <http://ams.confex.com/ams/pdfpapers/72110.pdf>

Luo, L. and E.F. Wood, 2007: Monitoring and predicting the 2007 U.S. drought. *Geophysical Research Letters*, **34**, L22702, doi:10.1029/2007GL031673.

Luo, L. and E.F. Wood, 2008: Use of Bayesian merging techniques in a multi-model seasonal hydrologic ensemble prediction system for the eastern U.S. *Journal of Hydrometeorology*, in press. Early online release doi:10.1175/2008JHM980.1

Mantua, N.J., S.R. Hare, Y. Zhang, J.M. Wallace, and R.C. Francis, 1997: A Pacific interdecadal climate oscillation with impacts on salmon production. *Bulletin of the American Meteorological Society*, **78(6)**, 1069-1079.

Maurer, E.P. and D.P. Lettenmaier, 2004: Potential effects of long-lead hydrologic predictability on Missouri River mainstem reservoirs. *Journal of Climate*, **17(1)**, 174-186.

McCabe, G.J., Jr., and M.D. Dettinger, 1999: Decadal variations in the strength of ENSO teleconnections with precipitation in the western United States. *International Journal of Climatology*, **19(13)**, 1399-1410.

McEnery, J., J. Ingram, Q. Duan, T. Adams, and L. Anderson, 2005: NOAA's Advanced Hydrologic Prediction Service: building pathways for better science in water forecasting. *Bulletin of the American Meteorological Society*, **86(3)**, 375-385.

McPhaden, M.J., A.J. Busalacchi, R. Cheney, J.R. Donguy, K.S. Gage, D. Halpern, M. Ji, P. Julian, G. Meyers, G.T. Mitchum, P.P. Niiler, J. Picaut, R.W. Reynolds, N. Smith, and K. Takeuchi, 1998: The Tropical Ocean-Global Atmosphere observing system: a decade of progress. *Journal of Geophysical Research*, **103(C7)**, 14169-14240.

McPhaden, M.J., S.E. Zebiak, and M.H. Glantz, 2006: ENSO as an integrating concept in earth science. *Science*, **314(5806)**, 1740-1745.

Miles, E.L., A.K. Snover, L.C. Whitely Binder, E.S. Sarachik, P.W. Mote, and N. Mantua, 2006: An approach to designing a national climate service. *Proceedings of the National Academy of Sciences*, **103(52)**, 19616-19623.

Milly, P.C.D., J. Betancourt, M. Falkenmark, R.M. Hirsch, Z.W. Kundzewicz, D.P. Lettenmaier, and R.J. Stouffer, 2008: Stationarity is dead: Whither water management? *Science*, **319(5863)**, 573-574.

Mitchell, K.E., D. Lohmann, P.R. Houser, E.F. Wood, J.C. Schaake, A. Robock, B.A. Cosgrove, J. Sheffield, Q.Y. Duan, L.F. Luo, R.W. Higgins, R.T. Pinker, J.D. Tarpley, D.P. Lettenmaier, C.H. Marshall, J.K. Entin, M. Pan, W. Shi, V. Koren, J. Meng, B.H. Ramsay, and A.A. Bailey, 2004: The multi-institution North American Land Data Assimilation System (NLDAS): utilizing multiple GCIP products and partners in a continental distributed hydrological modeling system. *Journal of Geophysical Research*, **109**, D07S90, doi:10.1029/2003JD003823.

Molotch, N. and R. Bales, 2005: Scaling snow observations from the point to the grid element: implications for observation network design. *Water Resources Research*, **41**, W11421, doi:10.1029/2005WR004229.

Molteni, F., R. Buizza, T.N. Palmer, and T. Petroliagis, 1996: The ECMWF ensemble prediction system: methodology and validation. *Quarterly Journal of the Royal Meteorological Society*, **122(529)**, 73-119.

Morss, R.E. and D.S. Battisti, 2004: Designing efficient observing networks for ENSO prediction. *Journal of Climate*, **17(16)**, 3074-3089.

Neelin, J.D., F.-F. Jin, and H.-H. Syu, 2000: Variations in ENSO phase locking. *Journal of Climate*, **13(14)**, 2570-2590.

NHWC (National Hydrologic Warning Council), 2002: *Use and Benefits of the National Weather Service River and Flood Forecasts*. National Hydrologic Warning Council, Silver Spring, MD. <http://www.nws.noaa.gov/oh/ahps/AHPS%20Benefits.pdf>

Ni-Meister, W., J.P. Walker, and P.R. Houser, 2005: Soil moisture initialization for climate prediction: characterization of model and observation errors. *Journal of Geophysical Research*, **110**, D13111, doi:10.1029/2004JD005745.

Nowlin, W.D., M. Briscoe, N. Smith, M.J. McPhaden, D. Roemmich, P. Chapman, and J.F. Grassle, 2001: Evolution of a sustained ocean observing system. *Bulletin of the American Meteorological Society*, **82(7)**, 1369-1376.

NRC (National Research Council), 2006: *Toward a New Advanced Hydrologic Prediction Service (AHPS)*. National Academies Press, Washington DC, 84 pp.

NWS (National Weather Service), 2006: *River Forecast Verification Plan*. Report of the Hydrologic Verification System Requirements Team at: <http://www.weather.gov/oh/rfcdev/docs/Final_Verification_Report.pdf>

OFCM (Office of the Federal Coordinator for Meteorology), 2007: *Interagency Strategic Research Plan for Tropical Cyclones: The Way Ahead*. FCM-P36-2007. Office of the Federal Coordinator for Meteorological Services and Supporting Research, Silver Spring, MD. <http://www.ofcm.gov/p36-isrtc/fcm-p36.htm>

Pagano, T.C. and D.C. Garen, 2005: Integration of climate information and forecasts into western US water supply forecasts. In: *Climate Variations, Climate Change, and Water Resources Engineering* [Garbrecht, J.D. and T.C. Peichota, (eds.)]. American Society of Civil Engineers, Reston, VA, pp. 86-102.

Pagano, T., D. Garen, and S. Sorooshian, 2004: Evaluation of official western US seasonal water supply outlooks, 1922-2002. *Journal of Hydrometerology*, **5(5)**, 896-909.

Palmer, T.N., A. Alessandri, U. Anderson, P. Canteluabe, M. Davey, P. Délécluse, M. Déqué, E. Díez, F.J. Doblas-Reyes, H. Feddersen, R. Graham, S. Gualdi, J.-F. Guérémy, R. Hagedorn, M. Hoshen, N. Keenlyside, M. Latif, A. Lazar, E. Maisonnave, V. Marletto, A.P. Morse, B. Orfila, P. Rogel, J.-M. Terres, and M.C. Thomson, 2004: Development of a European Multi-model Ensemble System for Season to Interannual Prediction (DEMETER). *Bulletin of the American Meteorological Society*, **85(6)**, 853-872.

Penland, C. and T. Magorian, 1993: Prediction of Niño 3 sea surface temperatures using linear inverse modeling. *Journal of Climate*, **6(6)**, 1067-1075.

Penland, C. and P.D. Sardeshmukh, 1995: The optimal growth of tropical sea surface temperature anomalies. *Journal of Climate*, **8(8)**, 1999-2024.

Pielke, R.A., Jr., 1999: Who decides? Forecasts and responsibilities in the 1997 Red River flood. *Applied Behavioral Science Review*, **7(2)**, 83-101.

Ploshay, J.J. and J.L. Anderson, 2002: Large sensitivity to initial conditions in seasonal predictions with a coupled ocean-atmosphere general circulation model. *Geophysical Research Letters*, **29(8)**, 1262, doi:10.1029/2000GL012710.

Pulwarty, R.S. and K.T. Redmond, 1997: Climate and salmon restoration in the Columbia River basin: the role and usability of seasonal forecasts. *Bulletin of the American Meteorological Society*, **78(3)**, 381-397.

Rajagopalan, B., U. Lall, and S.E. Zebiak, 2002: Categorical climate forecasts through regularization and optimal combination of multiple GCM ensembles. *Monthly Weather Review*, **130(7)**, 1792-1811.

Redmond, K.T. and R.W. Koch, 1991: Surface climate and streamflow variability in the western United States and their relationship to large-scale circulation indices. *Water Resources Research*, **27(9)**, 2381-2399.

Reichle, R.H., D.B. McLaughlin, and D. Entekhabi, 2002: Hydrologic data assimilation with the ensemble Kalman filter. *Monthly Weather Review*, **130(1)**, 103-114.

Rosati, A., K. Miyakoda, and R. Gudgel, 1997: The impact of ocean initial conditions on ENSO forecasting with a coupled model. *Monthly Weather Review*, **125(5)**, 754-772.

Saha, S., S. Nadiga, C. Thiaw, J. Wang, W. Wang, Q. Zhang, H.M. Van den Dool, H.L. Pan, S. Moorthi, D. Behringer, D. Stokes, M. Peña, S. Lord, G. White, W. Ebisuzaki, P. Peng, and P. Xie, 2006: The NCEP Climate Forecast System. *Journal of Climate*, **19(15)**, 3483-3517.

Sankarasubramanian, A. and U. Lall, 2003: Flood quantiles in a changing climate: seasonal forecasts and causal relations. *Water Resources Research*, **39(5)**, 1134, doi:10.1029/2002WR001593.

* **Schaake**, J.C., 1978: The National Weather Service extended streamflow prediction techniques: description and applications during 1977. In: *3rd Annual Climate Diagnostics Workshop*, Miami, FL, 31 Oct – 2 Nov 1978. National Oceanic and Atmospheric Administration, Silver Spring, MD, pp. 31.1-31.19.

* **Schaake**, J.C. and E.L. Peck, 1985: Analysis of water supply forecast accuracy. In: *Proceedings of the 53rd Annual Western Snow Conference*, Boulder, CO, pp. 44-53.

Schaake, J.C., T.M. Hamill, R. Buizza, and M. Clark, 2007: HEPEX: The Hydrological Ensemble Prediction Experiment. *Bulletin of the American Meteorological Society*, **88(10)**, 1541-1547.

Seo, D.-J., H.D. Herr, and J.C. Schaake, 2006: A statistical postprocessor for accounting of hydrologic uncertainties in short-range ensemble streamflow prediction. *Hydrologic and Earth System Science Discussions*, **3(4)**, 1987-2035.

Shukla, J., J. Anderson, D. Baumhefner, C. Brankovic, Y. Chang, E. Kalnay, L. Marx, T. Palmer, D. Paolino, J. Ploshay, S. Schubert, D. Straus, M. Suarez, and J. Tribbia, 2000: Dynamical seasonal prediction. *Bulletin of the American Meteorological Society*, **81(11)**, 2593-2606.

Stewart, I.T., D.R. Cayan, and M.D. Dettinger, 2004: Changes in snowmelt runoff timing in western North America under a 'business as usual' climate change scenario. *Climatic Change*, **62(1-3)**, 217-232.

Tippett, M.K., M. Barlow, and B. Lyon, 2003: Statistical correction of central southwest Asia winter precipitation simulations. *International Journal of Climatology*, **23(12)**, 1421-1433.

Tippett, M.K., A.G. Barnston, and A.W. Robertson, 2007: Estimation of seasonal precipitation tercile-based categorical probabilities from ensembles. *Journal of Climate*, **20(10)**, 2210-2228.

Trenberth, K.E., G.W. Branstator, D. Karoly, A. Kumar, N.-C. Lau, and C. Ropelewski, 1998: Progress during TOGA in understanding and modeling global teleconnections associated with tropical sea surface temperatures. *Journal of Geophysical Research*, **103 (C7)**, 14291-14324.

Van den Dool, H., 2007: *Empirical Methods in Short-Term Climate Prediction*. Oxford University Press, Oxford, 215 pp.

Vinnikov, K.Y., A. Robock, N.A. Speranskaya, and C.A. Schlosser, 1996: Scales of temporal and spatial variability of midlatitude soil moisture. *Journal of Geophysical Research*, **101(D3)**, 7163-7174.

Vrugt, J.A., C.G.H. Diks, H.V. Gupta, W. Bouten, and J.M. Verstraten, 2005: Improved treatment of uncertainty in hydrologic modeling: combining the strengths of global optimization and data assimilation. *Water Resources Research*, **41**, W01017, doi:10.1029/2004WR003059.

Wagener, T. and H.V. Gupta, 2005: Model identification for hydrological forecasting under uncertainty. *Stochastic Environmental Research and Risk Assessment*, **19(6)**, 378-387.

Wagner, W., G. Blöschl, P. Pampaloni, J.-C. Calvet, B. Bizzarri, J.-P. Wigneron, and Y. Kerr, 2007: Operational readiness of microwave soil moisture for hydrologic applications. *Nordic Hydrology*, **38(1)**, 1-20.

Welles, E., 2005: *Verification of River Stage Forecasts*. Ph.D. dissertation, Department of Hydrology and Water Resources. University of Arizona, Tucson, 157 pp.

Welles, E., S. Sorooshian, G. Carter, and B. Olsen, 2007: Hydrologic verification: a call for action and collaboration. *Bulletin of the American Meteorological Society*, **88(4)**, 503-511.

Western, A.W., S.-L. Zhou, R.B. Grayson, T.A. McMahon, G. Bloschl, and D.J. Wilson, 2004: Spatial correlation of soil moisture in small catchments and its relationship to dominant spatial hydrological processes. *Journal of Hydrology*, **286(1-4)**, 113-134.

* **Wiener**, J.D., C.W. Howe, D.S. Brookshire, C. Nunn Garcia, D. McCool, and G. Smoak, 2000: Preliminary results from Colorado for an exploratory assessment of the potential for improved water resources management by increased use of climate information in three western states and selected tribes. In: *25th Annual Climate Diagnostics and Prediction Workshop*, Palisades, NY, 23-27 October 2000. NOAA, Silver Spring, MD, pp. 231-234.

Wigmosta, M.S., L. Vail, and D.P. Lettenmaier, 1994: A distributed hydrology-vegetation model for complex terrain. *Water Resources Research*, **30(6)**, 1665-1679.

Wilks, D., 1995: *Statistical Methods in the Atmospheric Sciences: An Introduction.* Academic Press, San Diego, CA, 467 pp.

Wilson, D.J., A.W. Western, and R.B. Grayson 2004: Identifying and quantifying sources of variability in temporal and spatial soil moisture observations. *Water Resources Research*, **40**, W02507, doi:10.1029/2003WR002306.

Winter, T.C., J.W. Harvey, O.L. Franke, and W.M. Alley, 1998: *Ground Water and Surface Water: A Single Resource.* USGS circular 1139. U.S. Geological Survey, Denver, CO, 79 pp. <http://pubs.usgs.gov/circ/circ1139/>

Wood, A.W., 2008: The University of Washington surface water monitor: an experimental platform for national hydrologic assessment and prediction. In: *22nd Conference on Hydrology*, January 20-24, New Orleans, LA. American Meteorological Society, Boston, MA, paper 5.2. <http://ams.confex.com/ams/pdfpapers/134844.pdf>

Wood, A.W. and D.P. Lettenmaier, 2006: A test bed for new seasonal hydrologic forecasting approaches in the western U.S. *Bulletin of the American Meteorological Society*, **87(12)**, 1699-1712.

Wood, A.W. and J.C. Schaake, 2008: Correcting errors in streamflow forecast ensemble mean and spread. *Journal of Hydrometeorology*, **9(1)**, 132-148.

Wood, A.W., E.P. Maurer, A. Kumar, and D.P. Lettenmaier, 2002: Long range experimental hydrologic forecasting for the eastern U.S. *Journal of Geophysical Research*, **107**, 4429, doi:10.1029/2001JD000659.

Wood, A.W., A. Kumar, and D.P. Lettenmaier, 2005: A retrospective assessment of National Centers of Environmental Prediction climate model-based ensemble hydrologic forecasting in the western United States. *Journal of Geophysical Research*, **110**, D04105, doi:10.1029/2004JD004508.

Yun, W.T., L. Stefanova, A.K. Mitra, T.S.V. Vijaya Kumar, W. Dewar, and T.N. Krishnamurti, 2005: A multi-model superensemble algorithm for seasonal climate prediction using DEMETER forecasts. *Tellus A*, **57(3)**, 280-289.

Zheng F., J. Zhu, R.-H. Zhang, and G. Zhou, 2006: Improved ENSO forecasts by assimilating sea surface temperature observations into an intermediate coupled model. *Advances in Atmospheric Sciences*, **23(4)**, 615-624.

CHAPTER 3 REFERENCES

Adeel, Z. and M. Glantz, 2001: El Niño of the century: once burnt, twice shy? *Global Environmental Change*, **11(2)**, 171-174.

Allen, C.D. and D.D. Breshears, 1998: Drought-induced shift of a forest-woodland ecotone: rapid landscape response to climate variation. *Proceedings of the National Academy of Sciences*, **95(25)**, 14839-14842.

Allen Orchards, H.F., *et al.*, v. The United States, decision, United States Court of Claims, no. 113-80C, Washington, D.C. March 10, 1980.

Anderson, M.T. and L.H. Woosley Jr., 2005: *Water Availability for the Western United States: Key Scientific Challenges.* U.S. Geological Survey circular 1261. U.S. Geological Survey, Reston, VA, 85 pp. <http://pubs.usgs.gov/circ/2005/circ1261/>

Andrade, F.C., 1995: Modelo de Umidade do Solo para Atividades Agrícolas: Teoria e Prática. Fortaleza: FUNCEME.

Andrade, E.R. and W.D. Sellers, 1988: El Niño and its effect on precipitation in Arizona and Western New Mexico. *Journal of Climatology*, **8(4)**, 403-410.

Annear, T., I. Chisholm, H. Beecher, A. Locke, P. Aarrestad, N. Burkhardt, C. Comer, C. Estes, J. Hunt, R. Jacobson, G. Jobsis, J. Kaufman, J. Marshall, K. Mayes, C. Stalnaker, and R. Wentworth, 2002: *Instream Flows for Riverine Resource Stewardship.* Instream Flow Council, [Lansing, MI], 411 pp.

Attwood, D.W., T.C. Bruneau, and J.G. Galaty (eds.), 1988: *Power and Poverty: Development and Development Projects in the Third World.* Westview Press, Boulder, CO, 186 pp.

Bales, R.C., D.M. Liverman, and B.J. Morehouse, 2004: Integrated assessment as a step toward reducing climate vulnerability in the southwestern United States. *Bulletin of the American Meteorological Society*, **85(11)**, 1727-1734.

Blaikie, P., T. Cannon, I. Davies, and B. Wisner, 1994: *At Risk: Natural Hazards, People's Vulnerability, and Disasters.* Routledge, London and New York, 284 pp.

Bormann, B.T., P.G. Cunningham, M.H. Brookes, V.W. Maning, and M.W. Collopy, 1993: *Adaptive Ecosystem Management in the Pacific Northwest.* General technical report PNW 341. USDA Forest Service, Pacific Northwest Research Service, Portland, OR, 22 pp.

Brandon, D., B. Udall, and J. Lowrey, 2005: An overview of NOAA's Colorado Basin River Forecast Center. In: *Colorado River Basin Climate: Paleo, Present, Future.* California Department of Water Resources, Sacramento, pp. 30-35. <http://www.water.ca.gov/drought/docs/co_nov05.pdf>

Brown, G.E., Jr., 1997: Environmental science under siege in the U.S. Congress. *Environment,* **39(2),** 13-20, 29-30.

Brown, D.P. and A.C. Comrie, 2004: A winter precipitation 'dipole' in the western United States associated with multidecadal ENSO variability. *Geophysical Research Letters,* **31,** L09203, doi:10.1029/2003GL018726.

Cairo, R.A., 1997: Dealing with interstate water issues: the federal interstate compact experience. In: *Conflict and Cooperation on Trans-Boundary Water Resources* [Just, R.E. and S. Netanyahu (eds.)]. Kluwer, Boston, 432 pp.

Carbone, G.J. and K. Dow, 2005: Water resource management and drought forecasts in South Carolina. *Journal of the American Water Resources Association,* **41(1),** 145-155.

Cash, D.W., W.C. Clark, F. Alcock, N.M. Dickson, N. Eckley, D.H. Guston, J. Jäger, and R.B.H. Mitchell, 2003: Knowledge systems for sustainable development. *Proceedings of the National Academy of Sciences,* **100(14),** 8086-8091.

Cash, D.W., J.D. Borck, and A.G. Pratt, 2006: Countering the loading-dock approach to linking science and decision making: comparative analysis of El Niño/Southern Oscillation (ENSO) forecasting systems. *Science, Technology and Human Values,* **31(4),** 465-494.

Cayan, D.R. and R.H. Webb, 1992: El Niño/Southern Oscillation and streamflow in the western United States. In: *El Niño: Historical and Paleoclimatic Aspects of the Southern Oscillation* [Diaz, H.F. and V. Markgraf (eds.)]. Cambridge University Press, Cambridge, UK, and New York, pp. 29-68.

Cayan, D.R., K.T. Redmond, and L.G. Riddle, 1999: ENSO and hydrologic extremes in the western United States. *Journal of Climate,* **12(9),** 2881-2893.

Cayan, D.R., S.A. Kammerdiener, M.D. Dettinger, J.M. Caprio, and D.H. Peterson, 2001: Changes in the onset of spring in the western United States. *Bulletin of the American Meteorological Society,* **82(3),** 399-415.

CCSP (Climate Change Science Program), 2007: *Effects of Climate Change on Energy Production and Use in the United States.* [Wilbanks, T.J., V. Bhatt, D.E. Bilello, S.R. Bull, J. Ekmann, W.C. Horak, Y.J. Huang, M.D. Levine, M.J. Sale, D.K. Schmalzer, and M.J. Scott (eds.)]. Synthesis and assessment product 4.5. U.S. Climate Change Science Program, Washington, DC, 160 pp. <http://www.ornl.gov/sci/sap_4.5/energy_impacts/team.shtml>

* CDWR (California Department of Water Resources), 2007: *Climate Change Research Needs Workshop, May 2007.* California Department of Water Resources, [Sacramento], 51 pp. <http://wwwdwr.water.ca.gov/climatechange/docs/ClimateChangeReport-100307.pdf>

Chang, H., B.M. Evans, and D.R. Easterling, 2001: Potential effects of climate change on surface-water quality in North America. *Journal of the American Water Resources Association,* **37(4),** 973-985.

Changnon, S.A., 2002: Impacts of the midwestern drought forecasts of 2000. *Journal of Applied Meteorology,* **41(1),** 1042-1052.

Chen, C.C., D. Gillig, B.A. McCarl, and R.L. Williams, 2005: ENSO impacts on regional water management: a case study of the Edwards Aquifer (Texas, USA). *Climate Research,* **28(2),** 175-181.

Clark, M.P., L.E. Hay, G.J. McCabe, G.H. Leavesley, M.C. Serreze, and R.L. Wilby, 2003: The use of weather and climate information in forecasting water supply in the western United States. In: *Water and Climate in the Western United States* [Lewis, W.M. (ed.)]. University Press of Colorado, Boulder, pp. 69-92.

Cody, B.A., 1999: *Western Water Resource Issues.* A Congressional Research Service Brief for Congress. Congressional Research Service, Washington, DC, 13 pp.

Cook, E.R., C.A. Woodhouse, C.M. Eakin, and D.M. Meko, 2004: Long-term aridity changes in the western United States. *Science,* **306(5698),** 1015-1018.

Cortner, H.A. and M.A. Moote, 1994: Setting the political agenda: paradigmatic shifts in land and water policy. In: *Environmental Policy and Biodiversity* [Grumbine, R.E. (ed.)]. Island Press, Washington, DC, pp. 365-377.

Covello, V., E. Donovan, and J.E. Slavick, 1990: *Community Outreach.* Chemical Manufacturers Association, Washington, DC, 127 pp.

Cutter, S.L., 1996: Vulnerability to environmental hazards. *Progress in Human Geography,* **20(4),** 529-539.

Cutter, S.L., B.J. Boruff, and W.L. Shirley, 2003: Social vulnerability to environmental hazards. *Social Science Quarterly,* **84(2),** 242-261.

D'Arrigo, R.D. and G.C. Jacoby, 1991: A 1000-year record of winter precipitation from northwestern New Mexico, USA: a reconstruction from tree-rings and its relation to El Niño and the Southern Oscillation. *The Holocene,* **1(2),** 95-101.

Dettinger, M.D., D.R. Cayan, H.F. Diaz, and D.M. Meko, 1998: North-south precipitation patterns in western North America on interannual-to-decadal timescales. *Journal of Climate*, **11(12)**, 3095-3111.

Dewalle, D., J. Eismeier, and A. Rango, 2003: Early forecasts of snowmelt runoff using SNOTEL data in the upper Rio Grande Basin. In: *Proceedings of the Western Snow Conference*, April 21-25, 2003, Scottsdale, AZ, pp. 17-22.

DOE (U.S. Department of Energy), 2006: *Energy Demands on Water Resources.* Report to Congress on the Interdependency of Energy and Water. Department of Energy, Washington, DC, 80 pp. <http://www.sandia.gov/energy-water/congress_report. htm>

Dole, R.M., 2003: Predicting climate variations in the American West: What are our prospects? In: *Water and Climate in the Western United States* [Lewis, W.M. (ed.)]. University Press of Colorado, Boulder, pp. 9-28.

Dorner, S.M., W.B. Anderson, R.M. Slawson, N. Kouwen, and P.M. Huck, 2006: Hydrologic modeling of pathogen fate and transport. *Environmental Science and Technology*, **40(15)**, 4746-4753.

Dow, K., R.E. O'Connor, B.Yarnal, G.J. Carbone, and C.L. Jocoy, 2007: Why worry? Community water system managers' perceptions of climate vulnerability. *Global Environmental Change*, **17(2)**, 228-237.

Enfield, D.B., A.M. Mestas-Nuñez, and P.J. Trimble, 2001: The Atlantic Multidecadal Oscillation and its relationship to rainfall and river flow in the continental U.S. *Geophysical Research Letters*, **28(10)**, 2077-2080.

Engle, N.L., 2007: *Adaptive Capacity of Water Management to Climate Change in Brazil: A Case Study Analysis of the Baixo Jaguaribe and Pirapama River Basins.* Master's thesis, School of Natural Resources and Environment. University of Michigan, Ann Arbor, 86 pp. <http://hdl.handle.net/2027.42/50490>

Feldman, D.L., 1995: *Water Resources Management: In Search of an Environmental Ethic.* Johns Hopkins University Press, Baltimore, MD, 247 pp.

Feldman, D.L., 2007: *Water Policy for Sustainable Development.* Johns Hopkins University Press, Baltimore, MD, 371 pp.

FET (Forecast Evaluation Tool), 2008: <http://fet.hwr.arizona. edu/ForecastEvaluationTool/>

Formiga-Johnsson, R.M. and K.E. Kemper, 2005: *Institutional and Policy Analysis of River Basin Management – the Jaguaribe River Basin, Ceará, Brazil.* World Bank Policy research working paper 3649. World Bank, Washington, DC, 42 pp. <http://go.worldbank.org/06H9KDDFH0>

Freudenburg, W.R. and J.A. Rursch, 1994: The risks of putting the numbers in context. *Risk Analysis*, **14(6)**, 949-958.

Fulp, T., 2003: Management of Colorado River resources. In: *Water and Climate in the Western United States* [Lewis, W.M. (ed.)]. University Press of Colorado, Boulder, pp. 143-152.

Furlow, J., 2006: "The future ain't what it used to be": Climate change and water resources management. *Water Resources IMPACT*, **8(5)**, 5-7.

Garcia, P., P.F. Folliott, and D.G. Neary, 2005: Soil erosion following the Rodeo-Chediski wildfire: an initial assessment. *Hydrology and Water Resources in Arizona and the Southwest*, **34**, 51-56.

Garfin, G., M.A. Crimmins, and K.L. Jacobs, 2007: Drought, climate variability, and implications for water supply. In: *Arizona Water Policy: Management Innovations in an Urbanizing Arid Region* [Colby, B.G. and K.L. Jacobs (eds.)]. Resources for the Future, Washington, DC, pp. 61-78.

Garjulli, R., 2001: Experiência de Gestão Participativa de Recursos Hídricos: a Caso do Ceará. In: *Experiências de Gestão dos Recursos Hídricos.* [Alves, R.F.F. and G.B.B. de Carvalho (eds.)]. MMA/ANA, Brasília.

Garrick, D., K. Jacobs, and G. Garfin, 2008: Models, assumptions, and stakeholders: planning for water supply variability in the Colorado River basin. *Journal of the American Water Resources Association*, **44(2)**, 381-398.

GDTF (Governor's Drought Task Force), 2004: *Arizona Drought Preparedness Plan: Operational Drought Plan.* Arizona Department of Water Resources, Phoenix, 107 pp. <http://www. azwater.gov/dwr/drought/ADPPlan.html>

Georgakakos, A., 2006: Decision-support systems for integrated water resources management with an application to the Nile basin. In: *Topics on System Analysis and Integrated Water Resources Management* [Castelletti, A. and R. Soncini-Sessa (eds.)]. Elsevier, Amsterdam and London, chapter 6.

Georgakakos, K.P., N.E. Graham, T.M. Carpenter, A.P. Georgakakos, and H. Yao, 2005: Integrating climate-hydrology forecasts and multi-objective reservoir management for Northern California. *EOS Transactions of the American Geophysical Union*, **86(12)**, 122, 127.

Georgia DNR (Department of Natural Resources), 2003: *Georgia Drought Management Plan.* Georgia Department of Natural Resources, Atlanta, 23 pp. <http://www.gaepd.org/Files_PDF/ gaenviron/drought/drought_mgmtplan_2003.pdf>

Georgia Environmental Protection Division, 2002: *Progress on Interstate Water Compact.* Georgia Department of Natural Resources, Atlanta.

Georgia Forestry Commission, 2007: *Georgia Wildfires of 2007 Summary of Facts and Costs for Recovery.* Georgia Forestry Commission, Dry Branch, 1 p. <http://www.gfc.state.ga.us/GFCNews/documents/GFC2007FireFacts_000.pdf>

Gibbons, M., 1999: Science's new social contract with society. *Nature,* **402(6761 supp.)**, C81-C84.

Gillilan, D. and T.C. Brown, 1997: *Instream Flow Protection: Seeking a Balance in Western Water Use.* Island Press, Washington, DC, 417 pp.

Glantz, M., 1982: Consequences and responsibilities in drought forecasting: the case of the Yakima, 1977. *Water Resources Research,* **18(1)**, 3-13.

Gleick, P.H. (ed.), 2000: *Water: The Potential Consequences of Climate Variability and Change for the Water Resources of the United States.* Pacific Institute for Studies in Development, Environment, and Security, Oakland, CA, 151 pp. <http://www.gcrio.org/NationalAssessment/water/water.pdf>

Golnaraghi, M. and R. Kaul, 1995: The science of policymaking: responding to El Niño. *Environment,* **37(4)**, 16-44.

* **Goodman**, B., 2008: Georgia loses federal case in a dispute about water. *New York Times,* February 6. <http://www.nytimes.come/2008/02/06/us/06water.html>

Goodrich, G.B. and A.W. Ellis, 2006: Climatological drought in Arizona: an analysis of indicators for guiding the Governor's Drought Task Force. *Professional Geographer,* **58(4)**, 460-469.

Gunaji, N.N., 1995: Anatomy of the extraordinary drought at the U.S.–Mexico border. In: *Management of Water Resources in North America III: Anticipating the 21st Century* [Buras, N. (ed.)]. Proceedings of the Engineering Foundation Conference, Tucson, AZ, 4-8 September 1993. American Society of Civil Engineers, New York, pp. 139-148.

Gutzler, D.S., D.M. Kann, and C. Thornbrugh, 2002: Modulation of ENSO-based long-lead outlooks of southwestern U.S. winter precipitation by the Pacific decadal oscillation. *Weather and Forecasting,* **17(6)**, 1163-1172.

Handmer, J., 2004: Global flooding. In: *International Perspectives on Natural Disasters: Occurrence, Mitigation, and Consequences* [Stoltman, J.P., J. Lidstone, and L.M. DeChano (eds.)]. Kluwer Academic Publishers, Dordrecht, Netherlands, pp. 87-106.

Hanson, R.T. and M.D. Dettinger, 2005: Ground water/surface water responses to global climate simulations, Santa Clara-Calleguas Basin, Ventura, California. *Journal of the American Water Resources Association,* **43(3)**, 517-536.

Hartig, J.H., D.P. Dodge, L. Lovett-Doust, and K. Fuller, 1992: Identifying the critical path and building coalitions for restoring degraded areas of the Great Lakes. In: *Water Resources Planning and Management: Saving a Threatened Resource.* Proceedings of the water resources sessions at Water Forum '92. American Society of Civil Engineers, New York, pp. 823-830.

Hartmann, H., 2001: *Stakeholder Driven Research in a Hydroclimatic Context.* Ph.D. dissertation, Department of Hydrology and Water Resources. University of Arizona, Tucson, 256 leaves. <http://etd.library.arizona.edu/etd/>

Hartmann, H.C., T.C. Pagano, S. Sorooshian, and R. Bales, 2002: Confidence builders: evaluating seasonal climate forecasts from user perspectives. *Bulletin of the American Meteorological Society,* **83(5)**, 683-698.

Hawkins, T.W., A.W. Ellis, J.A. Skindlov, and D. Reigle, 2002: Intra-annual analysis of the North American snow cover-monsoon teleconnection: seasonal forecasting utility. *Journal of Climate,* **15(13)**, 1743-1753.

Hayes, M., O. Wilhelmi, and C. Knutson, 2004: Reducing drought risk: bridging theory and practice. *Natural Hazards Review,* **5(2)**, 106-113.

Hewitt, K., 1997: *Regions of Risk: A Geographical Introduction to Disasters.* Addison Wesley Longman, Harlow [UK], 389 pp.

Higgins, R.W., Y. Chen, and A.V. Douglass, 1999: Internannual variability of the North American warm season precipitation regime. *Journal of Climate,* **12(3)**, 653-680.

Hirschboeck, K.K. and D.M. Meko, 2005: *A Tree-Ring Based Assessment of Synchronous Extreme Streamflow Episodes in the Upper Colorado & Salt-Verde-Tonto River Basins.* University of Arizona Laboratory of Tree-Ring Research, Tucson, 31 pp. <http://fp.arizona.edu/kkh/srp.htm>

Holling, C.S. (ed.), 1978: *Adaptive Environmental Assessment and Management.* Wiley, New York, 377 pp.

Holmes, B.H., 1979: *A History of Federal Water Resources Programs and Policies, 1961-1970.* Economics, Statistics, and Cooperatives Service, U.S. Department of Agriculture, Washington, DC, 331 pp.

Holway, J., 2007: Urban growth and water supply. In: *Arizona Water Policy: Management Innovations in an Urbanizing Arid Region.* [Colby, B. and K. Jacobs (eds.)]. Resources for the Future, Washington, DC, pp. 157-172.

Homer-Dixon, T.F., 1999: *Environment, Scarcity, and Violence.* Princeton University Press, Princeton, NJ, 253 pp.

HRC-GWRI (Hydrologic Research Center-Georgia Water Resources Institute), 2006: *Integrated Forecast and Reservoir Management (INFORM) for Northern California: System Development and Initial Demonstration.* CEC-500-2006-109. California Energy Commission, PIER Energy-Related Environmental Research, Sacramento, 243 pp. + 9 Appendices. <http://www.energy.ca.gov/pier/project_reports/CEC-500-2006-109.html>

Hull, J.W., 2000: *The War Over Water.* Council of State Governments, Atlanta, GA, 12 pp.

Hurd, B.H., N. Leary, R. Jones, and J.B. Smith, 1999: Relative regional vulnerability of water resources to climate change. *Journal of the American Water Resources Association,* **35(6),** 1399-1410.

Hutson, S.S., N.L. Barber, J.F. Kenny, K.S. Linsey, D.S. Lumia, and M.A. Maupin, 2004: *Estimated Use of Water in the United States in 2000.* U.S. Geological Survey circular 1268. U.S. Geological Survey, Reston, VA, 46 pp.

Ingram, H. and L. Fraser, 2006: Path dependency and adroit innovation: the case of California water. In: *Punctuated Equilibrium and the Dynamics of U.S. Environmental Policy.* [Repetto, R. (ed.)]. Yale University Press, New Haven, CT, pp. 78-109.

International Joint Commission, 2000: *Protection of the Waters of the Great Lakes.* Final Report to the Governments of Canada and the United States. International Joint Commission, Ottawa, 69 pp.

IPCC (Intergovernmental Panel on Climate Change), 2007a: *Climate Change 2007: The Physical Science Basis.* Contribution of Working Group I to the Fourth Assessment Report of the Intergovernmental Panel on Climate Change [Solomon, S., D. Qin, M. Manning, Z. Chen, M. Marquis, K.B. Averyt, M. Tignor, and H.L. Miller (eds.)]. Cambridge University Press, Cambridge, UK, and New York, 996 pp.

IPCC (Intergovernmental Panel on Climate Change), 2007b: Summary for policymakers. In: *Climate Change 2007: Impacts, Adaptation and Vulnerability.* Contribution of Working Group II to the Fourth Assessment Report of the Intergovernmental Panel on Climate Change [Parry, M.L., O.F. Canziani, J.P. Palutikof, P.J. van der Linden, and C.E. Hanson (eds.)]. Cambridge University Press, Cambridge, UK, and New York, pp. 7-22.

Jacobs, K.L., 2002: *Connecting Science, Policy and Decision-Making: A Handbook for Researchers and Science Agencies.* National Oceanic and Atmospheric Administration, Office of Global Programs, Silver Spring, MD, 25 pp. <http://www.ogp.noaa.gov/mpe/sci/doc/hdbk.pdf>

Jacobs, K.L. and J.M. Holway, 2004: Managing for sustainability in an arid climate: lessons learned from 20 years of ground-water management in Arizona, USA. *Hydrogeology Journal,* **12(1),** 52-65.

Jacobs, K.L., G.M. Garfin, and M. Lenart, 2005: More than just talk: connecting science and decisionmaking. *Environment,* **47(9),** 6-22.

Jacoby, H.D., 1990: Water quality. In: *Climate Change and U.S. Water Resources.* [Wagonner, P.E. (ed.)]. John Wiley, New York, pp. 307-328.

Jasanoff, S., 1987: EPA's regulation of daminozide: unscrambling the messages of risk. *Science, Technology, and Human Values,* **12(3-4),** 116-124.

Kahneman, D., P. Slovic, and A. Tversky (eds.), 1982: *Judgment Under Uncertainty: Heuristics and Biases.* Cambridge University Press, Cambridge, UK, and New York, 555 pp.

Keeney, R.L., 1992: *Value-Focused Thinking: A Path to Creative Decision-making.* Harvard University Press, Cambridge, MA, 416 pp.

Kenney, D.S. and W.B. Lord, 1994: *Coordination Mechanisms for the Control of Interstate Water Resources: A Synthesis and Review of the Literature.* Report for the ACF-ACT Comprehensive Study. U.S. Army Corps of Engineers, Mobile District, Mobile, AL.

Kingdon, J., 1995: *Agendas, Alternatives, and Public Policies.* Harper Collins, New York, 2nd edition, 254 pp.

Kirby, K.W., 2000: *Beyond Common Knowledge: The Use of Technical Information in Policymaking.* Ph.D. dissertation, Civil and Environmental Engineering. University of California, Davis, 168 leaves.

Kirshen, P., M. Ruth, and W. Anderson, 2006: Climate's long-term impacts on urban infrastructures and services: the case of metro Boston. In: *Regional Climate Change and Variability: Impacts and Responses* [Ruth, M., K. Donaghy, and P.H. Kirshen (eds.)]. Edward Elgar Publishers, Cheltenham, England, and Northampton, MA, pp. 191-253.

Kundell, J.E. and D. Tetens, 1998: *Whose Water Is It? Major Water Allocation Issues Facing Georgia.* Carl Vinson Institute of Government, University of Georgia, Athens, 57 pp.

Kundell, J.E., T.A. DeMeo, and M. Myszewski, 2001: *Developing a Comprehensive State Water Management Plan: A Framework for Managing Georgia's Water Resources.* Research Atlanta, Inc., Atlanta, GA, 55 pp.

Landre, B.K. and B.A. Knuth, 1993: Success of citizen advisory committees in consensus based water resources planning in the Great Lakes basin. *Society and Natural Resources,* **6(3),** 229-257.

Leatherman, S.P. and G. White, 2005: Living on the edge: the coastal collision course. *Natural Hazards Observer,* **30(2),** 5-6.

Lee, K.N., 1993: *Compass and Gyroscope: Integrating Science and Politics for the Environment.* Island Press, Washington, DC, 243 pp.

Lemos, M.C., 2003: A tale of two policies: the politics of climate forecasting and drought relief in Ceará, Brazil. *Policy Sciences,* **36(2),**101-123.

Lemos, M.C. and J.L.F. Oliveira, 2004: Can water reform survive politics? Institutional change and river basin management in Ceará, Northeast Brazil. *World Development,* **32(12),** 2121-2137.

Lemos, M.C. and J.L.F. Oliveira, 2005: Water reform across the state/society divide: the case of Ceará, Brazil. *International Journal of Water Resources Development,* **21(1),** 93-107.

Lemos, M.C., D. Nelson, T. Finan, R. Fox, D. Mayorga, and I. Mayorga, 1999: *The Social and Policy Implications of Seasonal Forecasting: A Case Study of Ceará. Northeast Brazil.* NOAA Climate Program Office, Silver Spring, MD, 53 pp. <http://www.ogp.noaa.gov/mpe/sci/doc/NOAAReportFinal.pdf>

Lemos, M.C., T.J. Finan, R.W. Fox, D.R. Nelson, and J. Tucker, 2002: The use of seasonal climate forecasting in policymaking: lessons from Northeast Brazil. *Climatic Change,* **55(4),** 479-507.

Loáiciga, H.A., 2003: Climate change and ground water. *Annals of the Association of American Geographers,* **93(1),** 30-41.

Lyon, B., N. Christie-Blick, and Y. Gluzberg, 2005: Water shortages, development, and drought in Rockland County, New York. *Journal of the American Water Resources Association,* **41(6),** 1457-1469.

Mantua, N.J., 2004: Methods for detecting regime shifts in large marine ecosystems: a review with approaches applied to North Pacific data. *Progress in Oceanography,* **60(2-4),** 165-82.

Mantua, N.J., S. Hare, Y. Zhang, J.M. Wallace, and R.C. Francis, 1997: A Pacific interdecadal climate oscillation with impacts on salmon production. *Bulletin of the American Meteorological Society,* **78(6),** 1069-1079.

May, P.J., R.J. Burby, N.J. Ericksen, J.W. Handmer, J.E. Dixon, S. Michael, and S.D. Ingle, 1996: *Environmental Management and Governance: Intergovernmental Approaches to Hazards and Sustainability.* Routledge, New York, 254 pp.

McCabe, G.J., M.A. Palecki, and J.L. Betancourt, 2004: Pacific and Atlantic Ocean influences on multidecadal drought frequency in the United States. *Proceedings of the National Academy of Sciences,* **101(12),** 4136-4141.

McGinnis, M.V., 1995: On the verge of collapse: the Columbia River system, wild salmon, and the Northwest Power Planning Council. *Natural Resources Journal,* **35,** 63-92.

Meinke, H., R. Nelson, R. Stone, R. Selvaraju, and W. Baethgen, 2006: Actionable climate knowledge: from analysis to synthesis. *Climate Research,* **33(1),** 101-110.

Meko, D.M., C.W. Stockton, and W.R. Boggess, 1995: The tree-ring record of severe sustained drought. *Water Resources Bulletin,* **31(5),** 789-801.

Meko, D.M., C.A. Woodhouse, C.H. Baisan, T. Knight, J.J. Lukas, M.K. Hughes, and M.W. Salzer, 2007: Medieval drought in the upper Colorado River basin. *Geophysical Research Letters,* **34(10),** L10705, doi:10.1029/2007GL029988.

Merritt, R.H., 1979: *Creativity, Conflict, and Controversy: A History of the St. Paul District, U.S. Army Corps of Engineers.* U.S. Army Corps of Engineers, Washington, DC, 461 pp.

Miller, K. and D. Yates, 2005: *Climate Change and Water Resources: A Primer for Municipal Water Providers.* AWWA Research Foundation, Denver, CO, and University Corporation for Atmospheric Research, Boulder, CO, 83 pp.

Miller, K., S.L. Rhodes, and L.J. MacDonnell, 1996: Global change in microcosm: the case of U. S. water institutions. *Policy Sciences,* **29(4),** 271-272.

Milly, P.C.D., K.A. Dunne, and A.V. Vecchia, 2005: Global pattern of trends in streamflow and water availability in a changing climate. *Nature,* **438(7066),** 347-350.

Moody, J.A. and D.A. Martin, 2001: Post-fire, rainfall intensity-peak discharge relations for three mountainous watersheds in the western USA. *Hydrological Processes,* **15(15),** 2981-2993.

Morehouse, B.J., R.H. Carter, and P. Tschakert, 2002: Sensitivity of urban water resources in Phoenix, Tucson, and Sierra Vista, Arizona, to severe drought. *Climate Research,* **21(3),** 283-297.

Mumme, S.P., 1995: The new regime for managing U.S.-Mexican water resources. *Environmental Management,* **19(6),** 827-835.

Mumme, S.P., 2003: Strengthening binational drought management. *Utton Center Report: University of New Mexico School of Law,* **2(1),** 3-7.

Murdoch, P.S., J.S. Baron, and T.L. Miller, 2000: Potential effects of climate change on surface-water quality in North America. *Journal of the American Water Resources Association,* **36(2),** 347-366.

Naím, M., 2003: The five wars of globalization. *Foreign Policy,* **134,** 28-37.

NAST (National Assessment Synthesis Team), 2001: *Climate Change Impacts on the United States: The Potential Consequences of Climate Variability and Change.* Cambridge University Press, Cambridge, UK, and New York, 612 pp. <http://www.usgcrp.gov/usgcrp/Library/nationalassessment/>

Neary, D.G., K.C. Ryan, and L.F. DeBano (eds.), 2005: *Wildland Fire in Ecosystems: Effects of Fire on Soils and Water.* General technical report RMRS-GTR-42-vol.4. U.S. Department of Agriculture, Forest Service, Rocky Mountain Research Station, Ogden, UT, 250 pp. <http://purl.access.gpo.gov/GPO/LPS80060>

Nelson, R.R. and S.G. Winter, 1960: *Weather Information and Economic Decisions.* RM-2670-NASA. RAND Corporation, Santa Monica, CA 121 pp.

NMDTF (New Mexico Drought Task Force), 2006: *New Mexico Drought Plan: Update December 2006.* New Mexico Office of the State Engineer, Albuquerque, 37 pp. <http://www.nm-drought.state.nm.us/droughtplans.html>

NOAA Office of Hydrology, 1998: *Service Assessment and Hydraulic Analysis of the Red River of the North 1997 Floods.* Office of Hydrologic Development, Silver Spring, MD. <http://www.nws.noaa.gov/oh/Dis_Svy/RedR_Apr97/>

NRC (National Research Council), 1989: *Improving Risk Communication.* National Academy Press, Washington, DC, 332 pp.

NRC (National Research Council), 1996: *Understanding Risk: Informing Decisions in a Democratic Society.* [Stern, P.C. and H.V. Fineberg (eds.)]. National Academy Press, Washington, DC, 249 pp.

NRC (National Research Council), 1999: *Making Climate Forecasts Matter.* National Academy Press, Washington, DC, 175 pp.

NRC (National Research Council), 2005: *Decision Making for the Environment: Social and Behavioral Science Research Priorities.* [Brewer, G.D. and P.C. Stern (eds.)]. National Academies Press, Washington, DC, 281 pp.

NRC (National Research Council), 2006: *Toward a New Advanced Hydrologic Prediction Service (AHPS).* National Academies Press, Washington, DC, 84 pp.

NRC (National Research Council), 2007: *Colorado River Basin Water Management: Evaluating and Adjusting to Hydroclimatic Variability.* National Academies Press, Washington, DC, 222 pp.

NRC (National Research Council), 2008: *Research and Networks for Decision Support in the NOAA Sectoral Applications Research Program.* [Ingram, H.M. and P.C. Stern (eds.)]. National Academies Press, Washington DC, 85 pp.

Obeysekera, J., P. Trimble, C. Neidrauer, C. Pathak, J. VanArman, T. Strowd, and C. Hall, 2007: Appendix 2-2: Consideration of long-term climatic variability in regional modeling for SFWMD planning and operations. In: *2007 South Florida Environmental Report.* Florida Department of Environmental Protection and South Florida Water Management District, West Palm Beach, 47 pp. <http://www.sfwmd.gov/sfer/SFER_2007/Appendices/v1_app_2-2.pdf>

O'Connor, R.E., B. Yarnal, R. Neff, R. Bord, N. Wiefek, C. Reenock, R. Shudak, C.L. Jocoy, P. Pascale, and C.G. Knight, 1999: Weather and climate extremes, climate change, and planning: views of community water managers in Pennsylvania's Susquehanna River Basin. *Journal of the American Water Resources Association,* **35(6),** 1411-1419.

O'Connor, R.E., B. Yarnal, K. Dow, C.L. Jocoy, and G.J. Carbone, 2005: Feeling at-risk matters: water managers and the decision to use forecasts. *Risk Analysis,* **25(5),** 1265-1275.

Pagano, T.C., H.C. Hartmann, and S. Sorooshian, 2001: Using climate forecasts for water management: Arizona and the 1997-1998 El Niño. *Journal of the American Water Resources Association,* **37(5),** 1139-1153.

Pagano, T.C., H.C. Hartmann, and S. Sorooshian, 2002: Factors affecting seasonal forecast use in Arizona water management: a case study of the 1997-98 El Niño. *Climate Research,* **21(3),** 259-269.

Papadakis, E., 1996: *Environmental Politics and Institutional Change.* Cambridge University Press, Cambridge, UK and New York, 240 pp.

Payne, J., J. Bettman, and E. Johnson, 1993: *The Adaptive Decision Maker.* Cambridge University Press, Cambridge, UK, and New York, 33 pp.

Pfaff, A., K. Broad, and M. Glantz, 1999: Who benefits from climate forecasts? *Nature,* **397(6721),** 645-646.

Pielke, R.A., Jr., 1999: Who decides? Forecasts and responsibilities in the 1997 Red River flood. *Applied Behavioral Science Review,* **7(2),** 83-101.

Polsky, C., R. Neff, and B. Yarnal 2007: Building comparable global change vulnerability assessments: the vulnerability scoping diagram. *Global Environmental Change,* **17(3-4),** 472-485.

Pool, D.R., 2005: Variations in climate and ephemeral channel recharge in southeastern Arizona, United States. *Water Resources Research,* **41,** W11403, doi:10.1029/2004WR003255.

Power, S., B. Sadler, and N. Nicholls, 2005: The influence of climate science on water management in western Australia: lessons for climate scientists. *Bulletin of the American Meteorological Society,* **86(6),** 839-844.

Pringle, C.M., 2000: Threats to U.S. public lands from cumulative hydrologic alterations outside of their boundaries. *Ecological Applications*, **10(4)**, 971-989.

Pulwarty, R.S. and T.S. Melis, 2001: Climate extremes and adaptive management on the Colorado River: lessons from the 1997-1998 ENSO event. *Journal of Environmental Management*, **63(3)**, 307-324.

Rango, A., 2006: Snow: the real water supply for the Rio Grande basin. *New Mexico Journal of Science*, **44**, 99-118.

* **Ray**, A.J. and R.S. Webb, 2000: Demand-side perspective on climate services for reservoir management. In: *Proceedings, 25th Annual Climate Diagnostics and Prediction Workshop*, Palisades, NY, 23-27 October 2000. NOAA, Silver Spring, MD, pp. 219-222.

Rayner, S., D. Lach, and H. Ingram, 2005: Weather forecasts are for wimps: why water resource managers do not use climate forecasts. *Climatic Change*, **69(2-3)**, 197-227.

Redmond, K.T., 2003: Climate variability in the West: complex spatial structure associated with topography, and observational issues. In: *Water and Climate in the Western United States* [Lewis, W.M. (ed.)]. University Press of Colorado, Boulder, pp.29-48.

Restoring the Waters, 1997: Natural Resources Law Center, University of Colorado School of Law, Boulder, CO, 64 pp.

Rhodes, S.L., D. Ely, and J.A. Dracup, 1984: Climate and the Colorado River: the limits to management. *Bulletin of the American Meteorological Society*, **65(7)**, 682-691.

Rosenberg, D.M., P. McCully, and C.M. Pringle, 2000: Global-scale environmental effects of hydrological alterations: introduction. *BioScience*, **50(9)**, 746-751.

Roy, S.B., P.F. Ricci, K.V. Summers, C.F. Chung, and R.A. Goldstein, 2005: Evaluation of the sustainability of water withdrawals in the United States, 1995-2025. *Journal of the American Water Resources Association*, **41(5)**, 1091-1108.

Rubenstein, R.A., 1986: Reflections on action anthropology: some developmental dynamics of an anthropological tradition. *Human Organization*, **45(3)**, 270-279.

Rygel, L., D. O'Sullivan, and B. Yarnal, 2006: A method for constructing a social vulnerability index: an application to hurricane storm surges in a developed country. *Mitigation and Adaptations Strategies for Global Change*, **11(3)**, 741-764.

Sarewitz, D. and R.A. Pielke Jr., 2007: The neglected heart of science policy: reconciling supply of and demand for science. *Environmental Science and Policy*, **10(1)**, 5-16.

Schneider, S.H. and J. Sarukhan, 2001: Overview of impacts, adaptation, and vulnerability to climate change. In: *Climate Change 2001: Impacts, Adaptation and Vulnerability*. Contribution of Working Group 2 to the Third Assessment Report of the Intergovernmental Panel on Climate Change. [McCarthy, J.J., O.F. Canziani, N.A. Leary, D.J. Dokken, and K.S. White, (eds.)]. Cambridge University Press, Cambridge, UK, and New York, pp. 75-103.

Schröter, D., C. Polsky, and A.G. Patt, 2005: Assessing vulnerabilities to the effects of global change: an eight step approach. *Mitigation and Adaptation Strategies for Global Change*, **10(4)**, 573-596.

Shamir, E., D.M. Meko, N.E. Graham, and K.P. Georgakakos, 2007: Hydrologic model framework for water resources planning in the Santa Cruz River, southern Arizona. *Journal of the American Water Resources Association*, **43(5)**, 1155-1170.

Sheppard, P.R., A.C. Comrie, G.D. Packin, K. Angersbach and M.K. Hughes, 2002: The climate of the US southwest. *Climate Research*, **21(3)**, 219-238.

Slovic, P., B. Fischhoff, and S. Lichtenstein 1977: Behavioral decision theory. *Annual Review of Psychology*, **28**, 1-39.

Smith, S. and E. Reeves (eds.), 1988: *Human Systems Ecology: Studies in the Integration of Political Economy, Adaptation, and Socionatural Regions*. Westview Press, Boulder, CO, 233 pp.

Sonnett, J., B.J. Morehouse, T. Finger, G. Garfin, and N. Rattray, 2006: Drought and declining reservoirs: comparing media discourse in Arizona and New Mexico, 2002-2004. *Environmental Change*, **16(1)**, 95-113.

Steinemann, A.C. and L.F.N. Cavalcanti, 2006: Developing multiple indicators and triggers for drought plans. *Journal of Water Resources Planning and Management*, **132(3)**, 164-174.

Stewart, I.T., D.R. Cayan, and M.D. Dettinger, 2005: Changes toward earlier streamflow timing across western North America. *Journal of Climate*, **18(8)**, 1136-1155.

Stoltman, J.P., J. Lidstone, and L.M. DeChano, 2004: Introduction. In: *International Perspectives on Natural Disasters: Occurrence, Mitigation, and Consequences* [Stoltman, J.P., J. Lidstone, and L.M. DeChano (eds.)]. Kluwer Academic Publishers, Dordrecht, Netherlands, pp. 1-10.

Stone, D.A., 1997: *Policy Paradox: The Art of Political Decision Making*. W.W. Norton, New York, 394 pp.

Subcommittee on Disaster Reduction, 2005: Grand challenges for disaster reduction. *Natural Hazards Observer*, **30(2)**, 1-3.

Swetnam, T.W. and J.L. Betancourt, 1998: Mesoscale disturbance and ecological response to decadal climatic variability in the American southwest. *Journal of Climate*, **11(12)**, 3128-3147.

Taddei, R., 2005: *Of Clouds and Streams, Prophets and Profits: The Political Semiotics of Climate and Water in the Brazilian Northeast*. Ph.D. dissertation, Teachers College. Columbia University, New York, 405 leaves.

Torgerson, D., 2005: Obsolescent Leviathan: problems of order in administrative thought. In: *Managing Leviathan: Environmental Politics and the Administrative State* [Paehlke, R. and D. Torgerson (eds.)]. Broadview Press, Peterborough, Ontario, 2nd edition, pp. 11-24.

Trush, B. and S. McBain, 2000: Alluvial river ecosystem attributes. *Stream Notes*, **January**, 1-3.

Tversky, A. and D. Kahneman, 1974: Judgment under uncertainty: heuristics and biases. *Science*, **185(4157)**, 1124-1131.

Udall, B. and M. Hoerling, 2005: Seasonal forecasting: skill in the intermountain west? In *Colorado River Basin: Paleo, Present, Future*. California Department of Water Resources, Sacramento, pp. 23-29. <http://www.water.ca.gov/drought/docs/co_nov05.pdf>

Upendram, S. and J.M. Peterson, 2007: Irrigation technology and water conservation in the High Plains aquifer region. *Journal of Contemporary Water Research & Education*, **137**, 40-46.

U.S. District Court, 1945: Spokane, Washington, Consent decree in the District Court of the United States for the Eastern District of Washington, Southern Division: Civil action No. 21.

USGS (U.S. Geological Survey), 2004: *Climatic Fluctuations, Drought, and Flow in the Colorado River Basin*. USGS fact sheet 2004-3062 version 2. U.S. Geological Survey, Reston, VA, 4 pp. <http://water.usgs.gov/pubs/fs/2004/3062/>

Wagner, E.T., 1995: Canada-United States boundary waters management arrangements. In: *Management of Water Resources in North America III: Anticipating the 21st Century* [Buras, N. (ed.)]. Proceedings of the Engineering Foundation Conference, Tucson, AZ, 4-8 September 1993. American Society of Civil Engineers, New York, pp. 16-22.

Wahl, R.W., 1989: *Markets for Federal Water: Subsidies, Property Rights and the Bureau of Reclamation*. Resources for the Future, Washington DC, 308 pp.

Wallentine, C.B. and D. Matthews, 2003: Can climate predictions be of practical use in western water management? In: *Water and Climate in the Western United States* [Lewis, W.M. (ed.)]. University Press of Colorado, Boulder, pp. 161-168.

Walters, C.J., 1986: *Adaptive Management of Renewable Resources*. Macmillan, New York, 374 pp.

Water in the West: Challenge for the Next Century, 1998: Western Water Policy Review Advisory Commission, [Washington, DC]. <http://hdl.handle.net/1928/2788>

Weingart, P., A. Engels, and P. Pansegrau, 2000: Risks of communication: discourses on climate change in science, politics, and the mass media. *Public Understanding of Science*, **9(3)**, 261-283.

Weston, R.T., 1995: Delaware River basin: challenges and successes in interstate water management. Presented at *ASCE Water Resources Engineering Conference*, San Antonio, Texas, August 17.

Westphal, K.S., R.M. Vogel, P. Kirshen, and S.C. Chapra, 2003: Decision support system for adaptive water supply management. *Journal of Water Resources Planning and Management*, **129(3)**, 165-177.

***Wiener**, J.D., C.W. Howe, D.S. Brookshire, C.N. Garcia, D. McCool, and G. Smoak, 2000: Preliminary results for Colorado from an exploratory assessment of the potential for improved water resources management by increased use of climate information in three western states and selected tribes. In: *25th Annual Climate Diagnostcs and Prediction Workshop*, Palisades, NY, 23-27 October 2000. NOAA, Silver Spring, MD, pp. 231-234.

Wiener, J.D., K.A. Dwire, S.K. Skagen, R. Crifasi, and D. Yates, 2008: Riparian ecosystem consequences of water redistribution along the Colorado Front Range. *Water Resources IMPACT*, **10(3)**, 18-22.

Wilhite, D.A., 2004: Drought. In: *International Perspectives on Natural Disasters: Occurrence, Mitigation, and Consequences* [Stoltman, J.P., J. Lidstone, and L.M. DeChano (eds.)]. Kluwer Academic Publishers, Dordrecht, Netherlands, pp. 147-162.

Wilhite, D.A., M.J. Hayes, C. Knutson, and K.H. Smith, 2000: Planning for drought: moving from crisis to risk management. *Journal of the American Water Resources Association*, **36(4)**, 697-710.

Woodhouse, C.A. and J.J. Lukas, 2006: Drought, tree rings, and water resource management in Colorado. *Canadian Water Resources Journal*, **31(4)**, 1-14.

Woodhouse, C.A., S.T. Gray, and D.M. Meko, 2006: Updated streamflow reconstructions for the Upper Colorado River Basin. *Water Resources Research* **42**, W05415, doi:10.1029/2005WR004455.

Yarnal, B., R.E. O'Connor, K. Dow, G.J. Carbone, and C.L. Jocoy, 2005: Why don't community water system managers use weather and climate forecasts? In: *15th Conference on Applied Climatology*, Savannah, GA, June 2005. American Meteorological Society, Boston, paper 7.1. <http://ams.confex.com/ams/pdfpapers/93923.pdf>

Yarnal, B., A.L. Heasley, R.E. O'Connor, K. Dow, and C.L. Jocoy, 2006: The potential use of climate forecasts by community water system managers. *Land Use and Water Resources Research*, **6**, 3.1-3.8. <http://www.luwrr.com>

Zarriello, P.J. and K.G. Ries, 2000: *A Precipitation-Runoff Model for Analysis of the Effects of Water Withdrawals on Streamflow, Ipswich River Basin, Massachusetts*. Water resources investigations report 00-4029. U.S. Geological Survey, Northborough, MA, 99 pp. <http://purl.access.gpo.gov/GPO/LPS24844>

Zektser, S., H.A. Loaiciga, and J.T. Wolf, 2005: Environmental impacts of groundwater overdraft: selected case studies in the southwestern United States. *Environmental Geology*, **47(3)**, 396-404.

CHAPTER 4 REFERENCES

Atwater, R. and W. Blomquist, 2002: Rates, rights, and regional planning in the Metropolitan Water District of Southern California. *Journal of the American Water Resources Association*, **38(5)** 1195-1205.

Beecher, J.A., 1995: Integrated resource planning fundamentals. *Journal of the American Water Works Association*, **87(6)**, 34-48.

Bennis, W.G., 2003: *On Becoming a Leader*. De Capo Press, Cambridge, MA, 256 pp.

Bisson, P.A., B.E. Rieman, C. Luce, P.F. Hessburg, D.C. Lee, J.L. Kershner, G.H. Reeves, and R.E. Gresswell, 2003: Fire and aquatic ecosystems of the western USA: current knowledge and key questions. *Forest Ecology and Management*, **178(1-2)**, 213-229.

Bormann, B.T., P.G. Cunningham, M.H. Brookes, V.W. Maning, and M.W. Collopy, 1993: *Adaptive Ecosystem Management in the Pacific Northwest*. General technical report PNW 341. USDA Forest Service, Pacific Northwest Research Service, Portland, OR, 22 pp.

* **Bras**, R.L., 2006: *Summary of Reviews of "Consideration of Long-Term Climatic Variation in SFWMD Planning and Operations" by Obeysekera* et al. Report submitted to the South Florida Water Management District, 5 pp. <http://www.sfwmd.gov/pls/portal/docs/PAGE/PG_GRP_SFWMD_WEATHER/PORTLET_CLIMATE%20VARIABILITY/TAB2202367/BRAS_SUMMARY_REPORT.PDF>

Breuer, N., V.E. Cabrera, K.T. Ingram, K. Broad, and P.E. Hildebrand, 2007: AgClimate: a case study in participatory decision support system development. *Climate Change*, **87(3-4)**, 385-403.

Broad, K. and S. Agrawalla, 2000: The Ethiopia food crisis: uses and limits of climate forecasts. *Science*, **289(5485)**, 1693-1694.

Broad, K., A.S.P. Pfaff, and M.H. Glantz, 2002: Effective and equitable dissemination of seasonal-to-interannual climate forecasts: policy implications from the Peruvian fishery during El Niño 1997-98. *Climate Change*, **54(4)**, 415-438.

Bromberg, J.L., 2000: *NASA and the Space Industry*. Johns Hopkins University Press, Baltimore, MD, 247 pp.

Brown, G.E., Jr., 1997: Environmental science under siege in the U.S. Congress. *Environment*, **39(2)**, 13-20, 29-30.

Cabrera, V.E., N.E. Breuer, and P.E. Hildebrand, 2006: North Florida dairy farmer perceptions toward the use of seasonal climate forecast technology. *Climatic Change*, **78(2-4)**, 479-491.

Cabrera, V.E., N.E. Breuer, and P.E. Hildebrand, 2007: Participatory modeling in dairy farm systems: a method for building consensual environmental sustainability using seasonal climate forecasts. *Climatic Change*, **89(3-4)**, 395-409.

Cadavid, L.G., R. Van Zee, C. White, P. Trimble, and J.T.B. Obeysekera, 1999: Operational hydrology in south Florida using climate forecasts. In: *American Geophysical Union 19th Annual Hydrology Days*, Colorado State University, August 16-20. <https://my.sfwmd.gov/pls/portal/docs/page/pg_grp_sfwmd_hesm/portlet_opsplan_2/tab1340077/opln_hyd_web.pdf>

CALFED: Welcome to CALFED Bay-Delta Program website. <http://calwater.ca.gov/index.aspx>

Callahan, B., E. Miles, and D. Fluharty, 1999: Policy implications of climate forecasts for water resources management in the Pacific Northwest. *Policy Sciences*, **32(3)**, 269-293.

Carothers, S.W. and B.T. Brown, 1991: *The Colorado River Through Grand Canyon: Natural History and Human Change*. University of Arizona Press, Tucson, 235 pp.

Cash, D.W., 2001: In order to aid in diffusing useful and practical information: agricultural extension and boundary organizations. *Science, Technology, & Human Values*, **26(4)**, 431-453.

Cash, D.W. and J. Buizer, 2005: *Knowledge-Action Systems for Seasonal to Interannual Climate Forecasting: Summary of a Workshop*. National Academies Press, Washington, DC, 44 pp.

Cash, D.W., W.C. Clark, F. Alcock, N.M. Dickson, N. Eckley, D.H. Guston, J. Jäger, and R.B.H. Mitchell, 2003: Knowledge systems for sustainable development. *Proceedings of the National Academy of Sciences*, **100(14)**, 8086-8091.

* **CDWR** (California Department of Water Resources), 2007a: *DWR Signs Agreement with NOAA for Climate Research.* Press release. <http://www.publicaffairs.water.ca.gov/newsreleases/2007/090707summit.cfm>

* **CDWR** (California Department of Water Resources), 2007b: *Climate Change Research Needs Workshop, May 2007.* California Department of Water Resources, [Sacramento], 51 pp. <http://wwwdwr.water.ca.gov/climatechange/docs/ClimateChangeReport-100307.pdf>

Chambers, R., 1997: *Whose Reality Counts? Putting the First Last.* Intermediate Technology Publications, London, 297 pp.

Cody, B.A., 1999: *Western Water Resource Issues.* A Congressional Research Service Brief for Congress. Congressional Research Service, Washington, DC, 13 pp.

Corringham, T., A.L. Westerling, and B.J. Morehouse, 2008: Exploring use of climate information in wildland fire management: a decision calendar study. *Journal of Forestry,* **106(2)**, 71-77.

Cortner, H.A. and M.A. Moote, 1994: Setting the political agenda: paradigmatic shifts in land and water policy. In: *Environmental Policy and Biodiversity* [Grumbine, R.E. (ed.)]. Island Press, Washington DC, pp. 365-377.

Council of State Governments, 2003: *Water Wars.* Council of State Governments, Lexington, KY, 22 pp. <http://www.csg.org/pubs/Documents/TA0307WaterWars.pdf>

* **Crawford**, B., G. Garfin, R. Ochoa, R. Heffernan, T. Wordell, and T. Brown, 2006: *National Seasonal Assessment Workshop Proceedings.* CLIMAS, Institute for the Study of Planet Earth, University of Arizona, Tucson. <http://www.climas.arizona.edu/conferences/NSAW/publications/NSAWproceedings_06.pdf>

Delta Vision Blue Ribbon Task Force, 2008: *Delta Vision: Our Vision for the California Delta.* [State of California Resources Agency, Sacramento], 70 pp. <http://www.deltavision.ca.gov/BlueRibbonTaskForce/FinalVision/Delta_Vision_Final.pdf>

Desilets, S.L.E., B. Nijssen, B. Ekwurzel, and T.P.A. Ferré, 2006: Post-wildfire changes in suspended sediment rating curves: Sabino Canyon, Arizona. *Hydrologic Processes,* **21(11)**, 1413-1423.

Dickinson, J. (ed.), 2007: *Inventory of New York City Greenhouse Gas Emissions.* Mayor's Office of Long-Term Planning and Sustainability, New York City, 65 pp.

Donahue, J.M. and B.R. Johnston, 1998: *Water, Culture, and Power: Local Struggles in a Global Context.* Island Press, Washington, DC, 396 pp.

Dow, K., R.E. O'Connor, B. Yarnal, G.J. Carbone, and C.L. Jocoy, 2007: Why worry? Community water system managers' perceptions of climate vulnerability. *Global Environmental Change,* **17(2)**, 228-237.

Earles, T.A., K.R. Wright, C. Brown, and T.E. Langan, 2004: Los Alamos forest fire impact modeling. *Journal of the American Water Resources Association,* **40(2)**, 371-384.

Ekwurzel, B., 2004: Flooding during a drought? Climate variability and fire in the Southwest. *Southwest Hydrology,* **3(5)**, 16-17.

Enfield, D.B., A.M. Mestas-Nuñez, and P.J. Trimble, 2001: The Atlantic Multidecadal Oscillation and its relationship to rainfall and river flow in the continental U.S. *Geophysical Research Letters,* **28(10)**, 2077-2080.

Feldman, M. and A. Khademian, 2004: *Inclusive Management: Building Relationships with the Public.* Paper 0412. Center for the Study of Democracy, University of California, Irvine, 21 pp. <http://repositories.cdlib.org/csd/04-12>

Feldman, D.L. and C. Wilt, 1996: Evaluating the implementation of state-level global climate change programs. *Journal of Environment and Development,* **5(1)**, 46-72.

Feldman, D.L. and C.A. Wilt, 1999: Climate-change policy from a bioregional perspective: reconciling spatial scale with human and ecological impact. In: *Bioregionalism.* [McGinnis, M.V. (ed.)]. Routledge, London, pp. 133-154.

Fiske, G. and A. Dong, 1995: IRP: a case study from Nevada. *Journal of the American Water Works Association,* **87(6)**, 72-83.

Florida Department of Environmental Protection and South Florida Water Management District, 2007: *Executive Summary: 2007 South Florida Environmental Report.* South Florida Water Management District, West Palm Beach, 49 pp. <https://my.sfwmd.gov/pls/portal/url/ITEM/493263378434D93CE040E88D485237E8>

Fraisse, C., J. Bellow, N. Breuer, V. Cabrera, J. Jones, K. Ingram, G. Hoogenboom, and J. Paz, 2005: *Strategic Plan for the Southeast Climate Consortium Extension Program.* Southeast Climate Consortium technical report series. University of Florida, Gainesville, 12 pp. <http://secc.coaps.fsu.edu/pdfpubs/SECC05-002.pdf>

Fraisse, C.W., N.E. Breuer, D. Zierden, J.G. Bellow, J. Paz, V.E. Cabrera, A. Garcia y Garcia, K.T. Ingram, U. Hatch, G. Hoogenboom, J.W. Jones, and J.J. O'Brien, 2006: AgClimate: a climate forecast information system for agricultural risk management in the southeastern USA. *Computers and Electronics in Agriculture,* **53(1)**, 13-27.

Gallaher, B.M. and R.J. Koch, 2004: *Cerro Grande Fire Impacts to Water Quality and Streamflow Near Los Alamos National Laboratory: Results of Four Years of Monitoring.* LANL report no. LA-14177. Los Alamos National Laboratory, Los Alamos, NM, 195 pp.

Garfin, G.M., 2005: Fire season prospects split east of the Rockies. *Wildfire,* **14(2)**.

* **Garfin**, G., 2006: Arizona drought monitoring. Presented at: *North American Drought Monitor Forum,* Mexico City, Mexico, 18-19 October 2006, 7 pp. <http://www.ncdc.noaa.gov/oa/climate/research/2006/nadm-workshop/20061018/1161187800-abstract.pdf>

* **Garfin**, G. and R. Emanuel, 2006: *Arizona Weather & Climate.* Arizona watershed stewardship guide. University of Arizona Cooperative Extension, Tucson, 14 pp. <http://cals.arizona.edu/watershedsteward/resources/index.html>

* **Garfin**, G. and B. Morehouse (eds.), 2001: *2001 Fire Climate Workshops.* Proceedings of workshops held February 14-16 and March 28, 2001, Tucson, Arizona. CLIMAS, Institute for the Study of Planet Earth, University of Arizona, Tucson, 75 pp. <http://www.ispe.arizona.edu/climas/conferences/fire2001/fire2001.pdf>

GDTF (Governor's Drought Task Force), 2004: *Arizona Drought Preparedness Plan: Operational Drought Plan.* Arizona Department of Water Resources, Phoenix, 107 pp. <http://www.azwater.gov/dwr/drought/ADPPlan.html>

Gibbons, M., 1999: Science's new social contract with society. *Nature,* **402(6761 supp.)**, C81-C84.

Graham, M., 2002: *Democracy by Disclosure: The Rise of Technopopulism.* Brookings Institution Press, Washington, DC, 201 pp.

Grigg, N.S., 1996: *Water Resources Management: Principles, Regulations, and Cases.* McGraw Hill, New York, 540 pp.

Guston, D.H., 2001: Boundary organizations in environmental science and policy: an introduction. *Science, Technology, and Human Values,* **26(4)**, 399-408.

Hammer, G.L., J.W. Hansen, J.G. Philips, J.W. Mjelde, H. Hill, A. Love, and A. Potgieter, 2001: Advances in application of climate prediction in agriculture. *Agricultural Systems,* **70(2-3)**, 515-553.

Hartig, J.H., D.P. Dodge, L. Lovett-Doust, and K. Fuller, 1992: Identifying the critical path and building coalitions for restoring degraded areas of the Great Lakes. In: *Water Resources Planning and Management: Saving a Threatened Resource.* Proceedings of the water resources sessions at Water Forum '92. American Society of Civil Engineers, New York, pp. 823-830.

Hartmann, H., 2001: *Stakeholder Driven Research in a Hydroclimatic Context.* Ph.D. dissertation, Department of Hydrology and Water Resources. University of Arizona, Tucson, 256 leaves. <http://etd.library.arizona.edu/etd/>

Hartmann, H.C., T.C. Pagano, S. Sorooshian, and R. Bales, 2002: Confidence builders: evaluating seasonal climate forecasts from user perspectives. *Bulletin of the American Meteorological Society,* **83(5)**, 683-698.

Hayhoe, K., D. Cayan, C.B. Field, P.C. Frumhoff, E.P. Maurer, N.L. Miller, S.C. Moser, S.H. Schneider, K.N. Cahill, E.E. Cleland, L. Dale, R. Drapek, R.M. Hanemann, L.S. Kalkstein, J. Lenihan, C.K. Lunch, R.P. Neilson, S.C. Sheridan, and J.H. Verville, 2004: Emissions pathways, climate change, and impacts on California. *Proceedings of the National Academy of Sciences,* **101(34)**, 12422-12427.

Hildebrand, P.E., A. Caudle, V. Cabrera, M. Downs, M. Langholtz, A. Mugisha, R. Sandals, A. Shriar, and K. Veach, 1999: *Potential Use of Long Range Climate Forecasts by Agricultural Extension in Florida.* University of Florida, Gainesville, 25 pp.

HRC-GWRI (Hydrologic Research Center - Georgia Water Resources Institute), 2006: *Integrated Forecast and Reservoir Management (INFORM) for Northern California: System Development and Initial Demonstration.* CEC-500-2006-109. California Energy Commission, PIER Energy-Related Environmental Research, Sacramento, 243 pp. + 9 Appendices. <http://www.energy.ca.gov/pier/final_project_reports/CEC-500-2006-109.html>

Hundley, N., Jr., 2001: *The Great Thirst – Californians and Water: A History.* University of California Press, Berkeley, revised edition, 799 pp.

Hurd, B.H., N. Leary, R. Jones, and J.B. Smith, 1999: Relative regional vulnerability of water resources to climate change. *Journal of the American Water Resources Association,* **35(6)**, 1399-1410.

ICLEI (International Council on Local Environmental Initiatives), 2007: *ICLEI's Climate Resilient Communities Program Addresses Adaptation, Vulnerabilities.* ICLEI – Local Governments for Sustainability, News: April 11. <http://www.iclei.org/index.php?id=1487&no_cache=1&tx_ttnews[pointer]=42&tx_ttnews[tt_news]=969&tx_ttnews[backPid]=1556&cHash=6b2bb4ee62>

Ingram, H. and B. Bradley, 2006: Water sustainability: policy innovation and conditions for adaptive learning. Presented at: *Sustainability: Theory and Applications,* 2005 SMEP (Sustainable Michigan Endowment Project) Academy, November 18-19. <http://www.smep.msu.edu/documents/2005academy/inghram%20full%20paper.pdf>

IPCC (Intergovernmental Panel on Climate Change) 2007: *Climate Change 2007: The Physical Science Basis.* Contribution of Working Group I to the Fourth Assessment Report of the Intergovernmental Panel on Climate Change [Solomon, S., D. Qin, M. Manning, Z. Chen, M. Marquis, K.B. Averyt, M.Tignor, and H.L. Miller (eds.)]. Cambridge University Press, Cambridge, UK, and New York, 996 pp.

Jacobs, K., 2001: Risk increase to infrastructure due to sea level rise. In: *Climate Change and a Global City: The Potential Consequences of Climate Variability and Change – Metro East Coast.* [Rosenzweig, C. and W. Solecki, (eds.)]. Columbia Earth Institute, Columbia University, New York, [58 pp.] <http://metroeast_climate.ciesin.columbia.edu/reports/infrastructure.pdf>

Jacobs, K., 2003: *Connecting Science, Policy, and Decision-Making – A Handbook for Researchers and Science Agencies.* NOAA Office of Global Programs, Boulder, CO, 25 pp. <http://www.ogp.noaa.gov/mpe/csi/doc/hdbk.pdf>

Jacobs, K.L., G.M. Garfin, and M. Lenart, 2005: More than just talk: connecting science and decisionmaking. *Environment,* **47(9),** 6-22.

Jacobs, K., V. Gornitz, and C. Rosenzweig, 2007: Vulnerability of the New York City metropolitan area to coastal hazards, including sea level rise: inferences for urban coastal risk management and adaptation policies. In: *Managing Coastal Vulnerability* [McFadden, L., R.J. Nicholls, and E.E. Penning-Rowsell (eds.)]. Elsevier, Amsterdam and Oxford, pp. 141-158.

Jagtap, S.S., J.W. Jones, P. Hildebrand, D. Letson, J.J. O'Brien, G. Podestá, D. Zierden, and F. Zazueta, 2002: Responding to stakeholders' demands for climate information: from research to applications in Florida. *Agricultural Systems,* **74(3),** 415-430.

Jasanoff, S. and B. Wynne, 1998: Science and decision-making. In: *Human Choice and Climate Change. Volume 1: The Societal Framework* [Rayner, S. and E.L. Malone (eds.)]. Battelle Press, Columbus, OH, pp. 1-88.

Jones, J.W., J.W. Hansen, F.S. Royce, and C.D. Messina, 2000: Potential benefits of climate forecast to agriculture. *Agriculture, Ecosystems, and Environment,* **82(1-3),** 169-184.

Kame'enui, A., E.R. Hagen, and J.E. Kiang, 2005: *Water Supply Reliability Forecast for the Washington Metropolitan Area, Year 2025.* Report no. 05-06. Interstate Commission on the Potomac River Basin, Rockville, MD, [176 pp.] <http://www.potomacriver.org/cms/publicationspdf/ICPRB05-6.pdf>

Kaufman, H., 1967: *The Forest Ranger: A Study in Administrative Behavior.* Resources for the Future, Washington, DC, 259 pp.

Kenney, D.S. and W.B. Lord, 1994: *Coordination Mechanisms for the Control of Interstate Water Resources: A Synthesis and Review of the Literature.* Report for the ACF-ACT Comprehensive Study. U.S. Army Corps of Engineers, Mobile District, Mobile, AL.

Kitzberger, T., P.M. Brown, E.K. Heyerdahl, T.W. Swetnam, and T.T. Veblen, 2007: Contingent Pacific-Atlantic ocean influence on multicentury wildfire synchrony over western North America. *Proceedings of the National Academy of Sciences,* **104(2),** 543-548.

Knowles, N., M.D. Dettinger, and D.R. Cayan, 2006: Trends in snowfall versus rainfall in the western United States. *Journal of Climate,* **19(18),** 4545-4559.

* **Kreutz**, D., 2006: Sabino Canyon is 'forever changed.' *Arizona Daily Star,* August 12, 2006. <http://www.azstarnet.com/sn/printDS/141817>

Kuyumjian, G., 2004: The BAER Team: responding to post-fire threats. *Southwest Hydrology,* **3(5),** 14-15, 32.

Landre, B.K. and B.A. Knuth, 1993: Success of citizen advisory committees in consensus based water resources planning in the Great Lakes basin. *Society and Natural Resources,* **6(3),** 229.

Lee, K.N., 1993: *Compass and Gyroscope: Integrating Science and Politics for the Environment.* Island Press, Washington, DC, 243 pp.

Lee, K.N., 1999: Appraising adaptive management. *Conservation Ecology* **3(2),** 3.

Lemos, M.C. and L. Dilling, 2007: Equity in forecasting climate: Can science save the world's poor? *Science and Public Policy,* **34(2),** 109-116.

Lemos, M.C. and B. Morehouse, 2005: The co-production of science and policy in integrated climate assessments. *Global Environmental Change: Human and Policy Dimensions,* **15(1),** 57-68.

* **Lund**, J., E. Hanak, W. Fleenor, R. Howitt, J. Mount, and P. Moyle, 2007: *Envisioning futures for the Sacramento-San Joaquin Delta.* Public Policy Institute of California, San Francisco, 285 pp. <http://www.ppic.org/main/publication.asp?i=671>

May, P.J., R.J. Burby, N.J. Ericksen, J.W. Handmer, J.E. Dixon, S. Michael, and S.D. Ingle, 1996: *Environmental Management and Governance: Intergovernmental Approaches to Hazards and Sustainability.* Routledge, New York, 254 pp.

McGinnis, M.V., 1995: On the verge of collapse: the Columbia River system, wild salmon, and the Northwest Power Planning Council. *Natural Resources Journal,* **35(1),** 63-92.

McNie, E.C., 2007: Reconciling the supply of scientific information with user demands: an analysis of the problem and review of the literature. *Environmental Science and Policy*, **10(1)**, 17-38.

* **McNie**, E., R. Pielke Jr., and D. Sarewitz, 2007: *Climate Science Policy: Lessons from the RISAs*, 2005 SPARC Reconciling Supply and Demand Workshop, August 15-17, 2005, East-West Center Honolulu, Hawaii. SPARC, Boulder, CO, 110 pp.

MDBC (Murray-Darling River Basin), 2002: *About MDB Initiative*. [website] Canberra City, Australia. <http://www.mdbc.gov.au/about/governance/agreement_history.htm>

Meinke, H. and R.C. Stone, 2005: Seasonal and inter-annual climate forecasting: the new tool for increasing preparedness to climate variability and change in agricultural planning and operations. *Climatic Change*, **70(1-2)**, 221-253.

Meixner, T. and P. Wohlgemuth, 2004: Wildfire impacts on water quality. *Southwest Hydrology*, **3(5)**, 24-25.

Meko, D.M., C.A.Woodhouse, C.H. Baisan, T. Knight, J.J. Lukas, M.K. Hughes, and M.W. Salzer, 2007: Medieval drought in the upper Colorado River Basin. *Geophysical Research Letters*, **34**(10), L10705, doi:10.1029/2007GL029988.

Miller, K.A., S.L. Rhodes, and L.J. MacDonnell, 1996: Global change in microcosm: the case of U.S. water institutions. *Policy Sciences*, **29(4)**, 271-272.

Moody, T., 1997: Glen Canyon Dam: coming to an informed decision. *Colorado Plateau Advocate*, **Fall**.

* **Morehouse**, B. (ed.), 2000: *The Implications of La Niña and El Niño for Fire Management*. Proceedings of workshop held February 23-24, 2000, Tucson, Arizona. Climate Assessment for the Southwest, Institute for the Study of Planet Earth, University of Arizona, Tucson, 45 pp. <http://www.climas.arizona.edu/conferences/fire2000/index.html>

National Fire Plan, 2000: <http://www.forestsandrangelands.gov/NFP/index.shtml>

Newson, M.D., 1997: *Land, Water and Development: Sustainable Management of River Basin Systems*. Routledge, New York, 2nd ed., 423 pp.

Noss, R.F., J.F. Franklin, W.L. Baker, T. Schoennagel, and P.B. Moyle, 2006: Managing fire-prone forests in the western United States. *Frontiers in Ecology and the Environment*, **4(9)**, 481-487.

NRC (National Research Council), 1989: *Improving Risk Communication*. National Academy Press, Washington, DC, 332 pp.

NRC (National Research Council), 1996: *Understanding Risk: Informing Decisions in a Democratic Society*. [Stern, P.C. and H.V. Fineberg (eds.)]. National Academy Press, Washington, DC, 249 pp.

NRC (National Research Council), 1999: *Making Climate Forecasts Matter*. National Academy Press, Washington, DC, 175 pp.

NRC (National Research Council), 2004: *Adaptive Management for Water Resources Project Planning*. National Academies Press, Washington, DC, 123 pp.

NRC (National Research Council), 2005: *Decision Making for the Environment: Social and Behavioral Science Research Priorities*. [Brewer, G.D. and P.C. Stern (eds.)]. National Academies Press, Washington, DC, 281 pp.

NRC (National Research Council), 2006: *Toward a New Advanced Hydrologic Prediction Service (AHPS)*. National Academies Press, Washington, DC, 84 pp.

NRC (National Research Council), 2007: *Colorado River Basin Water Management: Evaluating and Adjusting to Hydroclimatic Variability*. National Academies Press, Washington, DC, 222 pp.

NRC (National Research Council), 2008: *Research and Networks for Decision Support in the NOAA Sectoral Applications Research Program*. [Ingram, H.M. and P.C. Stern (eds.)]. National Academies Press, Washington, DC, 85 pp.

Obeysekera, J., P. Trimble, C. Neidrauer, C. Pathak, J. VanArman, T. Strowd, and C. Hall, 2007: Appendix 2-2: Consideration of long-term climatic variability in regional modeling for SFWMD planning and operations. In: *2007 South Florida Environmental Report*. Florida Department of Environmental Protection and South Florida Water Management District, West Palm Beach, 47 pp. <http://www.sfwmd.gov/sfer/SFER_2007/Appendices/v1_app_2-2.pdf>

OFCM (Office of the Federal Coordinator for Meteorology), 2002: *Weather Information for Surface Transportation: National Needs Assessment Report*. FCM-R18-2002. Office of the Federal Coordinator for Meteorological Services and Supporting Research, Silver Spring, MD. <http://www.ofcm.gov/wist_report/pdf/entire_wist.pdf>

OFCM (Office of the Federal Coordinator for Meteorology), 2004: *Urban Meteorology: Meeting Weather Needs in the Urban Community*. FCM-R22-2004. Office of the Federal Coordinator for Meteorological Services and Supporting Research, Silver Spring, MD, 18 pp. <http://www.ofcm.gov/homepage/text/spc_proj/urban_met/Discussion-Framework.pdf>

OFCM (Office of the Federal Coordinator for Meteorology), 2007a: *Interagency Strategic Research Plan for Tropical Cyclones: The Way Ahead*. FCM-P36-2007. Office of the Federal Coordinator for Meteorological Services and Supporting Research, Silver Spring, MD. <http://www.ofcm.gov/p36-isrtc/fcm-p36.htm>

OFCM (Office of the Federal Coordinator for Meteorology), 2007b: *National Wildland Fire Weather: A Summary of User Needs and Issues*. Report to the Western Governors' Association. Office of the Federal Coordinator for Meteorological Services and Supporting Research, Silver Spring, MD, 42 pp. <http://www.ofcm.gov/jag-nwfwna/workingdocs/National-al%20Wildland%20Fire%20Weather%20-%20A%20Summary%20of%20User%20Needs%20and%20Issues.pdf>

Pagano, T.C., H.C. Hartmann, and S. Sorooshian, 2002: Factors affecting seasonal forecast use in Arizona water management: a case study of the 1997-98 El Niño. *Climate Research*, **21(3)**, 259-269.

Pfaff, A., K. Broad, and M. Glantz, 1999: Who benefits from climate forecasts? *Nature*, **397(6721)**, 645-646.

Pielke, R.A., Jr., D. Sarewitz, and R. Byerly Jr., 2000: Decision making and the future of nature: understanding and using predictions. In: *Prediction: Science, Decision Making, and the Future of Nature*. [Sarewitz, D., R.A. Pielke Jr., R. Byerly Jr. (eds.)]. Island Press, Washington, DC, pp. 361-387.

Power, S., B. Sadler, and N. Nicholls, 2005: The influence of climate science on water management in western Australia: lessons for climate scientists. *Bulletin of the American Meteorological Society*, **86(6)**, 839-844.

Preisler, H.K. and A.L. Westerling, 2007: Statistical model for forecasting monthly large wildfire events in western United States. *Journal of Applied Meteorology and Climatology*, **46(7)**, 1020-1030.

Rabe, B.G., 2004: *Statehouse and Greenhouse: The Emerging Politics of American Climate Change Policy*. Brookings Institution, Washington, DC, 212 pp.

Restoring the Waters, 1997: Natural Resources Law Center, University of Colorado School of Law, Boulder, CO, 64 pp.

Roads, J., F. Fujioka, S. Chen, and R. Burgan, 2005: Seasonal fire danger forecasts for the USA. *International Journal of Wildland Fire*, **14(1)**, 1-18.

Roncoli, C., J. Paz, N. Breuer, K. Ingram, G. Hoogenboom, and K. Broad, 2006: *Understanding Farming Decisions and Potential Applications of Climate Forecasts in South Georgia*. Southeast Climate Consortium, Gainesville, FL, 24 pp.

Rosenzweig, C. and W.D. Solecki (eds.), 2001: *Climate Change and a Global City: The Potential Consequences of Climate Variability and Change—Metro East Coast*. Columbia Earth Institute, Columbia University, New York, 224 pp. <http://metroeast_climate.ciesin.columbia.edu/>

Rosenzweig, C., D.C. Major, K. Demong, C. Stanton, R. Horton, and M. Stults, 2007: Managing climate change risks in New York City's water systems: assessment and adaptation planning. *Mitigation and Adaptation Strategies for Global Change*, **12(8)**, 1391-1409.

Sarewitz, D. and R.A. Pielke Jr., 2007: The neglected heart of science policy: reconciling supply of and demand for science. *Environmental Science and Policy*. **10(1)**, 5-16.

Schlobohm, P.M., B.L. Hall, and T.J. Brown, 2003: Using NDVI to determine green-up date for the National Fire Danger Rating System. In: *Fifth Symposium on Fire and Forest Meteorology*. American Meteorological Society, Boston, 15 pp. <http://ams.condex.com/ams/pdfpapers/65690.pdf>

South Florida Water Management District, 1996: *Climate Change and Variability: How Should The District Respond?* South Florida Water Management District, West Palm Beach, 27 pp. <http://www.sfwmd.gov/org/pld/hsm/pubs/ptrimble/solar/clim1.pdf>

SSCSE (Senate Standing Committee on Science and the Environment), 1979: *Continuing Scrutiny of Pollution: The River Murray*. Progress report, June 1979. Parliamentary paper 117/1979. Government Printer, Canberra, Australia, 38 pp.

Starling, G., 1989: *Strategies for Policy Making*. Dorsey Press, Chicago, 692 pp.

Steiner, R.C. and J.J. Boland, 1994: *Water Resources Management in the Potomac River Basin under Climate Uncertainty*. ICPRB 94-3. Interstate Commission on the Potomac River Basin, Rockville, MD, 84 pp.

Swetnam, T.W. and J.L. Betancourt, 1998: Mesoscale disturbance and ecological response to decadal climatic variability in the American southwest. *Journal of Climate*, **11(12)**, 3128-3147.

Tichy, N.M. and W.G. Bennis, 2007: *Judgment: How Winning Leaders Make Great Calls*. Penguin Group, New York, 392 pp.

TRACS (Transition of Research Applications to Climate Services Program), 2008: Web page describing program. NOAA Climate Program Office, Washington, DC <http://www.cpo.noaa.gov/cpo_pa/nctp/>

* **Trimble**, P.J. and B.M. Trimble, 1998: Recognition and predictability of climate variability within south-central Florida. In: *23rd Annual Climate Diagnostics and Prediction Workshop*, Rosenstiel School of Marine and Atmospheric Science, University of Miami, FL, October 26-30. National Oceanic and Atmospheric Administration, Silver Spring, MD, pp. 239-242. <http://www.sfwmd.gov/org/pld/hsm/pubs/ptrimble/solar/workshop/cpc_paper.htm>

* **Trimble**, P.J., E.R. Santee, and C.J. Neidrauer, 1997: Including the effects of solar activity for more efficient water management: an application of neural networks. In: *Second International Workshop on Artificial Intelligence Applications in Solar-Terrestrial Physics*, Sweden. 8 pp. <http://www.sfwmd.gov/org/pld/hsm/pubs/ptrimble/solar/final_dec3.pdf>

Trimble, P.J., E.R. Santee, and C.J. Neidrauer, 1998: *A Refined Approach to Lake Okeechobee Water Management: An Application of Climate Forecasts*. Special report. South Florida Water Management District, West Palm Beach, 73 pp. <http://www.sfwmd.gov/org/pld/hsm/pubs/ptrimble/solar/report/report.pdf>

UNCED (United Nations Conference on Environment and Development), 1992: *Nations of the Earth Report, volume 1: National Report Summaries*. Geneva, Switzerland, United Nations.

Wade, W.W., 2001: Least-cost water supply planning. Presentation to: *Eleventh Tennessee Water Symposium*, Nashville, Tennessee, April 15.

Warren, D.R., G.T. Blain, F.L. Shorney, and L.J. Klein, 1995: IRP: a case study from Kansas. *Journal of the American Water Works Association*, **87(6)**, 57-71.

Water in the West: Challenge for the Next Century, 1998: Western Water Policy Review Advisory Commission, [Washington, DC]. <http://hdl.handle.net/1928/2788>

Wells, A., 1994: *Up and Doing: A Brief History of the Murray Valley Development League, Now the Murray Darling Association, from 1944 to 1994*. Murray Darling Association, Sydney, [Australia], 97 pp.

Westerling, A.L., A. Gershunov, D.R. Cayan, and T.P. Barnett, 2002: Long lead statistical forecasts of area burned in western U.S. wildfires by ecosystem province. *International Journal of Wildland Fire*, **11(3&4)**, 257-266.

Westerling, A.L., H.G. Hidalgo, D.R. Cayan, and T.W. Swetnam, 2006: Warming and earlier spring increase western US forest wildfire activity. *Science*, **313(5789)**, 940-943.

* **Wiener**, J.D., 2004: Small agriculture needs and desires for weather and climate information in a case study in Colorado. In: *Second Annual User's Conference*, held in conjuction with 84th Annual Meeting, American Meteorological Society, 47 pp. <http://ams.confex.com/ams/pdfpapers/70298.pdf>

Woodhouse, C.A. and J.J. Lukas, 2006: Drought, tree rings, and water resource management in Colorado. *Canadian Water Resources Journal*, **31**, 297-310.

Woodhouse, C.A., S.T. Gray, and D.M. Meko, 2006: Updated streamflow reconstructions for the Upper Colorado River Basin. *Water Resources Research*, **42**, W05415, doi:10.1029/2005WR004455.

Wordell, T. and R. Ochoa, 2006: Improved decision support for proactive wildland fire management. *Fire Management Today*, **66(2)**, 25-28.

Yarnal, B., A.L. Heasley, R.E. O'Connor, K. Dow, and C.L. Jocoy, 2006: The potential use of climate forecasts by community water system managers. *Land Use and Water Resources Research*, **6**, 3.1-3.8. <http://www.luwrr.com>

Zhang, E. and P.J. Trimble, 1996: Predicting effects of climate fluctuations for water management by applying neural networks. *World Resource Review*, **8**, 334-348.

Zimmerman, R. and M. Cusker, 2001: Institutional decision-making. In: *Climate Change and a Global City: The Potential Consequences of Climate Variability and Change – Metro East Coast*. [Rosenzweig, C. and W. Solecki (eds.)]. Columbia Earth Institute, Columbia University, New York, [55 pp.] <http://metroeast_climate.ciesin.columbia.edu/reports/decision.pdf>

CHAPTER 5 REFERENCES

Agrawala, S., K. Broad, and D.H. Guston, 2001: Integrating climate forecasts and societal decision making: challenges to an emergent boundary organization. *Science, Technology & Human Values*, **26(4)**, 454-477.

Anderson, L.G., R. Bishop, M. Davidson, S. Hanna, M. Holliday, J. Kildow, D. Liverman, B.J. McCay, E.L. Miles, R. Pielke Jr., and R. Pulwarty, 2003: *Social Science Research within NOAA: Review and Recommendations*. NOAA Social Science Review Panel, Washington, DC, 98 pp. <http://www.economics.noaa.gov/ppi-economics/library/documents/social_science_initiative/social_science_research_within_noaa-review_recommend.doc>

Archer, E.R.M., 2003: Identifying underserved end-user groups in the provision of climate information. *Bulletin of the American Meteorological Society*, **84(11)**, 1525-1532.

Atwater, R. and W. Blomquist, 2002: Rates, rights, and regional planning in the Metropolitan Water District of Southern California. *Journal of the American Water Resources Association*, **38(5)**, 1195-1205.

Bäckstrand, K., 2003: Civic science for sustainability: reframing the role of experts, policy makers, and citizens in environmental governance. *Global Environmental Politics*, **3(4)**, 24-41.

Beecher, J.A., 1995: Integrated resource planning fundamentals. *Journal of the American Water Works Association*, **87(6)**, 34-48.

Beller-Simms, N., 2004: Planning for El Niño: the stages of natural hazard mitigation and preparation. *The Professional Geographer*, **56(2)**, 213-222.

Bennis, W.G., 2003: *On Becoming a Leader*. De Capo Press, Cambridge, MA, 256 pp.

Bharwani, S., M. Bithell, T.E. Downing, M. New, R. Washington, and G. Ziervogel, 2005: Multi-agent modeling of climate outlooks and food security on a community garden scheme in Limpopo, South Africa. *Philosophical Transactions of the Royal Society B: Biological Sciences*, **360(1463)**, 2183-2194.

Bormann, B.T., P.G. Cunningham, M.H. Brookes, V.W. Maning, and M.W. Collopy, 1994: *Adaptive Ecosystem Management in the Pacific Northwest*. General technical report PNW 341. USDA Forest Service, Pacific Northwest Research Service, Portland, OR, 22 pp.

Broad, K. and S. Agrawalla, 2000: The Ethiopia food crisis: uses and limits of climate forecasts. *Science*, **289(5485)**, 1693-1694.

Broad, K., A.S.P. Pfaff, and M.H. Glantz, 2002: Effective and equitable dissemination of seasonal-to-interannual climate forecasts: policy implications from the Peruvian fishery during El Niño 1997-98. *Climatic Change*, **54(4)**, 415-438.

Brunner, R.D., T.A. Steelman, L. Coe-Juell, C.M. Cromley, C.M. Edwards, and D.W. Tucker, 2005: *Adaptive Governance: Integrating Science, Policy, and Decision Making*. Columbia University Press, New York, 319 pp.

Carbone, G.J. and K. Dow, 2005: Water resource management and drought forecasts in South Carolina. *Journal American Water Resources Association*, **41(1)**, 145-155.

Cash, D.W. and J. Buizer, 2005: *Knowledge-Action Systems for Seasonal to Interannual Climate Forecasting: Summary of a Workshop*. National Academies Press, Washington, DC, 44 pp.

Cash, D.W., J.D. Borck, and A.G. Pratt, 2006: Countering the loading-dock approach to linking science and decision making: comparative analysis of El Niño/Southern Oscillation (ENSO) forecasting systems. *Science, Technology and Human Values*, **31(4)**, 465-494.

Cody, B.A., 1999: *Western Water Resource Issues*. A Congressional Research Service Brief for Congress. Congressional Research Service, Washington, DC, 13 pp.

Cortner, H.A. and M.A. Moote, 1994: Setting the political agenda: paradigmatic shifts in land and water policy. In: *Environmental Policy and Biodiversity* [Grumbine, R.E. (ed.)]. Island Press, Washington, DC, pp. 365-377.

Covello, V., E. Donovan, and J.E. Slavick, 1990: *Community Outreach*. Chemical Manufacturers Association, Washington, DC, 127 pp.

Dow, K., R.E. O'Connor, B. Yarnal, G.J. Carbone, and C.L. Jocoy, 2007: Why worry? Community water system managers' perceptions of climate vulnerability. *Global Environmental Change*, **17(2)**, 228-237.

Durodié, B., 2003: Limitations of public dialogue in science and the rise of new "experts." *Critical Review of International Social and Political Philosophy*, **6(4)**, 82-92.

Eakin, H., 2000: Smallholder maize production and climatic risk: a case study from Mexico. *Climatic Change*, **45(3-4)**, 19-36.

Eden, S., 1996: Public participation in environmental policy: considering scientific, counter-scientific, and non-scientific contributions. *Public Understanding of Science*, **5(3)**, 183-204.

Fischer, F., 2000: *Citizens, Experts, and the Environment: The Politics of Local Knowledge*. Duke University Press, Durham NC, 336 pp.

Fiske, G. and A. Dong, 1995: IRP: a case study from Nevada. *Journal of the American Water Works Association*, **87(6)**, 72-83.

Freudenburg, W.R. and J.A. Rursch, 1994: The risks of putting the numbers in context. *Risk Analysis*, **14(6)**, 949-958.

Georgia DNR (Georgia Department of Natural Resources), 2003: *Georgia Drought Management Plan*. Department of Natural Resources, Atlanta, GA, 23 pp. <http://www.gaepd.org/Files_PDF/gaenviron/drought/drought_mgmtplan_2003.pdf>

Gibbons, M., 1999: Science's new social contract with society. *Nature*, **402(6761 supp.)**, C81-C84.

Glantz, M.H., 1996: *Currents of Change: El Niño's Impact on Climate and Society*. Cambridge University Press, Cambridge, UK, and New York, 194 pp.

Gunderson, L., 1999: Resilience, flexibility and adaptive management – antidotes for spurious certitude? *Ecology and Society*, **3(1)**, article 7. <http://www.consecol.org/vol3/iss1/art7>

Hammer, G.L., J.W. Hansen, J.G. Philips, J.W. Mjelde, H. Hill, A. Love, and A. Potgieter, 2001: Advances in application of climate prediction in agriculture. *Agricultural Systems*, **70(2-3)**, 515-553.

Harding, S., 2000: Should philosophies of science encode democratic ideals? In: *Science, Technology, and Democracy* [Kleinmann, D.L. (ed.)]. State University of New York Press, Albany, pp. 121-138.

Hartig, J.H., D.P. Dodge, L. Lovett-Doust, and K. Fuller, 1992: Identifying the critical path and building coalitions for restoring degraded areas of the Great Lakes. In: *Water Resources Planning and Management: Saving a Threatened Resource*. Proceedings of the water resources sessions at Water Forum '92. American Society of Civil Engineers, New York, pp. 823-830.

Hartmann, H., 2001: *Stakeholder Driven Research in a Hydroclimatic Context.* Ph.D. dissertation, Department of Hydrology and Water Resources. University of Arizona, Tucson, 256 leaves. <http://etd.library.arizona.edu/etd/>

Hartmann, H.C., T.C. Pagano, S. Sorooshian, and R. Bales, 2002: Confidence builders: evaluating seasonal climate forecasts from user perspectives. *Bulletin of the American Meteorological Society,* **83(5)**, 683-698.

Holling, C.S. (ed.), 1978: *Adaptive Environmental Assessment and Management.* Wiley, New York, 377 pp.

Huda, A.K.S., R. Selvaraju, T.N. Balasubramanian, V. Geethalakshmi, D.A. George, and J.F. Clewett, 2004: Experiences of using seasonal climate information with farmers in Tamil Nadu, India. In: *Using Seasonal Climate Forecasting in Agriculture: A Participatory Decision-making Approach* [Huda, A.K.S. and R.G. Packham (eds.)]. ACIAR technical reports series 59. Australian Centre for International Agricultural Research, Canberra, pp. 22-30. <http://www.aciar.gov/au/publication/TR59>

Jasanoff, S., 1987: EPA's regulation of daminozide: unscrambling the messages of risk. *Science, Technology, and Human Values,* **12(3-4)**, 116-124.

Jasanoff, S., 1996: The dilemma of environmental democracy. *Issues in Science and Technology,* **Fall,** 63-70.

Jasanoff, S. (ed.), 2004a: *States of Knowledge: The Co-Production of Science and Social Order.* Routledge, London and New York, 317 pp.

Jasanoff, S., 2004b: Science and citizenship: a new synergy. *Science and Public Policy,* **31(2)**, 90-94.

Jasanoff, S. and B. Wynne, 1998: Science and decision making. In: *Human Choice and Climate Change. Volume 1: The Societal Framework* [Rayner, S. and E Malone (eds.)]. Battelle Press, Columbus, OH, pp. 1-88.

Klopper, E., 1999: The use of seasonal forecasts in South Africa during the 1997-1998 rainfall season. *Water SA,* **25(3)**, 311-316.

Klopper, E., C.H. Vogel, and W.A. Landman, 2006: Seasonal climate forecasts – potential agricultural-risk management tools? *Climatic Change,* **76(1-2)**, 73-90.

Landre, B.K. and B.A. Knuth, 1993: Success of citizen advisory committees in consensus based water resources planning in the Great Lakes basin. *Society and Natural Resources,* **6(3)**, 229.

Leatherman, S.P. and G. White, 2005: Living on the edge: the coastal collision course. *Natural Hazards Observer,* **30(2)**, 5-6.

Lee, K.N., 1993: *Compass and Gyroscope: Integrating Science and Politics for the Environment.* Island Press, Washington, DC, 243 pp.

Lemos, M.C., 2008: Whose water is it anyway? Water management, knowledge and equity in NE Brazil. In: *Water Place and Equity* [Whiteley, J.M., H. Ingram and R.W. Perry (eds.)]. MIT Press, Cambridge, MA, pp. 249-270.

Lemos, M.C. and L. Dilling, 2007: Equity in forecasting climate: Can science save the world's poor? *Science and Public Policy,* **34(2)**, 109-116.

Lemos, M.C. and B.J. Morehouse, 2005: The co-production of science and policy in integrated climate assessments. *Global Environmental Change: Human and Policy Dimensions,* **15(1)**, 57-68.

Lemos, M.C., T.J. Finan, R.W. Fox, D.R. Nelson, and J. Tucker, 2002: The use of seasonal climate forecasting in policymaking: lessons from Northeast Brazil. *Climatic Change,* **55(4)**, 479-507.

Letson, D., I. Llovet, G. Podestá, F. Royce, V. Brescia, D. Lema, and G. Parellada, 2001: User perspectives of climate forecasts: crop producers in Pergamino, Argentina. *Climate Research,* **19(1)**, 57-67.

Lusenso, W.K., J.G. McPeak, C.B. Barrett, P.D. Little, G. Gebru, 2003: Assessing the value of climate forecasts information for pastoralists: evidence from southern Ethiopia and northern Kenya. *World Development,* **31(9)**, 1477-1494.

McGinnis, M.V., 1995: On the verge of collapse: the Columbia River system, wild salmon, and the Northwest Power Planning Council. *Natural Resources Journal,* **35**, 63-92.

* McNie, E., R. Pielke Jr., and D. Sarewitz, 2007: *Climate Science Policy: Lessons from the RISAs,* 2005 SPARC Reconciling Supply and Demand Workshop, August 15-17, 2005, East-West Center Honolulu, Hawaii. SPARC, Boulder, CO, 110 pp.

McPhaden, M.J., S.E. Zebiak, and M.H. Glantz, 2006: ENSO as an integrating concept in earth science. *Science,* **314(5806)**, 1740-1745.

Meinke, H., R. Nelson, P. Kokic, R. Stone, R. Selvaraju, and W. Baethgen, 2006: Actionable climate knowledge: from analysis to synthesis. *Climate Research,* **33(1)**, 101-110.

Miles, E.L., A.K. Snover, L.C. Whitely Binder, E.S. Sarachik, P.W. Mote, and N. Mantua, 2006: An approach to designing a national climate service. *Proceedings of the National Academy of Sciences,* **103(52)**, 19616-19623.

Miller, K., S.L. Rhodes, and L.J. MacDonnell, 1996: Global change in microcosm: the case of U.S. water institutions. *Policy Sciences,* **29(4)**, 271-290.

Nicholls, N., 1999: Cognitive illusions, heuristics, and climate prediction. *Bulletin of the American Meteorological Society*, **80(7)**, 1385-1398.

NRC (National Research Council), 1989: *Improving Risk Communication*. National Academy Press, Washington, DC, 332 pp.

NRC (National Research Council), 2007: *Research and Networks for Decision Support in the NOAA Sectoral Applications Research Program*. [Ingram, H.M. and P.C. Stern (eds.)]. National Academies Press, Washington, DC, 85 pp.

Nowotny, H., P. Scott, and M. Gibbons, 2001: *Re-thinking Science: Knowledge and the Public in an Age of Uncertainty*. Polity, Cambridge, UK, 278 pp.

O'Brien, K. and C. Vogel (eds.), 2003: *Coping with Climate Variability: The Use of Seasonal Climate Forecasts in Southern Africa*. Ashgate Publishing, Aldershot, England, and Burlington, VT, 220 pp.

Pagano, T.C., H.C. Hartmann, and S. Sorooshian, 2002: Factors affecting seasonal forecast use in Arizona water management: a case study of the 1997-98 El Niño. *Climate Research*, **21(3)**, 259-269.

Papadakis, E., 1996: *Environmental Politics and Institutional Change*. Cambridge University Press, Cambridge, UK, and New York, 240 pp.

Patt, A. and C. Gwata, 2002: Effective seasonal climate forecast applications: examining constraints for subsistence farmers in Zimbabwe. *Global Environmental Change*, **12(3)**, 185-195.

Patt, A., P. Suarez, and C. Gwata, 2005: Effects of seasonal climate forecasts and participatory workshops among subsistence farmers in Zimbabwe. *Proceedings of the National Academy of Sciences*, **102(35)**, 12623-12628.

Pfaff, A., K. Broad, and M. Glantz, 1999: Who benefits from climate forecasts? *Nature*, **397(6721)**, 645-646.

Pulwarty, R.S. and T.S. Melis, 2001: Climate extremes and adaptive management on the Colorado River: lessons from the 1997-1998 ENSO event. *Journal of Environmental Management*, **63(3)**, 307-324.

Roncoli, C., K. Ingram., P. Kirshen, and C. Jost, 2004: Burkina Faso: integrating indigenous and scientific rainfall forecasting. In: *Indigenous Knowledge: Local Pathways to Global Development*. The World Bank, [Washingon, DC], pp. 197-200. <http://www.worldbank.org/afr/ik/ikcomplete.pdf>

Roncoli, C., J. Paz, N. Breuer, K. Ingram, G. Hoogenboom, and K. Broad, 2006: *Understanding Farming Decisions and Potential Applications of Climate Forecasts in South Georgia*. Southeast Climate Consortium, Gainesville, FL, 24 pp.

Sarewitz, D. and R.A. Pielke Jr., 2007: The neglected heart of science policy: reconciling supply of and demand for science. *Environmental Science and Policy*, **10(1)**, 5-16.

Subcommittee on Disaster Reduction, 2005: Grand challenges for disaster reduction. *Natural Hazards Observer*, **30(2)**, 1-3.

Tichy, N.M. and W.G. Bennis, 2007: *Judgment: How Winning Leaders Make Great Calls*. Penguin Group, New York, 392 pp.

Valdivia, C., J.L. Gilles, and S. Materer, 2000: Climate variability, a producer typology and the use of forecasts: experience from Andean semiarid small holder producers. In: *Proceedings of the International Forum on Climate Prediction Agriculture and Development*. International Research Institute for Climate Prediction, Palisades, NY, pp. 227-239.

Vogel, C., 2000: Usable science: an assessment of long-term seasonal forecasts amongst farmers in rural areas of South Africa. *South African Geographical Journal*, **82**, 107-116.

Wade, W.W., 2001: Least-cost water supply planning. Presentation to: *Eleventh Tennessee Water Symposium*, Nashville, Tennessee, April 15.

Warren, D.R., G.T. Blain, F.L. Shorney, and L.J. Klein, 1995: IRP: a case study from Kansas. *Journal of the American Water Works Association*, **87(6)**, 57-71.

Water in the West: Challenge for the Next Century, 1998: Western Water Policy Review Advisory Commission, [Washington, DC]. <http://hdl.handle.net/1928/2788>

Weingart, P., A. Engels, and P. Pansegrau, 2000: Risks of communication: discourses on climate change in science, politics, and the mass media. *Public Understanding of Science*, **9(3)**, 261-283.

Wiener, J., 2007: *The Climate of Uncertainty for Colorado's Agricultural Water: Management Responses and Coordination of Objectives*. Presentation at USDA CSREES Water Meeting, January. Presentation available from John.Wiener@Colorado.edu

Yarnal, B., A.L. Heasley, R.E. O'Connor, K. Dow, and C.L. Jocoy, 2006: The potential use of climate forecasts by community water system managers. *Land Use and Water Resources Research*, **6**, 3.1-3.8. <http://www.luwrr.com>

Ziervogel, G. and R. Calder, 2003: Climate variability and rural livelihoods: assessing the impact of seasonal climate forecasts in Lesotho. *Area*, **35(4)**, 403-417.

PHOTGRAPHY CREDITS

Cover/Title Page/Table of Contents
Image for Chapter 2, page 41, NOAA space satellite, NOAA

Chapter 1
Page 22, Image 1, (Downspout), ©iStockphotos.com/Don Wilkie
Page 29, Image 1, (Fish ladder), ©iStockphotos.com/Mark Jensen

Chapter 2
Page 41, Chapter heading, (NOAA space satellite), NOAA
Page 42, Image 1, (Low lake levels), Grant Goodge, STG Inc., Asheville, NC
Page 64, Image 1, (Weather instrument station), USGS
Page 65, Image 1, (Stream flow measurement), USGS
Page 66, Image 1, (Snow measuring equipment), USDA

Chapter 3
Page 77, Image 1, (Ship and buoy), NOAA
Page 78, Image 1, (Boulder Dam, Colorado), Grant Goodge, STG. Inc., Asheville, NC
Page 88, Image 1, (Flooded road and pick up truck), Weather Stock photo, Copyright ©1993, Warren Faidley/Weatherstock serial #000103
Page 90, Image 1, (Dried up Lake Lanier), Kent Frantz, SRH/NOAA
Page 95, Image 1, (Shasta Dam, California), ©iStockphotos.com/Andrew Zarivny
Page 96, Image 1, (Two lake levels at Lake Powell), John C. Dohrenwend
Page 100, Image 1, (Lake Okeechobee, Florida) , ©iStockphotos.com/Adventure Photo
Page 104, Image 1, (Apalachicola Estuary), ©iStockphotos.com/M. B. Cheatham

Chapter 4
Page 115, Image 1, (Everglades, Florida), ©iStockphotos.com/John Anderson
Page 125, Image 1, (San Franciso Bay area), ©iStockphotos.com/Joe Potato
Page 135, Image 1, (Tennessee River low levels), ©iStockphotos.com/Matt Tilghman
Page 139, Image 1, (Apache lake, Arizona), ©iStockphotos.com/Paul Hill
Page 142, Image 1, (Glen Canyon Dam), ©iStockphotos.com/Vlad Turchenko
Page 145, Image 1, (Washington DC, tidal basin), ©iStockphotos.com/Jim Pruitt
Page 150, Image 1, (Road builders/information users) ©iStock photos.com/Bart Coenders

Chapter 5
Page 155, Image 1, (Raging dust cloud), NOAA
Page 160, Image 1, (Buoy launch) Emily B. Christman, NOAA,
Page 165, Image 1, (Tree rings), ©iStockphotos.com/Ivan Mateev

Contact Information

Global Change Research Information Office
c/o Climate Change Science Program Office
1717 Pennsylvania Avenue, NW
Suite 250
Washington, DC 20006
202-223-6262 (voice)
202-223-3065 (fax)

The Climate Change Science Program incorporates the U.S. Global Change Research Program and the Climate Change Research Initiative.

To obtain a copy of this document, place an order at the Global Change Research Information Office (GCRIO) web site: http://www.gcrio.org/orders

Climate Change Science Program and the Subcommittee on Global Change Research

William Brennan, Chair
Department of Commerce
National Oceanic and Atmospheric Administration
Acting Director, Climate Change Science Program

Jack Kaye, Vice Chair
National Aeronautics and Space Administration

Allen Dearry
Department of Health and Human Services

Jerry Elwood
Department of Energy

Mary Glackin
National Oceanic and Atmospheric Administration

Patricia Gruber
Department of Defense

William Hohenstein
Department of Agriculture

Linda Lawson
Department of Transportation

Mark Myers
U.S. Geological Survey

Timothy Killeen
National Science Foundation

Patrick Neale
Smithsonian Institution

Jacqueline Schafer
U.S. Agency for International Development

Joel Scheraga
Environmental Protection Agency

Harlan Watson
Department of State

EXECUTIVE OFFICE AND OTHER LIAISONS

Stephen Eule
Department of Energy
Director, Climate Change Technology Program

Katharine Gebbie
National Institute of Standards & Technology

Stuart Levenbach
Office of Management and Budget

Margaret McCalla
Office of the Federal Coordinator for Meteorology

Rob Rainey
Council on Environmental Quality

Daniel Walker
Office of Science and Technology Policy

www.ingramcontent.com/pod-product-compliance
Lightning Source LLC
Chambersburg PA
CBHW081443170526
45166CB00008B/2294